Econophysics Approaches to Large-Scale Business Data and Financial Crisis

Misako Takayasu • Tsutomu Watanabe
Hideki Takayasu
Editors

Econophysics Approaches to Large-Scale Business Data and Financial Crisis

Proceedings of the Tokyo Tech–Hitotsubashi Interdisciplinary Conference + APFA7

Springer

Editors
Misako Takayasu
Associate Professor
Department of Computational
Intelligence and Systems Science
Interdisciplinary Graduate School
of Science and Engineering
Tokyo Institute of Technology
4259 Nagatsuta, Midori-ku
Yokohama 226-8502, Japan
takayasu@dis.titech.ac.jp

Hideki Takayasu
Senior Researcher
Fundamental Research Group
Sony Computer Science Laboratories
3-14-13 Higashigotanda, Shinagawa-ku
Tokyo 141-0022, Japan
takayasu@csl.sony.co.jp

Tsutomu Watanabe
Professor
Research Center for Price Dynamics
and Institute of Economic Research
Hitotsubashi University
2-1 Naka, Kunitachi
Tokyo 186-8603, Japan
tsutomu.w@srv.cc.hit-u.ac.jp

ISBN 978-4-431-54671-9 ISBN 978-4-431-53853-0 (eBook)
DOI 10.1007/978-4-431-53853-0
Springer Tokyo Dordrecht Heidelberg London New York

Softcover re-print of the Hardcover 1st edition 2010

Printed on acid-free paper

Springer is part of Springer Science+Business Media (www.springer.com)

Preface

In recent years, as part of the increasing "informationization" of industry and the economy, enterprises have been accumulating vast amounts of detailed data such as high-frequency transaction data in financial markets and point-of-sale information on individual items in the retail sector. Similarly, vast amounts of data are now available on business networks based on interfirm transactions and shareholdings. In the past, these types of information were studied only by economists and management scholars. More recently, however, researchers from other fields, such as physics, mathematics, and information sciences, have become interested in this kind of data and, based on novel empirical approaches to searching for regularities and "laws" akin to those in the natural sciences, have produced intriguing results.

This book is the proceedings of the international conference THIC+APFA7 that was titled "New Approaches to the Analysis of Large-Scale Business and Economic Data," held in Tokyo, March 1–5, 2009. The letters THIC denote the Tokyo Tech (Tokyo Institute of Technology)–Hitotsubashi Interdisciplinary Conference. The conference series, titled APFA (Applications of Physics in Financial Analysis), focuses on the analysis of large-scale economic data. It has traditionally brought physicists and economists together to exchange viewpoints and experience (APFA1 in Dublin 1999, APFA2 in Liège 2000, APFA3 in London 2001, APFA4 in Warsaw 2003, APFA5 in Torino 2006, and APFA6 in Lisbon 2007). The aim of the conference is to establish fundamental analytical techniques and data collection methods, taking into account the results from a variety of academic disciplines.

The workshop was supported by the Research Institute of Economy, Trade and Industry (RIETI); IAA; the Physical Society of Japan; the Japanese Economic Association; the Information Processing Society of Japan; the Japanese Society for Artificial Intelligence; and the Japan Association for Evolutionary Economics. We would like to acknowledge the following companies for their financial support: Monex Group, Inc.; Kao Corporation; Nikkei Digital Media, Inc.; Kakaku.com, Inc.; ASTMAX Co., Ltd.; Sony Computer Science Laboratories, Inc.; and CMD Laboratory Inc.

Misako Takayasu
Tsutomu Watanabe
Hideki Takayasu

Editors

Contents

Contributors

Peter Toke Heden Ahlgren Nykredit Asset Management, Otto Moensteds Plads 9, 1780 Copenhagen, Denmark, ahl@nykredit.dk

Masanao Aoki Department of Economics, University of California, 403 Hilgard Avenue, Los Angeles, CA 90095-1477, USA, aoki@econ.ucla.edu

Marcel Ausloos GRAPES, Université de Liège, B5 Sart-Tilman, 4000 Liège, Belgium, marcel.ausloos@ulg.ac.be

Mikhail I. Bogachev Radio System Department, St. Petersburg State Electrotechnical University, 197376, St. Petersburg, Russia, mikhail.bogachev@physik.uni-giessen.de

Armin Bunde Institut für Theoretische Physik III, Justus-Liebig-Universität Giessen, 35392 Giessen, Germany, bunde@uni-giessen.de

Henrik Dahl Nykredit Asset Management, Otto Mønsteds Plads9, 1780 Copenhagen, Denmark, heda@nykredit.dk

Ernesto Estrada Department of Mathematics, Department of Physics and Institute of Complex Systems, University of Strathclyde, 26 Richmond Street, Glasgow G11XQ, UK, ernesto.estrada@strath.ac.uk

Nuno Ferreira Department of Quantitative Methods, ISCTE-IUL and UNIDE, Avenida Forças Armadas, 1649-026 Lisbon, Portugal, nuno.ferreira@iscte.pt

Dariusz Grech Institute of Theoretical Physics, University of Wrocław, 50-204 Wrocław, Poland, dgrech@ift.uni.wroc.pl

Naomichi Hatano Institute of Industrial Science, The University of Tokyo, 4-6-1 Komaba, Meguro-ku, Tokyo 153-8505, Japan, hatano@iis.u-tokyo.ac.jp

Shlomo Havlin Minerva Center and Department of Physics, Bar-Ilan University, Ramat-Gan 52900, Israel, havlin@ophir.ph.biu.ac.il

Mogens Høgh Jensen Niels Bohr Institute, Blegdamsvej 17, 2100 Copenhagen, Denmark, mhjensen@nbi.dk

Taisei Kaizoji International Christian University, 3-10-2 Osawa, Mitaka, Tokyo 181-8585, Japan, kaizoji@icu.ac.jp

Genshiro Kitagawa The Institute of Statistical Mathematics;
Research Organization of Information and Systems, 10-3 Midori-cho, Tachikawa, Tokyo 190-8562, Japan, kitagawa@ism.ac.jp

Diana A. Mendes Department of Quantitative Methods, ISCTE-IUL and UNIDE, Avenida Forças Armadas, 1649-026 Lisbon, Portugal, diana.mendes@iscte.pt

Vivaldo M. Mendes Department of Economics, ISCTE-IUL and UNIDE, Lisbon, Portugal, vivaldo.mendes@iscte.pt

Rui Menezes Department of Quantitative Methods, ISCTE-IUL and UNIDE, Avenida Forças Armadas, 1649-026 Lisbon, Portugal, rui.menezes@iscte.pt

Takayuki Mizuno The Institute of Economic Research, Hitotsubashi University, 2-1 Naka, Kunitachi, Tokyo 186-8603, Japan, mizuno@ier.hit-u.ac.jp

Filippo Petroni GRAPES, Université de Liège, B5 Sart-Tilman, 4000 Liège, Belgium;
DIMADEFA, Facoltà di Economia, Università di Roma "La Sapienza", 00161 Rome, Italy, fpetroni@gmail.com

Tobias Preis Center for Polymer Studies, Department of Physics, Boston University, 590 Commonwealth Avenue, Boston, MA 02215, USA;
Institute of Physics, Johannes Gutenberg University Mainz, Staudinger Weg 7, 55128 Mainz, Germany;
Artemis Capital Asset Management GmbH, Gartenstr. 14, 65558 Holzheim, Germany, mail@tobiaspreis.de

Anatoly B. Schmidt Business Development and Research, ICAP Electronic Broking LLC, One Upper Pond Road, Building F, Parsippany, NJ 07054, USA, Alec.Schmidt@us.icap.com

Ingve Simonsen Department of Physics, Norwegian University of Science and Technology (NTNU), 7491 Trondheim, Norway, ingve.simonsen@ntnu.no

Didier Sornette Department of Management, Technology and Economics, ETH Zurich, Kreuzplatz 5, 8032 Zurich, Switzerland;
Swiss Finance Institute, c/o University of Geneva, 40 Blvd Du Pont d'Arve, 1211 Geneva 4, Switzerland, dsornette@ethz.ch

H. Eugene Stanley Center for Polymer Studies and Department of Physics, Boston University, 590 Commonwealth Avenue, Boston, MA 02215, USA, hes@bu.edu

Hideki Takayasu Fundamental Research Group, Sony Computer Science Laboratories, 3-14-13 Higashigotanda, Shinagawa-ku, Tokyo 141-0022, Japan, takayasu@csl.sony.co.jp

Misako Takayasu Department of Computational Intelligence and Systems Science, Interdisciplinary Graduate School of Science and Engineering, Tokyo Institute of Technology, 4259-G3-52 Nagatsuta-cho, Midori-ku, Yokohama 226-8502, Japan, takayasu@dis.titech.ac.jp

Fengzhong Wang Center for Polymer Studies and Department of Physics, Boston University, 590 Commonwealth Avenue, Boston, MA 02215, USA, fzwang@bu.edu

Kota Watanabe Department of Computational Intelligence and Systems Science, Interdisciplinary Graduate School of Science and Engineering, Tokyo Institute of Technology, 4259-G3-52 Nagatsuta-cho, Midori-ku, Yokohama 226-8502, Japan, watanabe@smp.dis.titech.ac.jp

Ryan Woodard Department of Management, Technology and Economics, ETH Zurich, Kreuzplatz 5, 8032 Zurich, Switzerland, rwoodard@ethz.ch

Kazuko Yamasaki Department of Environmental Sciences, Tokyo University of Information Sciences, 4-1 Onaridai, Wakaba-ku, Chiba 265-8501, Japan, yamasaki@edu.tuis.ac.jp

Hiroshi Yoshikawa Faculty of Economics, The University of Tokyo, 7-3-1 Hongo, Bunkyo-ku, Tokyo 113-0033, Japan, yoshikawa@e.u-tokyo.ac.jp

Part 1
Financial Market Properties

Trend Switching Processes in Financial Markets

Tobias Preis and H. Eugene Stanley

Abstract For an intriguing variety of switching processes in nature, the underlying complex system abruptly changes at a specific point from one state to another in a highly discontinuous fashion. Financial market fluctuations are characterized by many abrupt switchings creating increasing trends ("bubble formation") and decreasing trends ("bubble collapse"), on time scales ranging from macroscopic bubbles persisting for hundreds of days to microscopic bubbles persisting only for very short time scales. Our analysis is based on a German DAX Future data base containing 13,991,275 transactions recorded with a time resolution of 10^{-2} s. For a parallel analysis, we use a data base of all S&P500 stocks providing 2,592,531 daily closing prices. We ask whether these ubiquitous switching processes have quantifiable features independent of the time horizon studied. We find striking scale-free behavior of the volatility after each switching occurs. We interpret our findings as being consistent with time-dependent collective behavior of financial market participants. We test the possible universality of our result by performing a parallel analysis of fluctuations in transaction volume and time intervals between trades. We show that these financial market switching processes have features similar to those present in phase transitions. We find that the well-known catastrophic bubbles that occur on large time scales – such as the most recent financial crisis – are no outliers

---

T. Preis
Center for Polymer Studies, Department of Physics, Boston University, 590 Commonwealth Avenue, Boston, MA 02215, USA
and
Institute of Physics, Johannes Gutenberg University Mainz, Staudinger Weg 7, 55128 Mainz, Germany
and
Artemis Capital Asset Management GmbH, Gartenstr. 14, 65558 Holzheim, Germany
e-mail: mail@tobiaspreis.de

H.E. Stanley (✉)
Center for Polymer Studies and Department of Physics, Boston University, 590 Commonwealth Avenue, Boston, MA 02215, USA
e-mail: hes@bu.edu

M. Takayasu et al. (eds.), *Econophysics Approaches to Large-Scale Business Data and Financial Crisis*, DOI 10.1007/978-4-431-53853-0_1, © Springer 2010

but in fact single dramatic representatives caused by the formation of upward and downward trends on time scales varying over nine orders of magnitude from the very large down to the very small.

1 Introduction

In physics and in other natural sciences, it is often a successful strategy to analyze the behavior of a complex system by studying the smallest components of that system. For example, the molecule is composed of atoms, the atom consists of a nucleus and electrons, the nucleus consists of protons and neutrons, and so on. The fascinating point about analyses on steadily decreasing time and length scales is that one often finds that the complex system exhibits properties which cannot only be explained by the properties of its components alone. Instead, a complex behavior can emerge due to the interactions among these components [1]. In financial markets, these components are comprised by the market participants who buy and sell assets in order to realize their trading and investment decisions. The superimposed flow of all individual orders submitted to the exchange trading system initiated by market participants and, of course, its change in time generate a complex system with fascinating properties, similar to physical systems.

One of the key conceptual elements in modern statistical physics is the concept of scale invariance, codified in the scaling hypothesis that functions obey certain functional equations whose solutions are power laws [2–5]. The scaling hypothesis has two categories of predictions, both of which have been remarkably well verified by a wealth of experimental data on diverse systems. The first category is a set of relations, called *scaling laws*, that serve to relate the various critical-point exponents characterizing the singular behavior of functions such as thermodynamic functions. The second category is a sort of *data collapse*, where under appropriate axis normalization, diverse data "collapse" onto a single curve called a scaling function.

Econophysics research has been addressing a key question of interest: quantifying and understanding large stock market fluctuations. Previous work was focussed on the challenge of quantifying the behavior of the probability distributions of large fluctuations of relevant variables such as returns, volumes, and the number of transactions. Sampling the far tails of such distributions require a large amount of data. However, there is a truly gargantuan amount of pre-existing precise financial market data already collected, many orders of magnitude more than for typical complex systems. Accordingly, financial markets are becoming a paradigm of complex systems, and increasing numbers of scientists are analyzing market data [6–11, 13–19]. Empirical analyses have been focused on quantifying and testing the robustness of power-law distributions that characterize large movements in stock market activity. Using estimators that are designed for serially and cross-sectionally independent data, findings thus far support the hypothesis that the power law exponents that characterize fluctuations in stock price, trading volume, and the number of trades [20–27] are seemingly "universal" in the sense that they do not change their values significantly for different markets, different time periods, or different market conditions.

In contrast to these analyses of global financial market distributions, we focus on the temporal sequence of fluctuations in volatility, transaction volume, and inter-trade times before and after a trend switching point. Our analysis can provide insight into switching processes in complex systems in general and financial systems in particular. The study of dramatic crash events is limited by the fortunately rare number of such events. Increasingly, one seeks to understand the current financial crisis by comparisons with the depression of the 1930s. Here we ask if the smaller financial crises – trend switching processes on all time scales – also provide information of relevance for large crises. If this is so, then the large abundance of data on smaller crises should provide quantifiable statistical laws for *bubbles on all scales*.

2 Financial Market Data

To answer whether smaller financial crises also provide information of relevance to large crises, we perform parallel analyses of bubble formation and bursting using two different data bases on two quite different time scales: (1) from $\approx 10^1$ to $\approx 10^6$ ms, and (2) from $\approx 10^8$ to $\approx 10^{10}$ ms.

2.1 *German Market: DAX Future*

For the first analysis, we use a multivariate time series of the German DAX Future contract (FDAX) traded at the European Exchange (Eurex), which is one of the world's largest derivatives exchange markets. A so-called future contract is a contract to buy or sell at a specified price at a specific future date an underlying asset – in this case the German DAX index, which measures the performance of the 30 largest German companies in terms of order book volume and market capitalization.[1] The time series comprises $T_1 = 13{,}991{,}275$ transactions of three disjoint 3-month periods (see Table 1). Each end of the three disjoint periods corresponds to a last trading day of the FDAX contract, which is ruled to be the third Friday of one of the quarterly months March, June, September, and December, apart from the exceptions of national holidays. The data base we analyze contains the transaction prices, the volumes, and the corresponding time stamps [32–35], with a large liquidity and inter-trade times down to 10 ms, which allows us to perform an analysis of microtrends.

The time series analysis of future contracts has the advantage that the prices are created by trading decisions alone. In contrast, stock index data are derived from a weighted sum of a bunch of stock prices. Furthermore, systematic drifts by inflation

[1] More detailed information about German DAX index constituents and calculation principles can be found on http://www.deutsche-boerse.com.

Table 1 Three disjoint 3-month periods of the German DAX Future contract (FDAX) which we analyze. Additionally, the mean volume per transaction $\bar{v}$ and the mean inter-trade time $\bar{\tau}$ is given

Contract	Records	Time period	$\bar{v}$	$\bar{\tau}$ (s)
FDAX JUN 2007	3,205,353	16 March 2007 – 15 June 2007	3.628[a]	2.485[b]
FDAX SEP 2008	4,357,876	20 June 2008 – 19 September 2008	2.558[a]	1.828[b]
FDAX DEC 2008	6,428,046	19 September 2008 – 19 December 2008	2.011[a]	1.253[b]

[a] Measured in units of contract
[b] Including overnight gaps

are eliminated by construction. The theory of futures pricing based on arbitrage states that for an asset that can be stored at no cost and which does not yield any cash flows, the futures price F has to be equal to the spot price S plus the cost of financing the purchase of the underlying between the spot date and the expiry date [36, 37]. This theoretical futures price can be referred to as *fair value*. In the case of the German DAX index, the underlying purchase can be financed till expiry with a loan rate. Using a continuously compounded rate r, the *fair value* equation can be written as

$$F(t) = S(t)e^{rt}, \tag{1}$$

whereas t denotes the remaining time till expiry. The theoretical futures price expression – see (1) –, which simply reflects the *cost of carry*, compensates interest rate related effects of the underlying. At expiry time $t = 0$, the future's price and underlying price are identical.

2.2 US Market: S&P500 Stocks

For the second analysis, which focuses on macrotrends, we use price time series of daily closing prices of all stocks of the S&P500 index. This index consists of 500 large-cap common stocks, which are actively traded in the United States of America.[2] The time series comprises $T_2 = 2{,}592{,}531$ closing prices overall of those US stocks which were constituent of the S&P500 until 16 June 2009. Our oldest closing prices date back to 2 January 1962. The data base of closing prices we analyze contains the daily closing prices and the daily cumulative trading volume. As spot market prices undergo a significant shift by inflation over time periods of more than 40 years, we study the logarithm of stock prices instead of the raw closing prices. Thus, the results between the two different data bases on two quite different time scales become more comparable.

[2] More detailed information about S&P500 constituents and calculation principles can be found on http://www.standardandpoors.com.

3 Renormalization Method

Less studied than the large fluctuations of major national stock indices such as the S&P500 are the various jagged functions of time characterizing complex financial fluctuations down to time scales as short as a few milliseconds. These functions are not amenable to mathematical analysis at first sight, because they are characterized by sudden reversals between up and down microtrends (see Figs. 1 and 2a), which can also be referred to as microscopic *bubbles* on small time scales. On these small time scales, evidence can be found [11, 12] that the three major financial market quantities of interest – price, volume, and inter-trade times – are connected in a complex way, overburdening standard tools of time series analysis such as linear cross-correlation functions. Thus, more sophisticated methods are necessary in order to analyze these complex financial fluctuations creating complex financial market patterns.

We do not know how to characterize the sudden microtrend reversals. For example, the time derivative of the price $p(t)$ is discontinuous. This behavior is completely different from most real world trajectories, such as a thrown ball, for

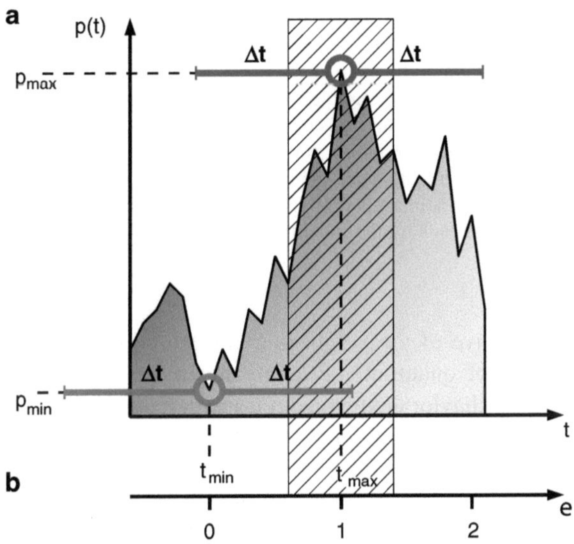

Fig. 1 Visualization of a *microtrend* in the price movement $p(t)$. (**a**) Positive microtrend starting at a local price minimum $p_{\min}$ of order Δt and ending at a local price maximum $p_{\max}$ of order Δt. The hatched region around $p_{\max}$ indicates the interval in which we find scale-free behavior of related quantities. This behavior is consistent with "self-organized" [28] macroscopic interactions among many traders [29], not unlike "tension" in a pedestrian crowd [30, 31]. The reason for tension among financial traders may be found in the risk aversions and profit targets of financial market participants. (**b**) Renormalized time scale ε between successive extrema, where $\varepsilon = 0$ corresponds to the start of a microtrend, and $\varepsilon = 1$ corresponds to the end. The hatched region is surprisingly large, starting at $\varepsilon = 0.6$ and ending at $\varepsilon = 1.4$

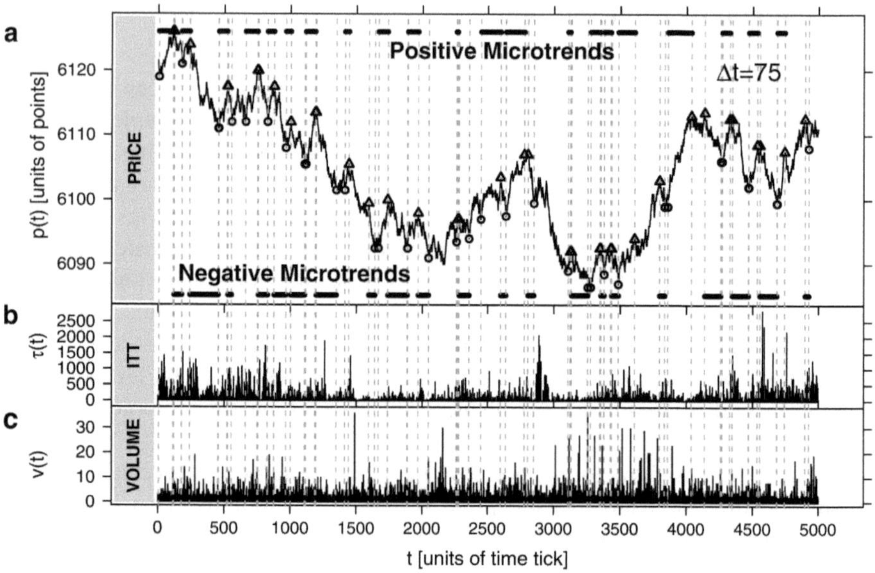

Fig. 2 Visualization of the quantities analyzed. (**a**) A small subset comprising 5,000 trades (0.04%) of the full $T_1 = 13{,}991{,}275$ trade data set analyzed, extracted from the German DAX future time series during part of one day. *Shown as circles and triangles* are the extrema of order Δt, defined to be the extremum in the interval $t - \Delta t \leq t \leq t + \Delta t$. We performed our analysis for $\Delta t = 1, 2, \ldots, 1000$ ticks; in this example, $\Delta t = 75$ ticks. Positive microtrends are indicated by *black bars on the top*, which start at a Δt-minimum and end at the next Δt-maximum. A negative microtrend (*black bars on the bottom*) starts at a Δt-maximum and ends at the consecutive Δt-minimum. (**b**) Time series of the corresponding inter-trade times $\tau(t)$ reflecting the *natural* time between consecutive trades in units of 10 ms, where $t = 1, 2, \ldots, 5000$ is the transactions index. (**c**) The volume $v(t)$ of each trade t in units of contracts

which the time derivative of the height is a smooth continuous function of time. Here, we find a way of quantitatively analyzing these sudden microtrend reversals, which exhibit a behavior analogous to transitions in systems in nature [2, 38], and we interpret these transitions in terms of the cooperative interactions of the traders involved. A wide range of examples of transitions exhibiting scale-free behavior ranges from magnetism in statistical physics to heartbeat intervals (sudden switching from heart contraction to heart expansion) [39], and to macroscopic social phenomena such as traffic flows (switching from a free to a congested traffic) [40].

To focus on switching processes of price movements down to a microscopic time scale, we first propose how a switching process can be quantitatively analyzed. Let $p(t)$ be the transaction price of trade t, which will be treated as a discrete variable $t = 1, \ldots, T$. Each transaction price $p(t)$ is defined to be a *local maximum* $p_{\max}(\Delta t)$ of order Δt, if there is no higher transaction price in the interval $t - \Delta t \leq t \leq t + \Delta t$. Thus, if $p(t) = p_{\max}(t, \Delta t)$, then $p(t)$ is a *local maximum* $p_{\max}(\Delta t)$, where

$$p_{\max}(t, \Delta t) = \max\{p(t)|t - \Delta t \leq t \leq t + \Delta t\}. \tag{2}$$

Analogously, each transaction price $p(t)$ is defined to be a *local minimum* $p_{\min}(\Delta t)$ of order Δt, if there is no lower transaction price in this interval. With

$$p_{\min}(t, \Delta t) = \min\{p(t)|t - \Delta t \leq t \leq t + \Delta t\}, \tag{3}$$

it follows that $p(t)$ is a *local minimum* $p_{\min}(\Delta t)$ if $p(t) = p_{\min}(t, \Delta t)$. In this sense, the two points in the time series in Fig. 1 marked by circles are a local minimum and a local maximum, respectively. Figure 2a shows a short subset of the FDAX time series for the case $\Delta t = 75$ ticks.

For the analysis of financial market quantities in dependence of trend fraction, we introduce a renormalized time scale ε between successive extrema as follows: Let $t_{\min}$ and $t_{\max}$ be the time (measured in units of ticks) at which the corresponding transactions take place of a successive pair of $p_{\min}(\Delta t)$ and $p_{\max}(\Delta t)$ (see Fig. 1). For a positive microtrend, the renormalized time scale is given by

$$\varepsilon(t) \equiv \frac{t - t_{\min}}{t_{\max} - t_{\min}}, \tag{4}$$

with $t_{\min} \leq t \leq t_{\max} + (t_{\max} - t_{\min})$, and for a negative microtrend by

$$\varepsilon(t) \equiv \frac{t - t_{\max}}{t_{\min} - t_{\max}}, \tag{5}$$

with $t_{\max} \leq t \leq t_{\min} + (t_{\min} - t_{\max})$. Thus, $\varepsilon = 0$ corresponds to the beginning of the microtrend and $\varepsilon = 1$ indicates the end of the microtrend. We analyze a range of ε for the interval $0 \leq \varepsilon \leq 2$, so we can analyze trend switching processes both before as well as after the critical value $\varepsilon = 1$ (Fig. 1). The renormalization is essential to assure that microtrends of various lengths can be aggregated and that all switching points have a common position in the renormalized time scale.

3.1 Volatility Analysis

First we analyze the fluctuations $\sigma^2(t)$ of the price time series during the short time interval of increasing microtrends from one price minimum to the next price maximum (see Fig. 3a) and decreasing microtrends from one price maximum to the next price minimum (see Fig. 3b). The quantity studied is given by squared price differences, $\sigma^2(t) = (p(t) - p(t-1))^2$ for $t > 1$, and can be referred to

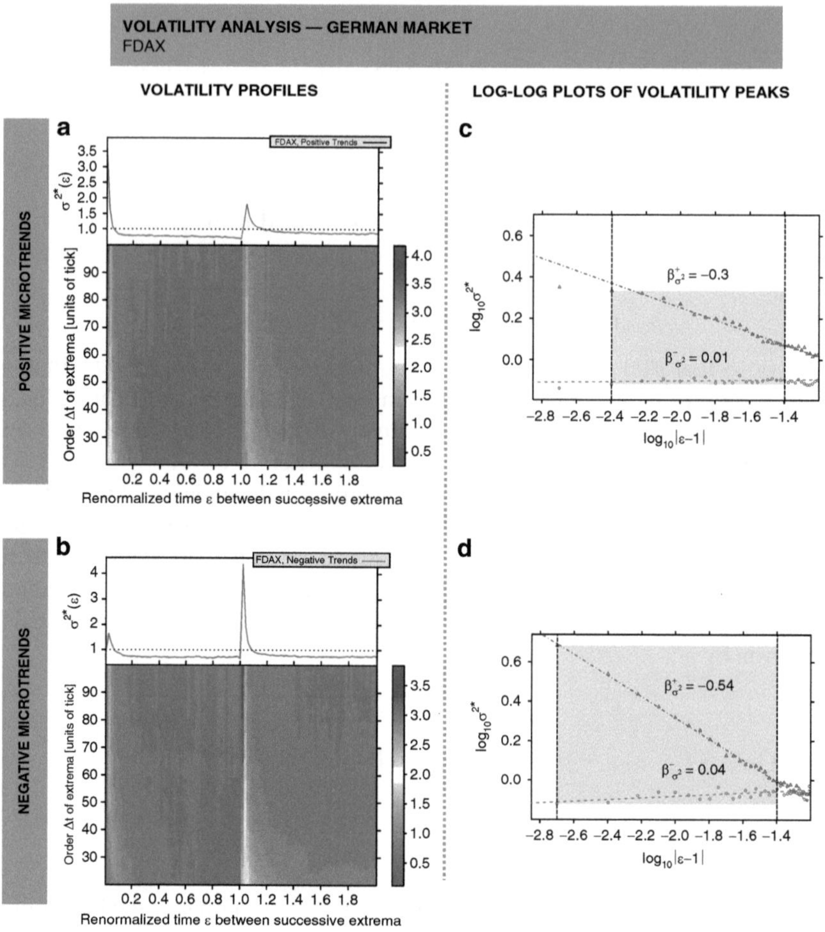

Fig. 3 Renormalization time analysis of volatility σ^2 for microtrends. (**a**) The greyscaled volatility profile, averaged over all positive microtrends in the German DAX future time series and normalized by the average volatility of all positive microtrends studied. The greyscale key gives the normalized mean volatility $\langle\sigma^2_{\text{pos}}\rangle(\varepsilon, \Delta t)/\bar{\sigma}_{\text{pos}}$. The greyscaled profile exhibits the clear link between mean volatility and price evolution. New maximum values of the price time series are reached with a significant sudden jump of the volatility, as indicated by the *vertical white regions* and the sharp maximum in the volatility aggregation $\sigma^{2*}(\varepsilon)$ shown in the *top panel*. Here, $\sigma^{2*}(\varepsilon)$ denotes the average of the volatility profile, averaged only for layers with $50 \leq \Delta t \leq 100$. After reaching new maximum values in the price the volatility decays and returns to the average value (*top panel*) for $\varepsilon > 1$. (**b**) Parallel analysis averaged over all negative microtrends in the time series. New minimum values of the price time series are reached with a pronounced sudden jump of the volatility, as indicated by the *vertical dark gray regions* in the volatility aggregation $\sigma^{2*}(\varepsilon)$ shown in the *top panel*. (**c**) The volatility (50 ticks $\leq \Delta t \leq$ 1000 ticks) before reaching a new maximum price value ($\varepsilon < 1$, *circles*) and after reaching a new maximum price value ($\varepsilon > 1$, *triangles*) aggregated for increasing microtrends. The *straight lines* correspond to power law scaling with exponents $\beta^+_{\sigma^2} = -0.30$ and $\beta^-_{\sigma^2} = 0.01$. The *shaded interval* marks the region in which this power law behavior is valid. (**d**) Log–log plot of $\sigma^{2*}(\varepsilon)$ for negative microtrends. The *straight lines* correspond to power law scaling with exponents $\beta^+_{\sigma^2} = -0.54$ and $\beta^-_{\sigma^2} = 0.04$. The *left border of the shaded region* is given by the first measuring point closest to the switching point

as local volatility. For the analysis of $\sigma^2(t)$ in dependence of the trend fraction, we use the renormalization time scale ε. In Fig. 3, the greyscale key represents the mean volatility $\langle\sigma^2\rangle(\varepsilon, \Delta t)$ in dependence of ε and Δt, normalized by the average volatility $\bar{\sigma}$, where the brackets denote the average over all increasing microtrends (see Fig. 3a) or all decreasing microtrends (see Fig. 3b) in the full time series of $T_1 = 13{,}991{,}275$ records. If one can find $N_{\text{pos}}(\Delta t)$ positive microtrends and $N_{\text{neg}}(\Delta t)$ negative microtrends, each of order Δt in the time series, and of $\sigma_i^2(\varepsilon)$ denotes the local volatility at position ε in the i-th positive or i-th negative microtrend, then the mean volatility is given by

$$\langle\sigma^2_{\text{pos}}\rangle(\varepsilon, \Delta t) = \frac{1}{N_{\text{pos}}(\Delta t)} \sum_{i=1}^{N_{\text{pos}}(\Delta t)} \sigma_i^2(\varepsilon) \tag{6}$$

for positive microtrends and

$$\langle\sigma^2_{\text{neg}}\rangle(\varepsilon, \Delta t) = \frac{1}{N_{\text{neg}}(\Delta t)} \sum_{i=1}^{N_{\text{neg}}(\Delta t)} \sigma_i^2(\varepsilon) \tag{7}$$

for negative microtrends. The mean volatility can be normalized by the average volatility $\bar{\sigma}$, which is determined by

$$\bar{\sigma}_{\text{pos}} = \frac{\varepsilon_{\text{bin}}}{\varepsilon_{\max}\Delta t_{\max}} \sum_{\varepsilon=0}^{\varepsilon_{\max}/\varepsilon_{\text{bin}}} \left(\sum_{\Delta t=0}^{\Delta t_{\max}} \langle\sigma^2_{\text{pos}}\rangle(\varepsilon, \Delta t) \right) \tag{8}$$

and

$$\bar{\sigma}_{\text{neg}} = \frac{\varepsilon_{\text{bin}}}{\varepsilon_{\max}\Delta t_{\max}} \sum_{\varepsilon=0}^{\varepsilon_{\max}/\varepsilon_{\text{bin}}} \left(\sum_{\Delta t=0}^{\Delta t_{\max}} \langle\sigma^2_{\text{neg}}\rangle(\varepsilon, \Delta t) \right), \tag{9}$$

where $\varepsilon_{\max}$ is the maximum value of the renormalization time scale ε studied, which is fixed to $\varepsilon_{\max} = 2$, and ε_{bin} denotes the bin size of the renormalization time scale. The maximum value of the extrema order Δt which we analyze is given by $\Delta t_{\max}$. The bin size is related to $\Delta t_{\max}$ by

$$\varepsilon_{\text{bin}} = \frac{\varepsilon_{\max}}{\Delta t_{\max}} \tag{10}$$

for reasons of convenience. The absence of changes of the greyscaled volatility profiles in Fig. 3 is consistent with a *data collapse* for Δt values larger than a certain cut-off value Δt_{cut}. Thus, we calculate the volatility aggregation $\sigma^{2*}(\varepsilon)$. This volatility aggregation $\sigma^{2*}(\varepsilon)$ is the average of the mean volatility $\langle\sigma^2_{\text{neg}}\rangle(\varepsilon, \Delta t)$, averaged only for layers with $\Delta t_{\text{cut}} \leq \Delta t \leq \Delta t_{\max}$. It is given by

$$\sigma^{2*}_{\text{pos}}(\varepsilon) = \frac{1}{\Delta t_{\max} - \Delta t_{\text{cut}}} \sum_{\Delta t=\Delta t_{\text{cut}}}^{\Delta t_{\max}} \frac{\langle\sigma^2_{\text{pos}}\rangle(\varepsilon, \Delta t)}{\bar{\sigma}_{\text{pos}}} \tag{11}$$

and the equivalent definition

$$\sigma^{2*}_{\mathrm{neg}}(\varepsilon) = \frac{1}{\Delta t_{\max} - \Delta t_{\mathrm{cut}}} \sum_{\Delta t = \Delta t_{\mathrm{cut}}}^{\Delta t_{\max}} \frac{\langle \sigma^2_{\mathrm{neg}} \rangle (\varepsilon, \Delta t)}{\bar{\sigma}_{\mathrm{neg}}} \tag{12}$$

for negative microtrends. Note that, in order to improve the readability, the subscripts "pos" and "neg" are removed if the context assures whether positive or negative microtrends are considered.

The greyscaled volatility profiles (see Fig. 3) provide the mean volatility $\langle \sigma^2 \rangle (\varepsilon, \Delta t)$ averaged over all increasing or all decreasing microtrends in the full time series of $T_1 = 13{,}991{,}275$ records, and are normalized by the average volatilities of microtrends studied in both cases. In order to remove outliers, only such microtrends are collected in which the time intervals between successive trades $\tau(t)$ [41] (Fig. 2b) are not longer than 1 min, which is roughly 60 times longer than the average inter-trade time ($\approx$0.94 s without overnight gaps), and in which the transaction volumes are not larger than 100 contracts (the average transaction volume is 2.55 contracts, see Table 1). This condition ensures that time t, which is measured in units of ticks, runs only over the working hours of the exchange – removing overnight gaps, weekends, and national holidays. Furthermore, the analysis is only based on those microtrends which provide a reasonable activity. The greyscale profiles exhibit a very clear link between volatility and price evolution. A new local price maximum is reached with a significant sudden jump of the volatility (top panel of Fig. 3a). After reaching new local maximum values in the price, the volatility decays and returns to the average value for $\varepsilon > 1$. The reaching of a maximum causes obviously tension among the market participants. A local price maximum can stoke the expectations that higher prices are possible and thus stimulate purchases. However, this development can also raise fears of traders to find an optimal price for selling their assets. Additionally, it is possible that market participants holding a short position, which means that they benefit from falling asset prices, have to cut their losses by reaching new maximum values. For negative microtrends, the reaching of local minimum values in the price coincides with a more pronounced sudden jump of the volatility (see Fig. 3b). A negative asset price evolution seems to create a situation in which market participants act in a more dramatic way after the end of a trend in comparison to the end of positive microtrends. One can conjecture that they are driven by tension at least or even by "panic" if they try to cut their losses. But of course, also the opposite situation should become relevant: A market participant who has no inventory is looking for entry opportunities. As asset prices are rising after reaching a local price minimum ($\varepsilon = 1$), a financial market actor, who has the intention to enter into the market, has to deal with the tension to find the "right" time – the optimal entry level is already missed at this time: the local price minimum.

This qualitative effect is intuitively understandable and should be obvious for market actors. In contrast, the shape of the volatility peak around extrema is surprising. The peak is characterized by asymmetric tails, which we analyze next. For this analysis, we use the volatility aggregation $\sigma^{2*}(\varepsilon)$, which is the mean volatility

$\langle\sigma^2\rangle(\varepsilon, \Delta t)$ averaged for layers from $\Delta t_{\text{cut}} = 50$ ticks to $\Delta t_{\text{max}} = 1000$ ticks. Figure 3c shows the aggregated average volatility $\sigma^{2*}(\varepsilon)$ for positive microtrends on a log–log plot. Surprisingly, the evolution of the volatility before and after reaching a maximum shows up as straight lines and thus are consistent with a power law scaling behavior

$$\sigma^{2*}(|\varepsilon - 1|) \sim |\varepsilon - 1|^{\beta_{\sigma^2}} \tag{13}$$

within the range indicated by the vertical dashed lines. Over one order of magnitude, we find distinct exponents, $\beta^-_{\sigma^2} = 0.01$ before a price maximum and $\beta^+_{\sigma^2} = -0.30$ after. Figure 3d shows the aggregated average volatility $\sigma^{2*}(\varepsilon)$ for negative microtrends on a log–log plot. Over more than one order of magnitude, we find for negative microtrends a qualitatively consistent behavior to positive microtrends with distinct exponents, $\beta^-_{\sigma^2} = 0.04$ before a price minimum and $\beta^+_{\sigma^2} = -0.54$ after.

Next we test the possible universality of our results by performing a parallel analysis for trends on long time scales, using the daily closing price data base of S&P500 stocks. In this sense, universality means that our renormalized market quantities do not change their values significantly for different markets, different time periods, or different market conditions.

Note that for our parallel analysis on macroscopic time scales, the order of a extremum Δt is measured in units of days, and that $\langle\sigma^2\rangle(\varepsilon, \Delta t)$ is averaged additionally over all closing price time series of all S&P500 components. In order to avoid biased contributions for the rescaled averaging caused by inflation based drifts over more than 47 years as described in Sect. 2.2, the analyzed price time series $p(t)$ contains the logarithm of the daily closing prices. Figure 4a shows the mean volatility $\langle\sigma^2\rangle(\varepsilon, \Delta t)$ for positive microtrends averaged for layers from $\Delta t_{\text{cut}} = 10$ days to $\Delta t_{\text{max}} = 100$ days. Figure 4b shows the mean volatility $\langle\sigma^2\rangle(\varepsilon, \Delta t)$ for negative microtrends averaged for the same layers' range. As already uncovered for microtrends, the sudden volatility rise is more dramatic for negative macrotrends than for positive macrotrends. The aggregated average volatilities $\sigma^{2*}(\varepsilon)$ for positive and negative macrotrends on a log–log plot show surprisingly again distinct tail exponents around the switching point $\varepsilon = 1$. For positive macrotrends, we obtain $\beta^-_{\sigma^2} = -0.05$ before a price maximum and $\beta^+_{\sigma^2} = -0.40$ after. For negative microtrends, we obtain $\beta^-_{\sigma^2} = -0.09$ before a price minimum and $\beta^+_{\sigma^2} = -0.50$ after, which is both similar to the values obtained for our study of positive and negative microtrends.

3.2 *Volume Analysis*

To test the possible universality of these results obtained for volatility, we perform a parallel analysis of the corresponding volume fluctuations $v(t)$, the numbers of contracts traded in each individual transaction (see Fig. 2c) in case of microtrends for the German market and the cumulative number of traded stocks per day in case of macrotrends for the US market. In Fig. 5a, the greyscaled volume profile provides the mean volume averaged over all increasing microtrends in the time

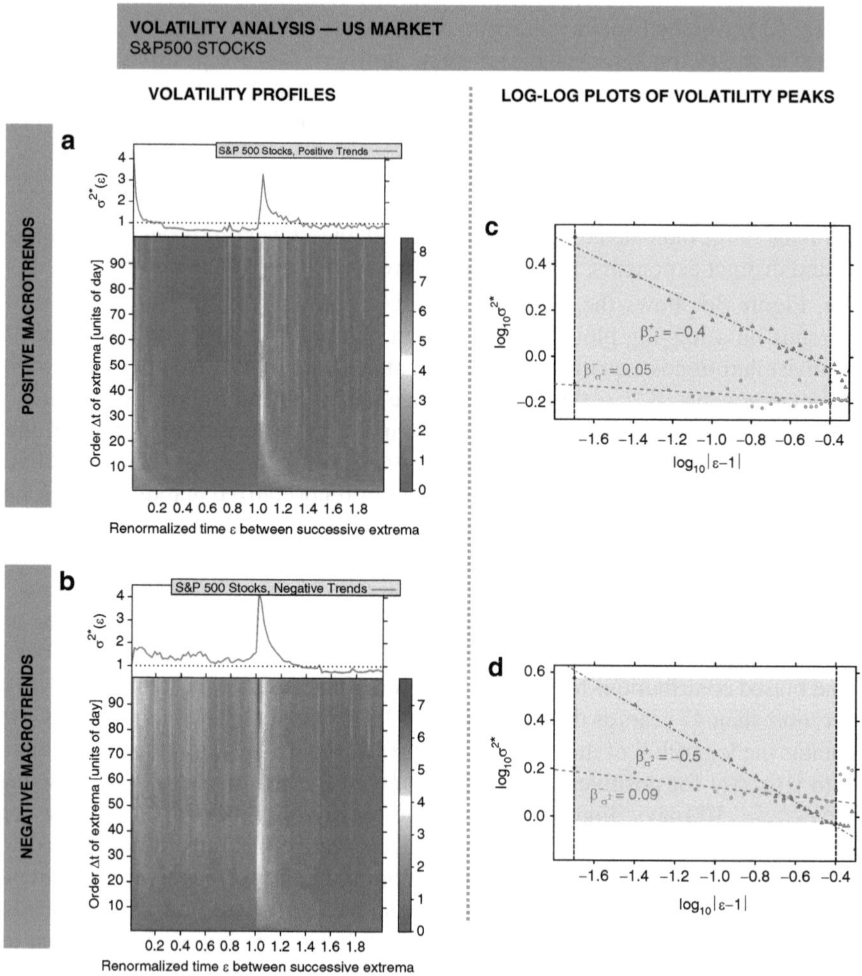

Fig. 4 Renormalization time analysis of volatility σ^2 for macrotrends. (**a**) The greyscaled volatility profile, averaged over all positive macrotrends in the daily closing price time series of all S&P500 stocks and normalized by the average volatility of all positive macrotrends studied. The stylized fact that new maximum values of the price time series are reached with a significant sudden jump of the volatility can also be found for macrotrends. Note that Δt is measured in units of day for macrotrends. (**b**) Parallel analysis performed for all negative macrotrends in the daily closing price time series of all S&P500 stocks. As already uncovered for microtrends (see Fig. 3), the sudden jump of the volatility at $\varepsilon = 1$ is more pronounced for negative trends than for positive microtrends. (**c**) The volatility ($10\,\text{days} \leq \Delta t \leq 100\,\text{days}$) before reaching a new maximum price value ($\varepsilon < 1$, *circles*) and after reaching a new maximum price value ($\varepsilon > 1$, *triangles*) aggregated for increasing macrotrends. The *straight lines* correspond to power law scaling with exponents $\beta^{+}_{\sigma^2} = -0.40$ and $\beta^{-}_{\sigma^2} = -0.05$. (**d**) Log–log plot of $\sigma^{2*}(\varepsilon)$ for negative macrotrends. The *straight lines* correspond to power law scaling with exponents $\beta^{+}_{\sigma^2} = -0.50$ and $\beta^{-}_{\sigma^2} = -0.09$

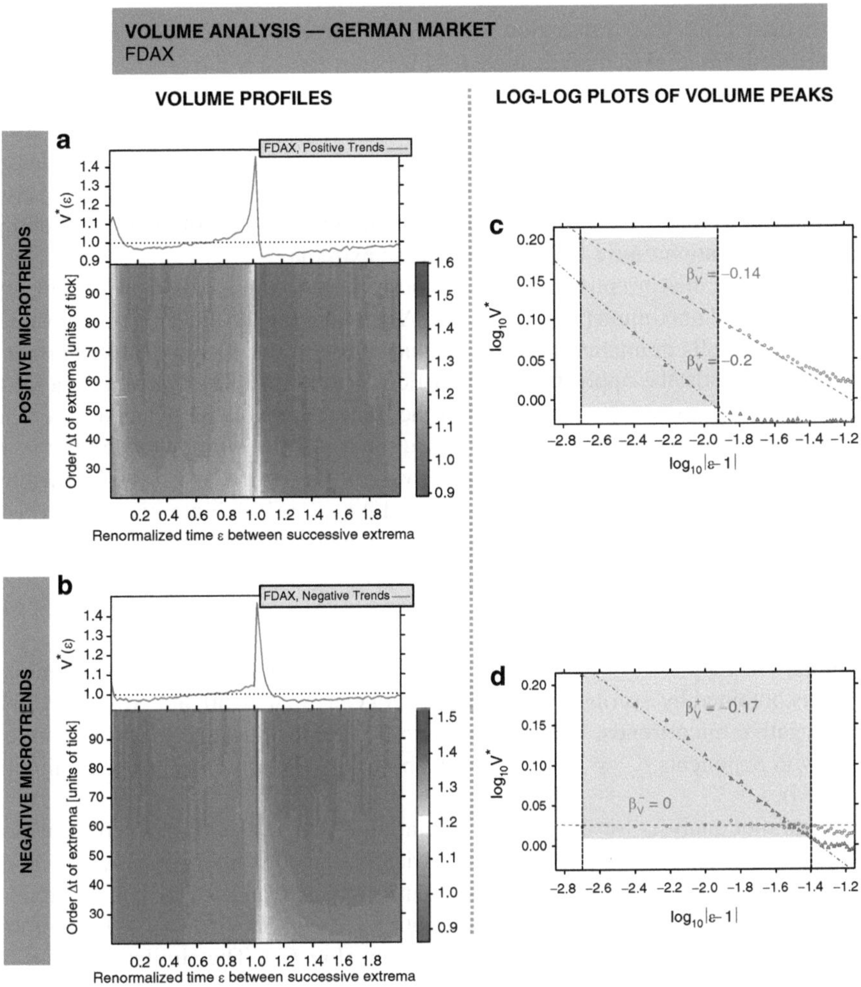

Fig. 5 Renormalization time analysis of volume for microtrends. (**a**) The greyscaled volume profile, averaged over all positive microtrends in the FDAX time series and normalized by the average volume of all positive microtrends studied. New maximum values of the price time series coincide with peaks in the volume. (**b**) Parallel analysis performed for all negative microtrends in FDAX time series. (**c**) The volume (50 ticks $\leq \Delta t \leq$ 1000 ticks) before reaching a new maximum price value ($\varepsilon < 1$, *circles*) and after reaching a new maximum price value ($\varepsilon > 1$, *triangles*) aggregated for increasing microtrends. The *straight lines* correspond to power law scaling with exponents $\beta_v^+ = -0.20$ and $\beta_v^- = -0.14$. (**d**) Log–log plot of $v^*(\varepsilon)$ for negative microtrends. The *straight lines* correspond to power law scaling with exponents $\beta_v^+ = -0.17$ and $\beta_v^- = 0$

series of the German market and is also normalized by the average volume of all microtrends studied. Analogously, Fig. 5b shows the mean volume averaged over all decreasing microtrends in the time series. In order to remove outliers in this analysis, we consider only those microtrends which include inter-trade times not longer than 1 min and transaction volumes not larger than 100 contracts. As expected, new price extrema are linked with peaks in the volume time series but, surprisingly, we find that the usual cross-correlation function between price changes and volumes vanishes. Thus, one can conjecture that the tendency towards increased volumes occurring at the end of positive microtrends is counteracted by the tendency towards increased volumes occurring at the end of negative microtrends. The crucial issue is to distinguish between positive and negative microtrends, realized by the renormalization time ε between successive extrema.

For positive microtrends, a significant increase of volumes can be found already before the local maximum price is reached. After reaching the local maximum value, the volatility falls dramatically back to values close to the average value. For negative microtrends, the opposite characteristic is observable. The reaching of a local price minimum causes a sudden jump of the transaction volume, whereas after the local price minimum the volume decays and returns to the average value for $\varepsilon > 1$. In the top panel of Figs. 5a,b, we show the volume aggregations $v^*(\varepsilon)$, obtained by averaging Δt "slices" between $\Delta t_{\text{cut}} = 50$ and $\Delta t_{\max} = 100$. Figure 5c shows $v^*(\varepsilon)$ versus $|\varepsilon - 1|$ as a log–log histogram supporting a power law behavior of the form

$$v^*(|\varepsilon - 1|) \sim |\varepsilon - 1|^{\beta_v} \tag{14}$$

with exponents $\beta_v^- = -0.14$ before, and $\beta_v^+ = -0.20$ after a price maximum – $v^*(\varepsilon)$ is obtained by averaging Δt "slices" between $\Delta t_{\text{cut}} = 50$ and $\Delta t_{\max} = 1000$. For negative microtrends, straight lines can be fitted to the log–log histogram as well with exponents $\beta_v^- = -0.17$ before, and $\beta_v^+ = 0$ after a local price minimum (Fig. 5d).

A parallel analysis for the US market on large time scales (Figs. 6a,b) provides evidence that the volume peaks are symmetrically shaped around the switching point $\varepsilon = 1$ and that the characteristics are similar for positive and negative macrotrends. The power law exponents for positive microtrends are given by $\beta_v^- = -0.04$ before, and $\beta_v^+ = -0.08$ after a local price maximum (Fig. 6c). The similar behavior of negative microtrends is supported by exponents $\beta_v^- = -0.05$ before, and $\beta_v^+ = -0.15$ after a local price minimum as shown in Fig. 6d.

3.3 *Inter-trade Time Analysis*

In order to verify a possible universality, we analyze additionally the behavior of the inter-trade times $\tau(t)$ of the German market during the short time interval from one price extremum to the next (see Fig. 2b). The linear cross-correlation function between price changes and inter-trade times as standard tool of time series analysis exhibits no significant correlation values as well. Thus, one can again conjecture that the tendency towards decreased inter-trade times at the end of positive microtrends is counteracted by the tendency towards decreased inter-trade times for the end of negative microtrends. It is of crucial importance to distinguish between

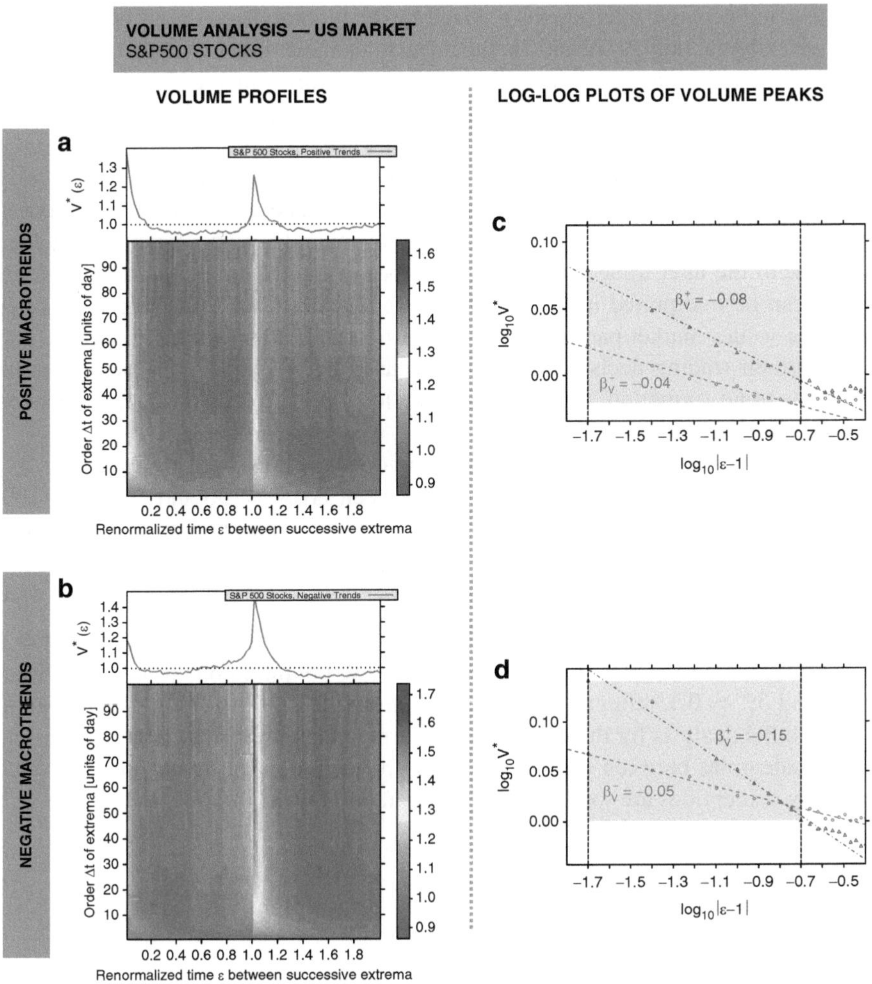

Fig. 6 Renormalization time analysis of volume for macrotrends. (**a**) The greyscaled volume profile, averaged over all positive macrotrends in the daily closing price time series of all S&P500 stocks and normalized by the average volatility of all positive macrotrends studied. Consistent with our results for microtrends maximum values of the price time series are reached with a peak of the volume. Note that Δt is measured in units of day for macrotrends. (**b**) Parallel analysis performed for all negative macrotrends in the daily closing price time series of all S&P500 stocks. Minimum values of the price time series coincide with peaks of volume as for positive macrotrends. (**c**) The volume (10 days $\leq \Delta t \leq$ 100 days) before reaching a new maximum price value ($\varepsilon < 1$, *circles*) and after reaching a new maximum price value ($\varepsilon > 1$, *triangles*) aggregated for increasing macrotrends. The straight lines correspond to power law scaling with exponents $\beta_v^+ = -0.08$ and $\beta_v^- = -0.04$. (**d**) Log–log plot of $v^*(\varepsilon)$ for negative macrotrends. The *straight lines* correspond to power law scaling with exponents $\beta_v^+ = -0.15$ and $\beta_v^- = -0.05$

positive and negative microtrends realized by the renormalized time ε between successive extrema. In Figs. 7a and 7b, the mean inter-trade time $\langle\tau\rangle(\varepsilon, \Delta t)/\bar{\tau}$ is shown for positive and negative microtrends, respectively, mirroring the clear link between inter-trade times and price extrema. Far away from the critical point $\varepsilon = 1$ the mean inter-trade time starts to decrease. After the formation of a new local price maximum the mean inter-trade times increase and return to the average value in a very symmetrical way. Negative microtrends obey the same behavior with one exception. The reaching of a local price minimum ($\varepsilon = 1$) coincides with a temporary sudden increase of the inter-trade times. For both types of trends, the dip of the inter-trade times can be interpreted in terms of "panic". Already before reaching local price extreme values market participants try to participate in the forming trend or try to correct their trading decision which was caused by the hope to participate in an opposite trend formation. After reaching the local price extreme value, the tension persists but becomes steadily smaller.

In the top panels of Figs. 7a,b, the aggregation of the inter-trade time profile $\tau^*(\varepsilon)$ is shown calculated for all values of Δt between $\Delta t_{\mathrm{cut}} = 50$ and $\Delta t_{\max} = 100$. Figure 7c shows $\tau^*(\varepsilon)$ versus $|\varepsilon - 1|$ as a log–log histogram supporting a power law behavior of the form

$$\tau^*(|\varepsilon - 1|) \sim |\varepsilon - 1|^{\beta_\tau} \tag{15}$$

for positive microtrends with exponents $\beta_\tau^- = 0.10$ before, and $\beta_\tau^+ = 0.12$ after a local price maximum. For negative microtrends, we obtain exponents $\beta_\tau^- = 0.09$ before, and $\beta_\tau^+ = 0.15$ after a local price minimum (see Fig. 7d). A log–log histogram of a parallel analysis for the US market on large time scales is not obtainable as the inter-trade times between successive closing prices are given by the constant value of 1 day (exceptions are weekends and general holidays).

3.4 *Random Shuffling*

To confirm that our results are a consequence of the exact time series sequence and thus sensitive to the time ordering of the original time series of volumes and inter-trade times, we randomly shuffle γT pairs of data points of both the volume time series and inter-trade time series in order to weaken their connection with the price evolution. We find that the clear link between volumes fluctuations and price evolution (see Fig. 8a) and between inter-trade times and price evolution (see Fig. 8b) disappears with increasing γ and entirely vanishes for $\gamma \geq 1$ for microtrends. The dip of the inter-trade times at $\varepsilon = 1$ becomes less pronounced with increasing γ and, correspondingly, the peak of the volume maximum decreases. For the S&P500 data set (Fig. 8c), the volume peak disappears with increasing γ obeying the same characteristics. These shuffling induced processes can also be characterized by power law relationships which support our result that a fluctuating price time series passes through a sequence of distinct transitions with scale-free properties. The disappearance phenomenon follows a power law behavior. The maximum value of $v^*(\varepsilon)_\gamma$ at $\varepsilon = 1$ scales with exponent $\beta_v^s = -0.12$ for microtrends (Fig. 8d).

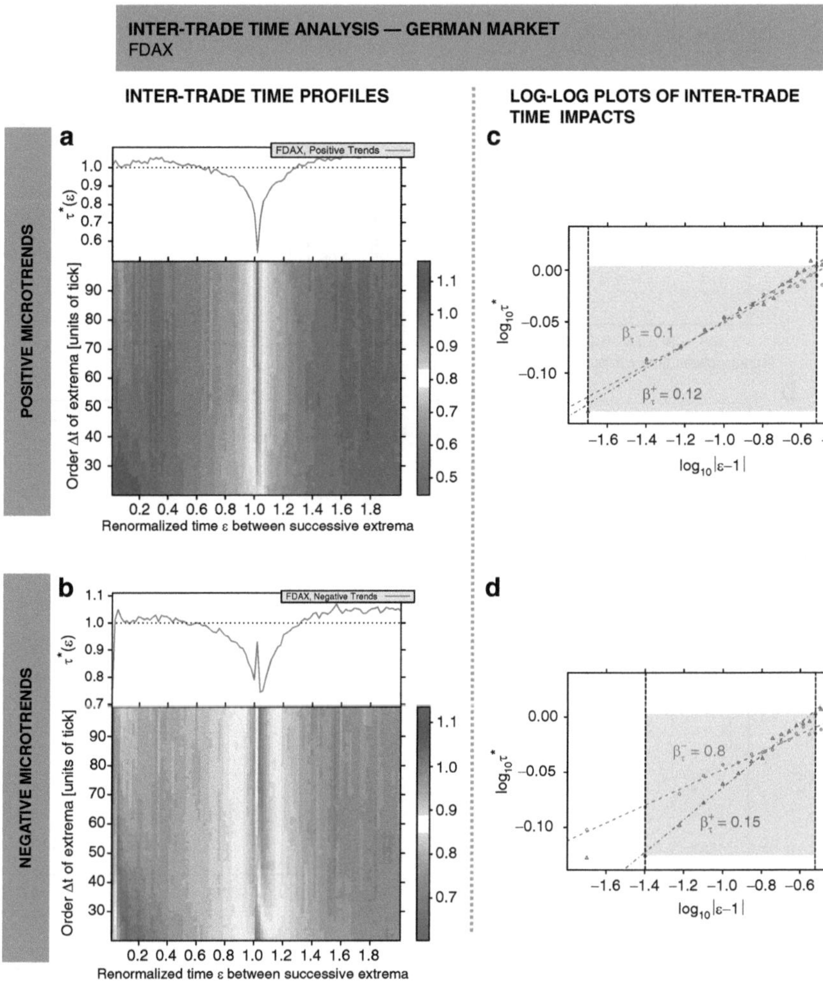

Fig. 7 Renormalization time analysis of inter-trade times for microtrends. (**a**) The greyscaled inter-trade time profile – averaged over all increasing microtrends in the German DAX Future time series and normalized by the average inter-trade times of all positive microtrends studied – is performed analogously to our study of volatility and volume. New maximum values of the price time series are reached with a significant decay of the inter-trade times. (**b**) Parallel analysis performed for all negative microtrends in the FDAX price time series. Minimum values of the price time series coincide with a dip of inter-trade times. In contrast to increasing trends, we observe for exactly $\varepsilon = 1$ an interim increase of the inter-trade times. (**c**) Inter-trade times (50 ticks $\leq \Delta t \leq$ 100 ticks) before reaching a new maximum price value ($\varepsilon < 1$, *circles*) and after reaching a new maximum price value ($\varepsilon > 1$, *triangles*) aggregated for increasing microtrends. The *straight lines* correspond to power law scaling with exponents $\beta_\tau^+ = 0.12$ and $\beta_\tau^- = 0.10$. (**d**) Log–log plot of $\tau^*(\varepsilon)$ for negative microtrends. The *straight lines* correspond to power law scaling with exponents $\beta_\tau^+ = 0.15$ and $\beta_\tau^- = 0.08$

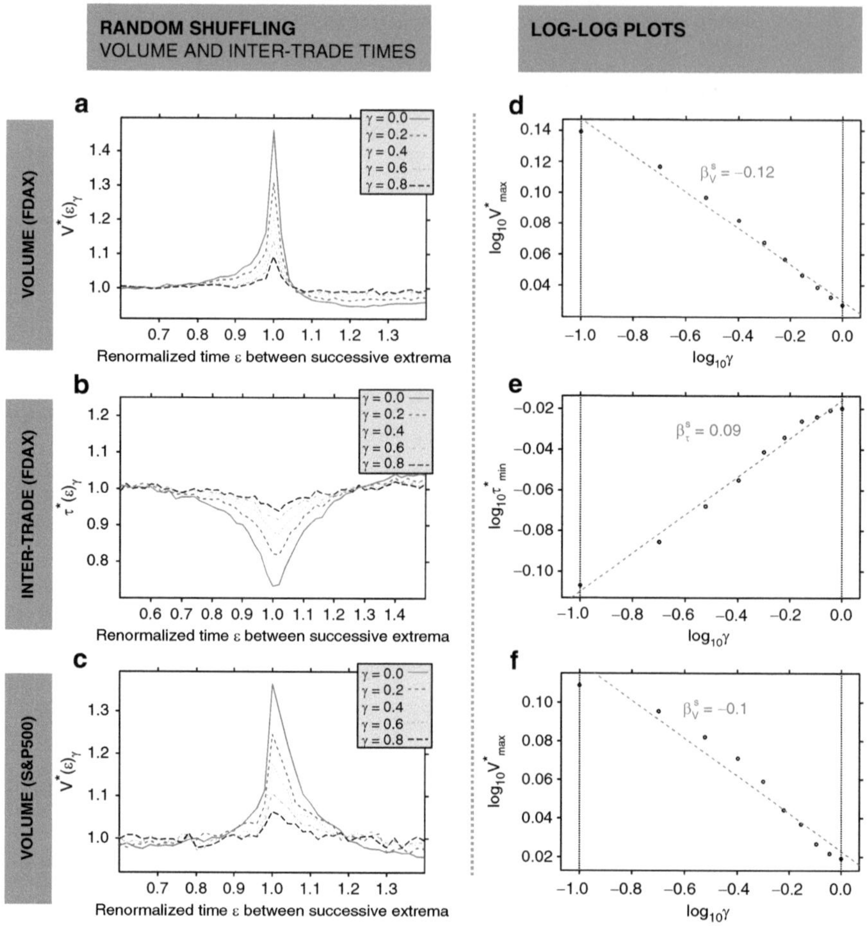

Fig. 8 Stability test of power law dependence. (**a**) If one shuffles randomly γT pairs of volume entries in the multivariate time series, the significant link between volume and price evolution starts to disappear as γ increases. (**b**) If γT pairs of inter-trade time entries are randomly shuffled the inter-trade time dip starts to disappear. (**c**) We find an identical behavior for the volume peak on long time scales using daily closing prices of S&P500 stocks. (**d**) The disappearance phenomenon also follows a power law behavior. The maximum value of $v^*(\varepsilon)_\gamma$ at $\varepsilon = 1$ scales with exponent $\beta_v^s = -0.115 \pm 0.005$. (**e**) The minimum value of $\tau^*(\varepsilon)_\gamma$ at $\varepsilon = 1$ scales with exponent $\beta_v^s = 0.094 \pm 0.004$. (**f**) In the case of the maximum of $v^*(\varepsilon)_\gamma$ at $\varepsilon = 1$ for the S&P500 stocks, the plot provides a power law with exponent $\beta_v^s = -0.095 \pm 0.008$

The minimum value of $\tau^*(\varepsilon)_\gamma$ at $\varepsilon = 1$ scales with exponent $\beta_v^s = 0.09$ as shown in Fig. 8e. In the case of the maximum of $v^*(\varepsilon)_\gamma$ at $\varepsilon = 1$ on large time scales, the log–log plot provides a straight line with a power law exponent $\beta_v^s = -0.10$ for the S&P500 stocks which is consistent with the underlying data set. In fact, deviations can be observed for macrotrends which are caused by the limited number of closing prices in the S&P500 data base ($T_2 \ll T_1$).

3.5 Universality of Power Law Exponents

Thus far, we distinguished between positive and negative trends on small and large time scales. In order to emphasize the possible universality of our results we present in this section a direct comparison of microtrends and macrotrends for our three financial market quantities of interest – volatility, volume, and inter-trade times. Figure 9 shows the renormalization time analysis of volatility σ^2, trade volumes v, and inter-trade times τ for all increasing and decreasing microtrends in the German DAX Future time series. The greyscaled volatility profile exhibits the clear link between mean volatility and price evolution. New extreme values of the price time series are reached with a significant sudden jump of the volatility, as indicated by the vertical dark gray regions and the sharp maximum in the volatility aggregation. After reaching new extreme values in the price, the volatility decays and returns to the average value for $\varepsilon > 1$ as observed in Sect. 3.1 for positive and negative microtrends, respectively. The greyscaled volume profile exhibits that the volume is clearly connected to the price evolution: new extreme values of the price coincide

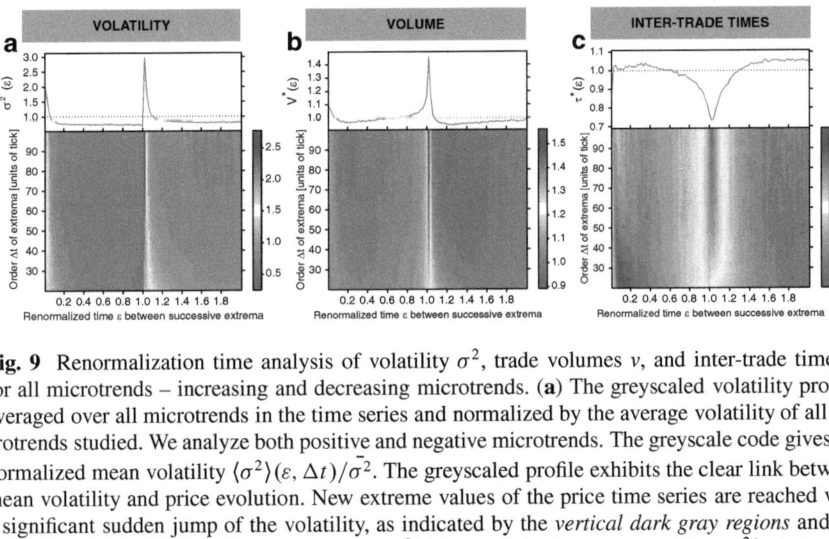

Fig. 9 Renormalization time analysis of volatility σ^2, trade volumes v, and inter-trade times τ for all microtrends – increasing and decreasing microtrends. (**a**) The greyscaled volatility profile, averaged over all microtrends in the time series and normalized by the average volatility of all microtrends studied. We analyze both positive and negative microtrends. The greyscale code gives the normalized mean volatility $\langle\sigma^2\rangle(\varepsilon, \Delta t)/\bar{\sigma^2}$. The greyscaled profile exhibits the clear link between mean volatility and price evolution. New extreme values of the price time series are reached with a significant sudden jump of the volatility, as indicated by the *vertical dark gray regions* and the sharp maximum in the volatility aggregation $\sigma^{2*}(\varepsilon)$ shown in the *top panel*. Here, $\sigma^{2*}(\varepsilon)$ denotes the average of the volatility profile, averaged only for layers with $50 \leq \Delta t \leq 100$. After reaching new extreme values in the price the volatility decays and returns to the average value (*top panel*) for $\varepsilon > 1$. (**b**) The greyscaled volume profile, averaged over all microtrends in the time series and normalized by the average volume of all microtrends studied. The greyscale code gives the normalized mean volume $\langle v\rangle(\varepsilon, \Delta t)/\bar{v}$. The volume is connected to the price evolution: new extreme values of the price coincide with peaks in the volume time series, as indicated by the *vertical dark gray regions* close to $\varepsilon = 1$. The *top panel* shows the volume aggregation $v^*(\varepsilon)$, where $v^*(\varepsilon)$ is the average over layers with $50 \leq \Delta t \leq 100$. The sharp maximum in $v^*(\varepsilon)$ is shown in the *top panel*. (**c**) The greyscaled inter-trade time profile – averaged over all microtrends in the time series and normalized by the average inter-trade times of all microtrends studied – is performed analogously to our study of volatility and volume. New extreme values of the price time series are reached with a significant decay of the inter-trade times

with peaks in the volume time series, as indicated by the vertical dark gray regions close to $\varepsilon = 1$. The greyscaled inter-trade time profile shows that new extreme values of the price time series are reached with a significant decay of the inter-trade times. The log–log plots of all these quantities can be found in Fig. 10. Additionally, the time scales which we study are visualized for both the German market and the US market. For the analysis of microtrends, we use the German DAX future data base which enables us to analyze microtrends starting at roughly 10^6 ms down to

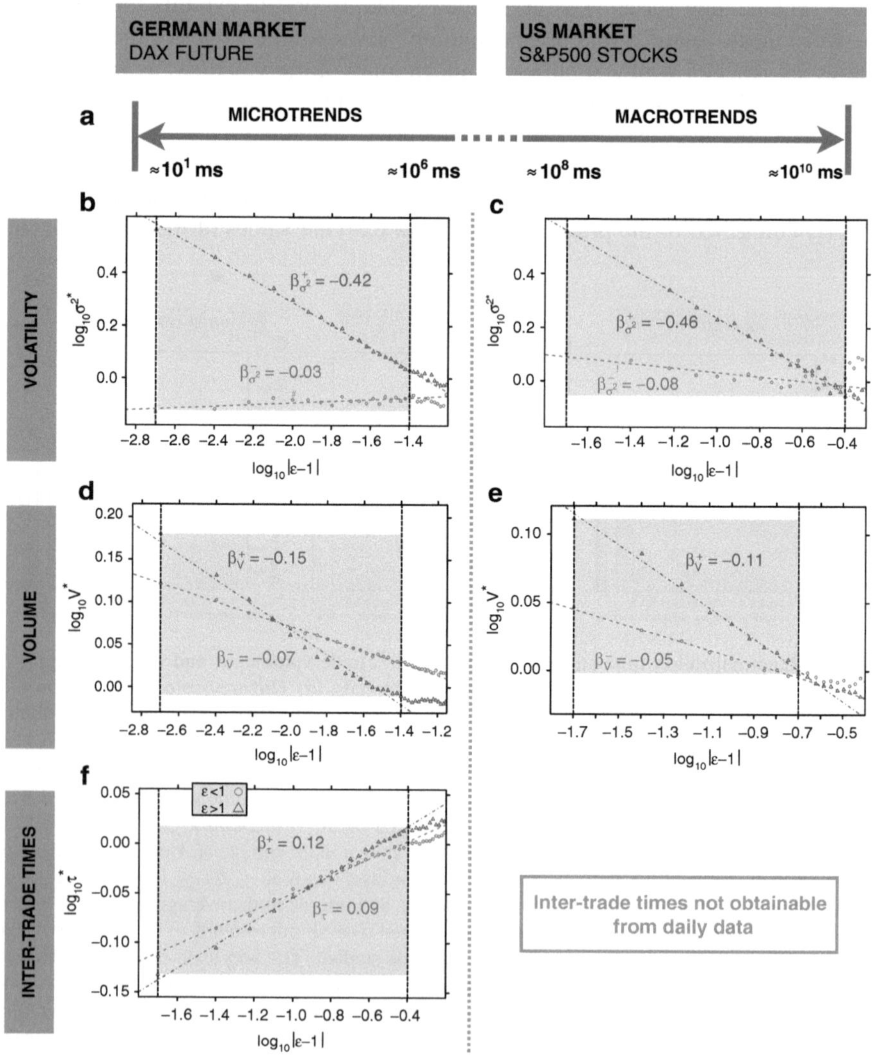

Fig. 10 Overview of time scales studied and log–log plots of quantities with scale-free properties. (**a**) Visualization of time scales studied for both the German market and the US market. For the analysis of microtrends, we use the German DAX future data base which enables us to analyze

the smallest possible time scale of individual transactions measured in multiples of 10 ms. The log–log plots of quantities with scale-free behavior on short time scales are shown in the left column. For the analysis of macrotrends, we use the data base of daily closing prices of all S&P500 stocks which enables us to perform an equivalent analysis of macrotrends on long time scales which are shown the right column. Thus, our analysis of switching processes ranges over nine orders of magnitude from 10 to 10^{10} ms. Surprisingly, the region around an extreme value in which the power law scaling can be found is large, especially for inter-trade waiting times on small time scales (see Fig. 10f) and volumes on long time scales (see Fig. 10c). This range around a local extreme price value is marked as hatched region in Fig. 1. Far away from the switching point a tension among market participants is established and propagates steadily until the critical point is reached – the switching point changing from an upward to a downward or from a downward to an upward trend.

4 Summary and Conclusions

The straight lines in Fig. 10 offer insight into financial market fluctuations: (1) a clear connection between volatility, volumes, inter-trade times, and price fluctuations on the path from one extremum to the next extremum, and (2) the underlying law, which describes the tails of volatility, volumes, and inter-trade times around extrema varying over nine orders of magnitude starting from the smallest possible time

Fig. 10 (continued) microtrends starting at roughly 10^6 ms down to the smallest possible time scale of individual transactions measured in multiples of 10 ms. The log–log plots of quantities with scale-free behavior on short time scales are shown in the left column. For the analysis of macrotrends, we use the data base of daily closing prices of all S&P500 stocks which enables us to perform equivalent analysis of macrotrends on long time scales which are shown the *right column*. Thus, our analysis of switching processes ranges over nine orders of magnitude from 10 to 10^{10} ms. (**b**) The volatility (50 ticks $\leq \Delta t \leq$ 1000 ticks) before reaching a new extreme price value ($\varepsilon < 1$, *circles*) and after reaching a new extreme price value ($\varepsilon > 1$, *triangles*) aggregated for microtrends. The *straight lines* correspond to power law scaling with exponents $\beta^+_{\sigma^2} = -0.42 \pm 0.01$ and $\beta^-_{\sigma^2} = 0.03 \pm 0.01$. The *shaded interval* marks the region in which this power law behavior is valid. The *left border of the shaded region* is given by the first measuring point closest to the switching point. (**c**) The volatility aggregation of macrotrends determined for the US market on long time scales (10 days $\leq \Delta t \leq$ 100 days). The *straight lines* correspond to power law scaling with exponents $\beta^+_{\sigma^2} = -0.46 \pm 0.01$ and $\beta^-_{\sigma^2} = -0.08 \pm 0.02$ which are consistent with the exponents determined for the German market on short time scales. (**d**) Log–log plot of the volume aggregation on short time scales (50 ticks $\leq \Delta t \leq$ 1000 ticks) exhibits a power law behavior with exponents $\beta^+_v = -0.146 \pm 0.005$ and $\beta^-_v = -0.072 \pm 0.001$. (**e**) Log–log plot of the volume aggregation on long time scales (10 days $\leq \Delta t \leq$ 100 days) exhibits a power law behavior with exponents $\beta^+_v = -0.115 \pm 0.003$ and $\beta^-_v = -0.050 \pm 0.002$ which are consistent with our results for short time scales. (**f**) Log–log plot of the inter-trade time aggregation on short time scales (50 ticks $\leq \Delta t \leq$ 100 ticks) exhibits a power law behavior with exponents $\beta^+_\tau = 0.120 \pm 0.002$ and $\beta^-_\tau = 0.087 \pm 0.002$. An equivalent analysis on long time scales is not possible as daily closing prices are recorded with equidistant time steps scale, is a power law with a unique exponents which quantitatively characterize the region around the trend switching point. As a direct consequence of the existence of power law tails, the behavior does not depend on the scale. Thus, we find identical behavior for other sub-intervals of $50 \le \Delta t \le 1000$. With a decreasing value of Δt, the number of local minima and maxima increases (see Fig. 1), around which we find scale-free behavior, for exactly the same ε interval $0.6 \le \varepsilon \le 1.4$. The peaks in $\sigma^2(\varepsilon)$ and $v(\varepsilon)$ around $\varepsilon = 1$ and the dip of $\tau(\varepsilon)$ around $\varepsilon = 1$ offer a challenge for multi-agent based financial market models [42–48] to reproduce these empirical facts. The characterization of volatility, volume, and inter-trade times by power law relationships in the time domain supports our hypothesis that a fluctuating price time series passes through a sequence of "phase transitions" [49, 50].

Before concluding, we may ask "what kind of phase transition" could the end of a microtrend or macrotrend correspond to, or is the end of a trend an altogether different kind of phase transition that resembles all phase transitions by displaying a regime of scale free behavior characterized by a critical exponent. It may be premature to speculate on possible analogies, so we will limit ourselves here to describe what seems to be a leading candidate. Consider a simple Ising magnet characterized by one-dimensional spins that can point North or South. Each spin interacts with some (or even with all) of its neighbors with positive interaction strength J, such that when J is positive neighboring spins lower their energy by being parallel. The entire system is bathed in a magnetic field that interacts with all the spins equally with a strength parametrized by H, such that when H is positive the field points North and when H is negative the field points South. Thus, when H is positive, the system lowers its energy by each spin pointing North. Thus, there are two competing control parameters J and H. If, e.g., the system is prepared in a state with the majority of spins pointing North yet the field H points South, the competition will be between the relative effects of J and H: the J interaction motivates the spins to point North but the H interaction motivates the spin to point South. Such a system is termed *metastable* since if each North-pointing spin suddenly flips its state to point South, the system can achieve a lower total energy. This flipping will occur in time in a fashion not unlike the trading frequency near the end of a trend: first one or two spins will randomly switch their state, then more, and suddenly in an "avalanche" the majority of spins will point South. The phase transition is termed a spinodal singularity, characterized by its own set of exponents. Why should the end of microtrends or macrotrends have a parallel with the metastable physical system? Presumably near the end of a positive trend, all the market participants watching the market begin to sense that the market is metastable and that if they do not sell soon, it could be too late to make any profit because the price will drop. First a few traders sell, pushing the market imperceptibly lower. Then additional traders, sensing this microscopic downturn, may decide that now is the time to sell and they too sell. Then an "avalanche" of selling begins, with traders all hoping to protect their profits by selling before the market drops. Thus, the set of N market participants "holding their position" are in this sense analogous to the set of N mostly North-pointing spins, bathed in a South-pointing magnetic field.

The above analogy may not be the best and it will be future challenge to find a coherently convincing explanation for why the end of a microtrend or macrotrend displays such striking parallels to a phase transition. In any case, the set of interacting spins surely is analogous to the set of interacting traders.

The end of the negative microtrend or macrotrend is the same mechanism but with everything reversed. The N Ising spins point mostly South, the magnetic field is North, and the spins flip from South to North one by one and the conclude in an avalanche corresponding to the spinodal singularity. Analogously, the N traders begin to suspect that the market is becoming metastable, so they one by one start to buy and as all the traders witness the price increasing, they jump in to buy before the price becomes too high.

In summary we have seen that each trend – microtrend and macrotrend – in a financial market starts and ends with a unique switching process, and each extremum shares properties of macroscopic cooperative behavior. We have seen that the mechanism of bubble formation and bubble bursting has no scale for time scales varying over nine orders of magnitude down to the smallest possible time scale – the scale of single transactions measured in units of 10 ms. On large time scales, histograms of price returns provide the same scale-free behavior. Thus, the formation of positive and negative trends on all scales is a fundamental principle of trading, starting on the smallest possible time scale, which leads to the non-stationary nature of financial markets as well as to crash events on large time scales. Thus, the well-known catastrophic bubbles occurring on large time scales – such as the most recent financial crisis – may not be outliers but in fact single dramatic events caused by the scale-free behavior of the forming of increasing and decreasing trends on time scales from the very large down to the very small.

Acknowledgements The authors thank K. Binder, S.V. Buldyrev, C. De Grandi, S. Havlin, D. Helbing, U. Krey, H.-G. Matuttis, M.G. Mazza, I. Morgenstern, W. Paul, J.J. Schneider, R.H.R. Stanley, T. Vicsek, and G.M. Viswanathan for discussions, and we also thank the German Research Foundation (DFG), the Gutenberg Academy, and the NSF for financial support.

References

1. Anderson PW (1972) Science 177:393
2. Stanley HE (1971) Introduction to phase transitions and critical phenomena. Oxford University Press, London
3. Stanley HE (1999) Rev Mod Phys 71:358
4. Stanley HE, Amaral LAN, Gabaix X, Gopikrishnan P, Plerou V, Rosenow B (1999) Physica A 301:126
5. Mantegna RN, Stanley HE (2000) Introduction to econophysics correlations and complexity in finance. Cambridge University Press, Cambridge, MA
6. Axtell RL (2001) Science 293:1818
7. Takayasu H (ed) (2006) Practical fruits of econophysics. Springer, Berlin
8. Kiyono K, Struzik ZR, Yamamoto Y (2006) Phys Rev Lett 96:068701
9. Watanabe K, Takayasu H, Takayasu M (2007) Physica A 383:120
10. Gabaix X, Gopikrishnan P, Plerou V, Stanley HE (2003) Nature 423:267

11. Preis T, Paul W, Schneider JJ (2008) Europhys Lett 82:68005
12. Preis T, Virnau P, Paul W, Schneider JJ (2009) New J Phys 11:093024
13. Lillo F, Farmer JD, Mantegna RN (2003) Nature 421:129
14. Plerou V, Gopikrishnan P, Gabaix X, Stanley HE (2002) Phys Rev E 66:027104
15. Cont R, Bouchaud JP (2000) Macroecon Dyn 4:170
16. Krawiecki A, Holyst JA, Helbing D (2002) Phys Rev Lett 89:158701
17. O'Hara M (1995) Market microstructure theory. Blackwell, Cambridge, MA
18. Vandewalle N, Ausloos M (1997) Physica A 246:454
19. Eisler Z, Kertész J (2006) Phys Rev E 73:046109
20. Mandelbrot B (1963) J Business 36:394
21. Fama EF (1963) J Business 36:420
22. Lux T (1996) Appl Finan Econ 6:463
23. Guillaume DM, Dacorogna MM, Davé RR, Müller UA, Olsen RB, Pictet OV (1997) Fin Stochastics 1:95
24. Gopikrishnan P, Meyer M, Amaral L, Stanley HE (1998) Eur J Phys B 3:139
25. Plerou V, Gopikrishnan P, Rosenow B, Amaral LAN, Stanley HE (1999) Phys Rev Lett 83:1471
26. Gopikrishnan P, Plerou V, Amaral LAN, Meyer M, Stanley HE (1999) Phys Rev E 60:5305
27. Gopikrishnan P, Plerou V, Gabaix X, Stanley HE (2000) Phys Rev E 62:4493
28. Krugman P (1996) The self-organizing economy. Blackwell, Cambridge, MA
29. Shleifer A (2000) Inefficient markets: an introduction to behavioral finance. Oxford University Press, Oxford
30. Helbing D, Farkas I, Vicsek T (2000) Nature 407:487
31. Bunde A, Schellnhuber HJ, Kropp J (eds) (2002) The science of disasters: climate disruptions, heart attacks, and market crashes. Springer, Berlin
32. Jones CM, Kaul G, Lipson ML (1994) Rev Fin Stud 7:631
33. Chan L, Fong WM (2000) J Fin Econ 57:247
34. Politi M, Scalas E (2008) Physica A 387:2025
35. Jiang ZQ, Chen W, Zhou WX (2009) Physica A 388:433
36. Dubil R (2004) An arbitrage guide to financial markets. Wiley, Chichester
37. Deutsch HP (2001) Derivate und interne modelle: modernes risk management. Schaefer-Poeschel, Stuttgart
38. Binder K (1987) Rep Prog Phys 50:783
39. Peng CK, Mietus J, Hausdorff JM, Havlin S, Stanley HE, Goldberger AL (1993) Phys Rev Lett 70:1343
40. Helbing D, Huberman BA (1998) Nature 396:738
41. Ivanov PC, Yuen A, Podobnik B, Lee Y (2004) Phys Rev E 69:056107
42. Smith E, Farmer JD, Gillemot L, Krishnamurthy S (2003) Quant Finance 3:481
43. Lux T, Marchesi M (1999) Nature 397:498
44. Preis T, Golke S, Paul W, Schneider JJ (2006) Europhys Lett 75:510
45. Preis T, Golke S, Paul W, Schneider JJ (2007) Phys Rev E 76:016108
46. Bouchaud JP, Matacz A, Potters M (2001) Phys Rev Lett 87:228701
47. Haerdle W, Kleinow T, Korostelev A, Logeay C, Platen E (2008) Quant Fin 8:81
48. Halla AD, Hautsch N (2007) J Fin Markets 10:249
49. Preis T, Stanley HE (2009) J Stat Phys (Article in press) doi: 10.1007/s10955-009-9914-y
50. Preis T, Stanley HE (2009) APCTP Bulletin 23–24:18

Nonlinear Memory and Risk Estimation in Financial Records

Armin Bunde and Mikhail I. Bogachev

Abstract It is well known that financial data sets are multifractal and governed by nonlinear correlations. Here we are interested in the daily returns of a financial asset and in the way the occurrence of large gains or losses is triggered by the nonlinear memory. To this end, we study the statistics of the return intervals between gains (or losses) above a certain threshold Q. In the case of i.i.d. random numbers the probability density function (pdf) of the return intervals decays exponentially and the return intervals are uncorrelated. Here we show that the nonlinear correlations lead to a power law decay of the pdf and linear long-term correlations between the return intervals that are described by a power-law decay of the corresponding autocorrelation function. From the pdf of the return intervals one obtains the risk function $W_Q(t;\Delta t)$, which is the probability that within the next Δt units of time at least one event above Q occurs, if the last event occurred t time units ago. We propose an analytical estimate of W_Q and show explicitly that the proposed method is superior to the conventional precursory pattern recognition technique widely used in signal analysis, which requires considerable fine-tuning and is difficult to implement. We also show that the estimation of the Value at Risk, which is a standard tool in finances, can be improved considerably compared with previous estimates.

1 Introduction

In the past years, the occurrence of rare (extreme) events has attracted much attention [1–3]. Usually, rare events with magnitudes considerably exceeding the average magnitude, have been considered as independent, since the typical time span

A. Bunde (✉)
Institut für Theoretische Physik III, Justus-Liebig-Universität Giessen, 35392 Giessen, Germany
e-mail: bunde@physik.uni-giessen.de

M.I. Bogachev
Radio System Department, St. Petersburg State Electrotechnical University,
197376 St. Petersburg, Russia
e-mail: mikhail.bogachev@physik.uni-giessen.de

M. Takayasu et al. (eds.), *Econophysics Approaches to Large-Scale Business Data and Financial Crisis*, DOI 10.1007/978-4-431-53853-0_2, © Springer 2010

between them is very large. In recent years, however, there is growing evidence, that this assumption is not always true. In particular, in financial markets large volatilities seem to cluster [4], and in paleoclimate records a clustering of large river flows or high temperatures has also been observed [5]. To quantify the occurrence of rare events one usually considers the time interval between successive events above (or below) some threshold Q. One is interested in the probability distribution function (PDF) of these return intervals as well as in their long-term-dependencies (autocorrelation function, conditional return periods, etc.). In numerical treatments, one usually considers not too large thresholds Q where the statistics of the return intervals is good, and then tries to extrapolate the results towards very large threshold where the statistics, by definition is poor.

For independent data sets, the return intervals are independent and (according to the Poisson statistics) exponentially distributed. Clustering of rare events indicates a certain memory in the return intervals, and indeed, recent studies have shown that this kind of memory is a consequence of long-term dependencies in the time series itself [5–8], which occur, for example in climate [5, 9] and physiological records [10–12] as well as in time series demonstrating human behavior, including economic records [13–15], teletraffic in large networks [16], and crowd behavior [17].

Long-term memory can be either (1) linear, (2) nonlinear, or both (3) linear and nonlinear. In the first case, which is often referred to as "monofractal" the (linear) autocorrelation function $C_x(s)$ of the data decays with time s by a power law, $C_x(s) \sim s^{-\gamma}, 0 < \gamma < 1$, and the exponent γ fully describes the correlations within the record. In this case, the return intervals are long-term correlated in the same way as the original record, and its distribution is characterized at large scales by a stretched exponential with exponent γ, and at short scales by a power law with exponent $\gamma - 1$ [5–7]. It was shown that those features can be observed in long climate records [5] as well as in the volatility of financial records [4], even though the volatility also contains nonlinear memory and thus belongs to the case (3).

In the second case, where the record is "multifractal", the linear autocorrelation function $C_x(s)$ vanishes for $s > 0$ and nonlinear (multifractal) correlations that cannot be described by a single exponent, characterize the record. In this article we are solely interested in this second case, since financial data fall into this category. Since the asset prizes P_i are usually nonstationary, one considers usually the returns R_i after the i-th unit trading period (which might differ from minutes to years) defined as

$$R_i = \frac{P_i - P_{i-1}}{P_{i-1}}. \quad (1)$$

We will concentrate on daily closing prices P_i, where i denotes subsequent bank days. By definition, positive returns characterize gains, negative returns characterize losses. Figure 1 shows representative examples of arithmetic returns of the daily closing prices for (1) indices, (2) stocks, (3) exchange rates, and (4) commodities obtained from Yahoo! Finance (http://finance.yahoo.com/), The Federal Reserve Bank

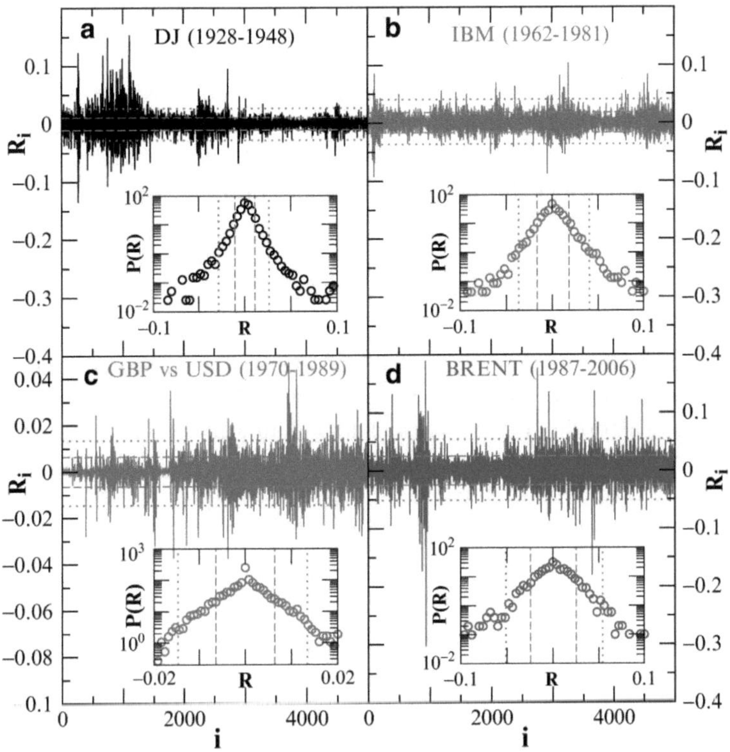

Fig. 1 Examples of the arithmetic returns between daily closing prices: (**a**) Dow Jones index (1928–1948), (**b**) IBM stock price 1962–1981, (**c**) British Pound versus US Dollar exchange rate (1970–1989), and (**d**) Brent crude oil price (1987–2006). Selected fragments of 5,000 data points starting from the beginning of the available data set are shown. Estimates of distribution densities for each of the data series are shown in the relevant *insets*. *Dashed lines* show the positive and the negative thresholds Q corresponding to the mean return times $R_Q = 10$ and *dotted lines* for $R_Q = 70$, in particular (listed in ascending order): Dow Jones (−0.0284, −0.0106, 0.011, 0.0266), IBM (−0.0365, −0.0166, 0.0183, 0.0405), GBP (−0.0147, −0.0064, 0.0064, 0.0134) and Brent (−0.0512, −0.0243, 0.0254, 0.0543). The R_Q values have been estimated from the whole record length (from the beginning of the record by 2007), while the examples contain only the first 20 years of each record

of St. Louis (http://research.stlouisfed.org/), and US Department of Energy, Energy Information Administration (http://www.eia.doe.gov/).

The figure displays the bursty behavior characteristic for multifractal records with nonlinear memory and shows also the distribution of the data.

In order to learn what are the consequences of the nonlinear memory on the statistics of the return intervals and how it can be used for risk estimation, we first describe two multifractal models that are able to capture the essential features of financial data sets.

2 Generation of Multifractal Data Series

The first algorithm is a variant of the multiplicative random cascade (MRC) process, described, e.g., in [18–21]. In this process [8], the data set is obtained in an iterative way, where the length of the record doubles in each iteration. We start with the zero-th iteration $n = 0$, where the data set (x_i) consist of one value, $x_1^{(n=0)} = 1$. In the n-th iteration, the data $x_i^{(n)}$, $i = 1, 2, \ldots, 2^n$, is obtained from

$$x_{2l-1}^{(n)} = x_l^{(n-1)} m_{2l-1}^{(n)} \quad \text{and} \quad x_{2l}^{(n)} = x_l^{(n-1)} m_{2l}^{(n)}, \tag{2}$$

where the multipliers m are independent and identically distributed (i.i.d.) random numbers with zero mean and unit variance (see Fig. 2).

The second algorithm is the multifractal random walk (MRW) proposed in [22]. In this algorithm, first we generate a record a_i, $i = 1, \ldots, N$ whose power spectrum decays as $1/f$ ("$1/f$-noise"). Next, we exponentiate these numbers and multiply them by Gaussian random numbers b_i, providing the resulting multifractal series x_i (see Fig. 3)

$$x_i = (e^{a_i}) b_i. \tag{3}$$

Both models create data series with a symmetric distribution characterized by lognormal tails (see Fig. 4), and both kinds of data sets are characterized by a vanishing autocorrelation function, i.e., $C_x(s) = 0$ for $s > 0$ [8].

In the following, we will focus on the MRC model, but like to note that similar results can be obtained also for the MRW model [23].

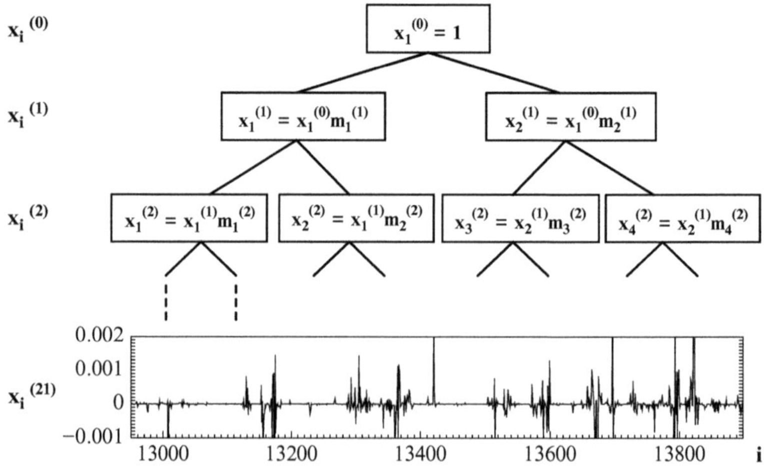

Fig. 2 Illustration of the iterative random cascade process. After each iteration the length of the generated records is doubled and after $n = 21$ iterations the multifractal set consists of $L = 2^{21}$ numbers. A subset is shown in the *bottom panel*

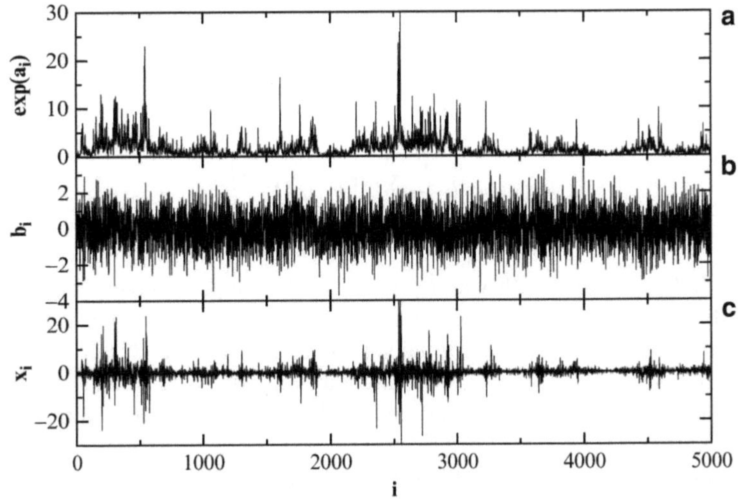

Fig. 3 Illustration of the multifractal random walk generation procedure: the exponentiated $1/f$ noise a_i (**a**) is multiplied by Gaussian random numbers b_i (**b**), resulting in the multifractal series x_i (**c**)

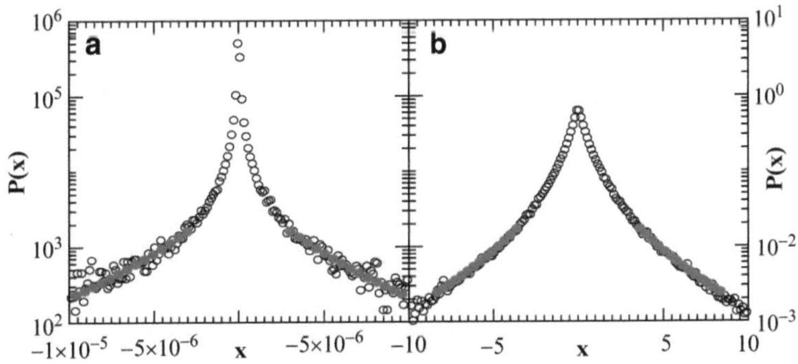

Fig. 4 PDFs of the simulated data series, created by (**a**) MRC and (**b**) MRW models. The *open symbols* provide a numerical estimate, the *full symbols* are the best lognormal fits for the tails of the distribution

3 Multifractal Analysis

There are several ways to characterize multifractal data sets. Here we chose the multifractal detrended fluctuation analysis (MF-DFA), introduced by Kantelhardt et al. [24]. In the MF-DFA one considers the profile, i.e., the cumulated data series $Y_j = \sum_{i=1}^{j}(x_i - \langle x \rangle)$, and splits the record into N_s (non-overlapping) segments of size s. In each segment k a local polynomial fit $y_k(j)$ of, e.g., second order is estimated. Then one determines the variance $F_k^2(s) = (1/s)\sum_{j=1}^{s}(Y_{[(k-1)s+j]} - y_k(j))^2$ between the local trend and the profile in each segment k and calculates a generalized fluctuation function $F_q(s)$,

$$F_q(s) \equiv \left\{ \frac{1}{N_s} \sum_{k=1}^{N_s} \left[F_k^2(s)\right]^{q/2} \right\}^{1/q}. \tag{4}$$

In general, $F_q(s)$ scales with s as $F_q(s) \sim s^{h(q)}$. The generalized Hurst exponent $h(q)$ is directly related to the scaling exponent $\tau(q)$ defined by the standard partition function-based multifractal formalism, via $\tau(q) = qh(q) - 1$. For a monofractal record, $h(q)$ is independent of q. For stationary records, $h(2)$ is related to the autocorrelation function $C_x(s)$. In the absence of linear correlations (where $C_x(s) = 0$ for $s > 0$), $h(2) = 1/2$.

In general, $h(q)$ depends on both, the distribution of the data and their correlation structure [24]. To eliminate the dependence on the log-normal tailed distribution, and thus to elucidate the contribution of the non-linear correlations on $h(q)$, we have first ranked the N numbers in the multifractal data set, and then exchanged them rankwise by a set of N numbers from a Gaussian distribution. Now the deviations of the resulting $h(q)$ from $h(2)$ depend only on the non-linear correlations and therefore can be used to characterize them. Figure 5 displays $F_q(s)/s^{1/2}$ for $q = 1, 2$ and 5. The double-logarithmic plot shows that $F_q(s)$ follows the anticipated power-law scaling, with different exponents for the different values of q. For $q = 2$, $F_q(s)/s^{1/2}$ has reached a plateau, i.e., $h(2) = 1/2$, indicating the absence

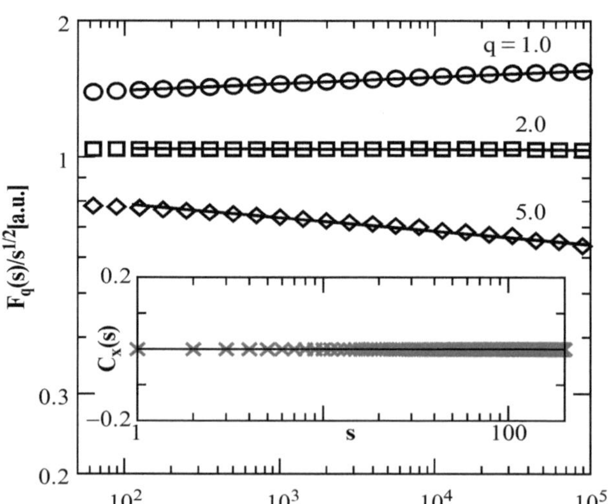

Fig. 5 Analysis of the multifractal cascade model: MF-DFA fluctuation function for $q = 1$ (*circles*), 2 (*squares*), and 5 (*diamonds*). The autocorrelation function $C_x(s)$ of the data is shown in the *inset*

of linear correlations in the data. To show this feature explicitly, we also calculated directly the autocorrelation function $C_x(s)$. The inset in Fig. 5 shows that (as expected) $C_x(s)$ fluctuates around zero for all $s \geq 1$.

4 Return Intervals in the MRC Record

In the following, we are interested in the statistics of the interoccurrence times, or return intervals r_i, between events above some threshold Q both in the MRC and in the MRW models. For the illustration of the procedure of extracting the return interval series from a data series, see Fig. 6.

For a given record, there is a one-by-one correspondence between the threshold Q and the mean return interval (or return period) R_Q, $R_Q = 1/\int_Q^\infty P(x)dx$, where $P(x)$ is the distribution of the data. By fixing R_Q instead of Q, return interval statistics remain unchanged, when the rankwise exchange procedure described above is applied. Accordingly, return interval statistics depend solely on the memory inherent in the data, and hence can be used as an effective instrument of quantifying such memory, independently of a multifractal analysis of the data.

Figure 7a shows the pdf of the return intervals for $R_Q = 10, 70$, and 500. For all return periods, we find a pronounced power-law behavior

$$P_Q(r) \sim (r/R_Q)^{-\delta(Q)}, \tag{5}$$

in marked contrast to the uncorrelated or long-term correlated monofractal data sets. The exponent δ (shown in Fig. 7b) depends explicitly on R_Q and seems to converge to a limiting curve for large data sets. When shuffling the multifractal data set, the non-linear correlations are destroyed and the scaled pdfs collapse (as expected) to a single exponential (also shown in Fig. 7a).

Next, we study the way the return intervals are arranged in time. Figure 7c shows the autocorrelation function $C_Q(s)$ [25] of the return intervals for $R_Q = 10, 70$, and 500. While $C_x(s) \equiv 0$ for $s \geq 1$, $C_Q(s)$ decays by a power law,

$$C_Q(s) \sim s^{-\beta(Q)}, \tag{6}$$

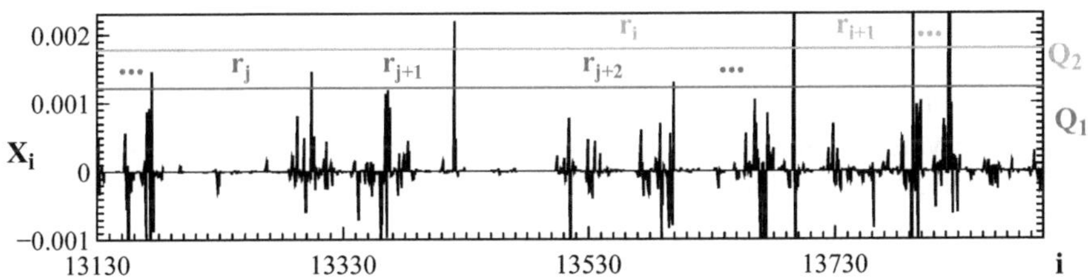

Fig. 6 Extraction of the return interval sequences of events above thresholds Q_1 and Q_2 from a data series

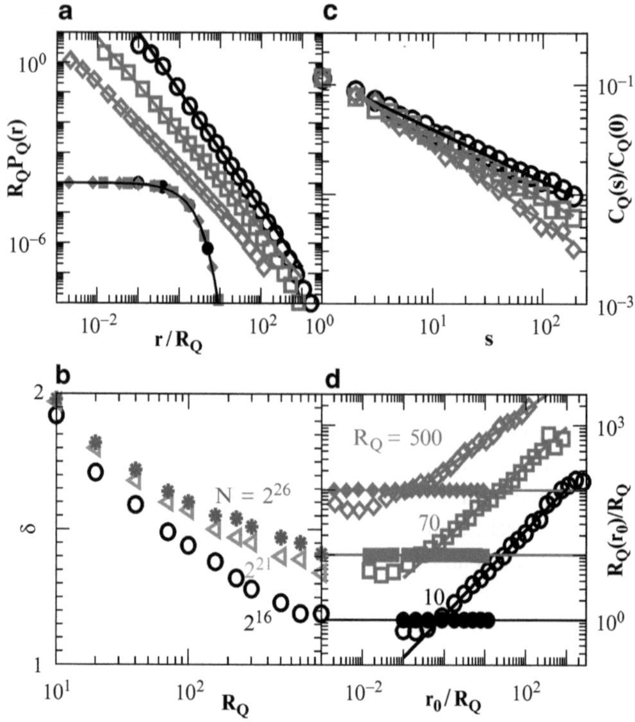

Fig. 7 Return interval statistics for the MRC cascade: (**a**) Scaled pdfs $P_Q(r)$ of the return intervals for return periods $R_Q = 10$ (*circles*), 70 (*squares*), and 500 (*diamonds*). To avoid overlapping, symbols were shifted downwards by a factor of 10 (*squares*), and 100 (*diamonds*). The relevant *filled symbols* show the corresponding pdfs for the shuffled data, shifted downwards by a factor of 10^4. (**b**) Exponents $\delta(Q)$ versus R_Q, for different systems sizes $N = 2^{16}$ (*circles*), 2^{21} (*triangles*), and 2^{26} (*asterisks*). (**c**) Return intervals autocorrelation function $C_Q(s)$ for the same R_Q values as in (**b**). (**d**) Conditional return periods $R_Q(r_0)$ in units of R_Q versus r_0/R_Q for the same R_Q values as in (**b**) (*circles*), The *filled symbols* are for the shuffled data. The curves for $R_Q = 70$ and 500 were raised by a factor of 10 and 100, respectively, to avoid overlapping symbols. The results are based on data sets of length $N = 2^{21}$ and averaged over 150 configurations

indicating long-term correlations among the return intervals. Figure 7c shows that the exponent β increases monotonically with R_Q, $\beta = 0.47$, 0.56, and 0.7 for $R_Q = 10$, 70, and 500, respectively. Obviously, these long-term correlations have been induced by the nonlinear correlations in the multifractal data set. Extracting the return interval sequence from a data set is a nonlinear operation, and thus the return intervals are influenced by the nonlinear correlations in the original data set. Accordingly, the return intervals in data sets without linear correlations are sensitive indicators for nonlinear correlations in the data sets.

To further quantify the memory among the return intervals, we consider the conditional return intervals, i.e., we regard only those intervals whose preceding interval is of a fixed size r_0. In Fig. 7d the conditional return period $R_Q(r_0)$, which is the average of all conditional return intervals for a fixed threshold Q, is plotted versus r_0/R_Q (in units of R_Q). The figure demonstrates that, as a consequence of the memory, large return intervals are rather followed by large ones, and small intervals by small ones. In particular, for r_0 values exceeding the return period R_Q, $R_Q(r_0)$ increases by a power law,

$$R_Q(r_0) \sim r_0^{\nu(Q)} \quad \text{for} \quad r_0 > R_Q, \tag{7}$$

where the exponent ν approximately decreases logarithmically with increasing value of R_Q. Note that only for an infinite record the value of $R_Q(r_0)$ can increase infinitely with r_0. For real (finite) records, there exists a maximum return interval which limits the values of r_0, and therefore $R_Q(r_0)$. As well as for $P_Q(r)$ and $C_Q(s)$, there is no scaling, and accordingly, the occurrence of extreme events cannot be deducted straightforwardly from the occurrence of smaller events. When shuffling the original data, the memory vanishes and $R_Q(r_0) \equiv R_Q$, indicated by the filled symbols.

5 Return Intervals in Financial Records

To show that the effects found in the MRC model can be also observed in financial data sets, we have analyzed several stocks (IBM, GM, GE, Boeing), several exchange rates versus the US Dollar (Danish crone, British pound, Deutsche mark, and Swiss frank), several commodities (Rotterdam gasoline, Singapore gasoline, Brent, and WTI oil crude prices) and integral market indices (Nasdaq, FTSE, Dow Jones and S&P 500), with qualitatively identical results.

As representative examples we focus on the Dow Jones index, the IBM record, and the GBP versus USD exchange rate. First, to compare the MF-DFA results with the simulated data, we exchange the returns by Gaussian data by conserving the rank ordering, as we did for the model data. For the IBM record, the result is displayed in Fig. 8a. The dashed lines correspond to the simulated data, with a record length comparable to the length of the IBM data set. The inset shows the (vanishing) autocorrelation function for the IBM returns.

The figure shows that the nonlinear correlations measured by $F_q(s)$, are modeled quite well by (2). Thus we expect that the return intervals will show a similar behavior, too. This is shown in Figs. 8b–d, where for several values of R_Q the pdf $P_Q(r)$, the autocorrelation function $C_Q(r)$ and the conditional return periods $R_Q(r_0)$ of the return intervals are shown and compared with the model data of similar length (shown as dashed lines) averaged over 500 configurations each. The agreement between observed and model data is striking. As in Fig. 7a the pdfs decay approximately by a power law, with different exponents for different values of R_Q. Due to the comparably short data set of the IBM data set, finite size effects are considerably stronger than for the model data, leading to deviations from the power law at smaller return intervals than in Fig. 7a. Also the autocorrelation function $C_Q(s)$

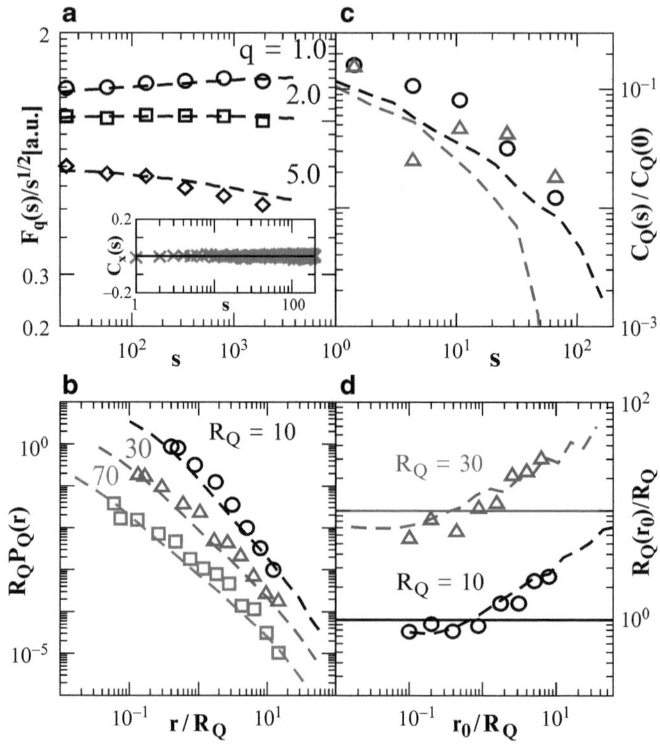

Fig. 8 Analysis of the daily arithmetic returns of IBM stock closing prices (*open symbols*), and simulated multifractal data of similar length $N = 2^{14}$ (*dashed lines*), averaged over 500 configurations: (**a**) MF-DFA fluctuation function for three moments $q = 1$ (*circles*), 2 (*squares*), and 5 (*diamonds*). The autocorrelation function $C_x(s)$ of the IBM returns is shown in the *inset*. (**b**) Scaled pdfs $P_Q(r)$ of the return intervals versus r/R_Q for $R_Q = 10$ (*circles*), 30 (*triangles*), and 70 (*squares*). To avoid overlapping, symbols were shifted downwards by a factor of 10 (*triangles*), and 100 (*squares*). (**c**) Autocorrelation function $C_Q(s)$ of return intervals (r_j) for $R_Q = 10$ and 30 (symbols correspond to those in (**b**)). (**d**) Conditional return periods $R_Q(r_0)$ in units of R_Q versus r_0/R_Q for the same R_Q values as in (**c**). The curve for $R_Q = 30$ was raised by a factor of 10 to avoid overlapping symbols

behaves qualitatively the same as the model data (see Fig. 7c) but due to less statistics finite-size effects are more pronounced. The conditional return periods shown in Fig. 8d agree very well with the model data, but due to less statistics very large values of r_0/R_Q cannot be tested. A similar behavior can be observed for all records we analyzed. Two more examples (Dow Jones index and the GBP versus USD exchange rate) are shown in Figs. 9 and 10. We like to note that $P_Q(r)$ and $R_Q(r_0)$ for several financial data sets for negative thresholds Q have been studied by Yamasaki et al. [26], but different conclusions regarding scaling and functional forms have been drawn.

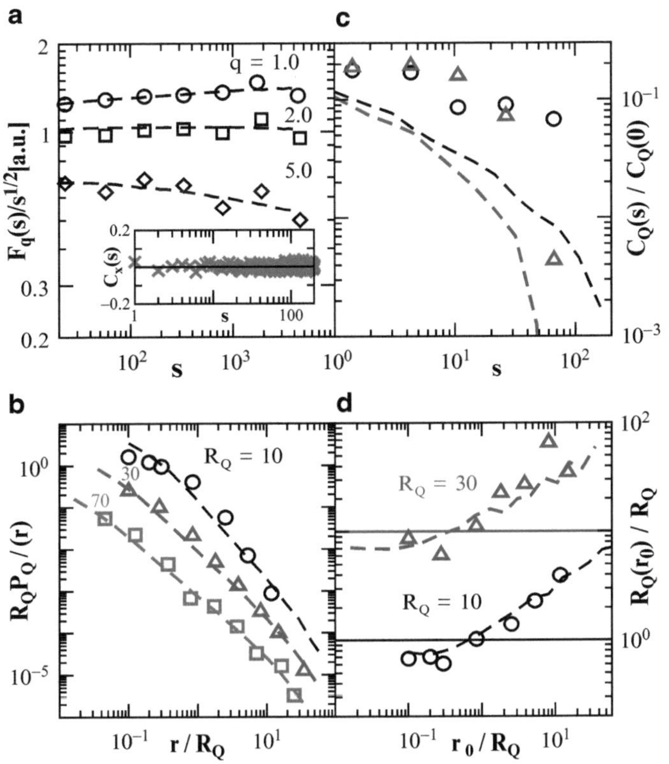

Fig. 9 Analysis of the daily arithmetic returns of Dow Jones index (*open symbols*), and simulated multifractal data of similar length $N = 2^{14}$ (*dashed lines*), averaged over 500 configurations: (**a**) MF-DFA fluctuation function for three moments $q = 1$ (*circles*), 2 (*squares*), and 5 (*diamonds*). The autocorrelation function $C_x(s)$ of the IBM returns is shown in the *inset*. (**b**) Scaled pdfs $P_Q(r)$ of the return intervals versus r/R_Q for $R_Q = 10$ (*circles*), 30 (*triangles*), and 70 (*squares*). To avoid overlapping, symbols were shifted downwards by a factor of 10 (*triangles*), and 100 (*squares*). (**c**) Autocorrelation function $C_Q(s)$ of return intervals for $R_Q = 10$ and 30 (symbols correspond to those in (**b**)). (**d**) Conditional return periods $R_Q(r)$ in units of R_Q versus r/R_Q for the same R_Q values as in (**c**). The curve for $R_Q = 30$ was raised by a factor of 10 to avoid overlapping symbols

6 Risk Estimation

6.1 Return Interval Approach

In the return interval approach (RIA), the central quantity for risk evaluation is the probability $W_Q(t; \Delta t)$ that within the next Δt units of time at least one extreme event (above Q) occurs, if the last extreme event occurred t time units ago [27]. This quantity is related to the probability density function (pdf) $P_Q(r)$ of the return intervals by

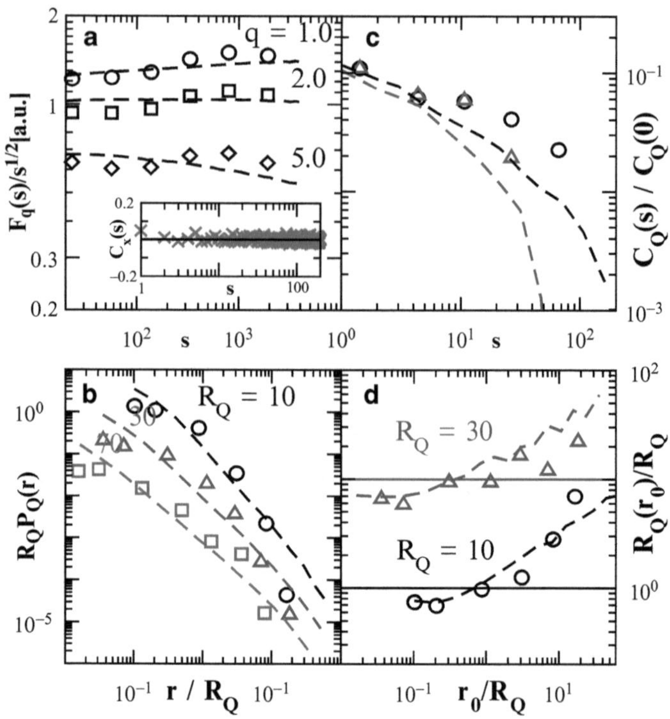

Fig. 10 Analysis of the daily arithmetic returns of the GBP vs USD exchange rate (*open symbols*), and simulated multifractal data of similar length $N = 2^{14}$ (*dashed lines*), averaged over 500 configurations: (**a**) MF-DFA fluctuation function for three moments $q = 1$ (*circles*), 2 (*squares*), and 5 (*diamonds*). The autocorrelation function $C_x(s)$ of the IBM returns is shown in the *inset*. (**b**) Scaled pdfs $P_Q(r)$ of the return intervals versus r/R_Q for $R_Q = 10$ (*circles*), 30 (*triangles*), and 70 (*squares*). To avoid overlapping, symbols were shifted downwards by a factor of 10 (*triangles*), and 100 (*squares*). (**c**) Autocorrelation function $C_Q(s)$ of return intervals for $R_Q = 10$ and 30 (symbols correspond to those in (**b**)). (**d**) Conditional return periods $R_Q(r)$ in units of R_Q versus r/R_Q for the same R_Q values as in (**c**). The curve for $R_Q = 30$ was raised by a factor of 10 to avoid overlapping symbols

$$W_Q(t;\Delta t) = \int_t^{t+\Delta t} P_Q(r)dr \Big/ \int_t^{\infty} P_Q(r)dr. \tag{8}$$

As a consequence of the algebraical decay of $P_Q(r)$, $W_Q(t;\Delta t) = (\delta(Q)-1)\,\Delta t/t$ for $\Delta t \ll t$. To obtain by numerical simulation a more general expression for $W_Q(t;\Delta t)$ valid for all arguments, we have employed the MRC model.

Figures 11a,b show, for the MRC record, the risk function $W(Q;\Delta t)$, for $R_Q = 10$ and 70, respectively. In the inset, the related pdf of the return intervals $P_Q(r)$ is shown. Since $W(t;\Delta t)$ is bounded by one for $t/R_Q \to 0$, the power law behavior can only be valid for $t/R_Q > (\delta(Q)-1)\Delta t/R_Q$. For large t/R_Q, strong finite size effects occur in $P_Q(r)$ which become more pronounced for large R_Q values. Since these finite size effects decrease with decreasing R_Q and increasing data

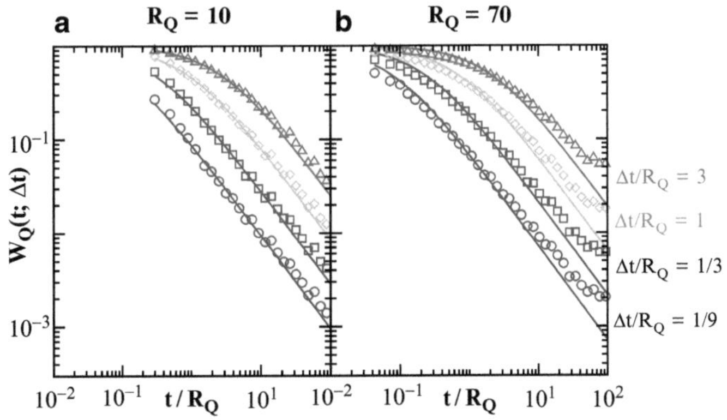

Fig. 11 The risk functions $W_Q(t; \Delta t)$ for the MRC record for (**a**) $R_Q = 10$ and (**b**) $R_Q = 70$. The *symbols* show the numerical estimates for $\Delta t/R_Q = 1/9$ (*circles*), 1/3 (*squares*), 1 (*diamonds*) and 3 (*triangles*) for an average over 150 records of length $L = 2^{21}$. The corresponding analytical approximations according to (9) are shown by *full lines*. The pdfs of the return intervals $P_Q(r)$, for the same records, are shown in the *insets*

length L, they underestimate the denominator in (8) and thus lead to an artificial overestimation of W_Q. To account for the proper short and large time behavior, we are thus led to the ansatz

$$W_Q(t; \Delta t) = \frac{(\delta(Q) - 1)\Delta t/R_Q}{(t/R_Q) + (\delta(Q) - 1)\Delta t/R_Q}, \tag{9}$$

which for the MRC yields the correct behavior for both small and large arguments t/R_Q (shown in Fig. 11 with full lines). We will use (9) for the risk estimation.

Figure 12 shows $W_Q(t; 1)$ for the price returns of three representative financial records (Dow Jones index, IBM stock, British Pound vs. US Dollar exchange rate) for (a) $R_Q = 10$ and (b) $R_Q = 70$. The corresponding pdfs of the return intervals are shown, as in Fig.1, in the insets. Since the data is short, finite size effects are more pronounced than in the simulated data of Figs. 11a,b. For comparison, the results for the simulated data ($\langle m \rangle = 0$ and $\sigma_m = 1$) of comparable system size $L = 2^{14}$ are also shown (dashed lines). The model pdfs represent slightly better the observational data for $R_Q = 70$ than for $R_Q = 10$, in agreement with the conclusions from [23]. The risk functions from the model and from the observational data agree remarkably well. Due to the short record length and no averaging, the finite size effects are comparable in the simulated and in the observational data.

6.2 Pattern Recognition Technique

In conventional signal analysis, the standard strategies are based on finding precursory patterns $y_{n,k} : y_{n-k}, y_{n-k+1}, \ldots, y_{n-1}$ of k events that typically precede an

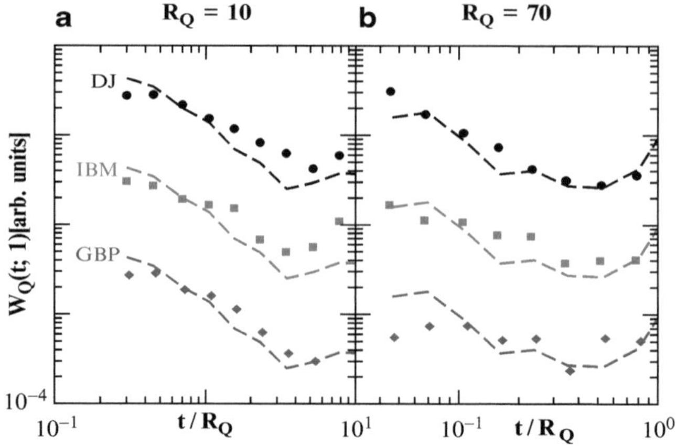

Fig. 12 Risk functions $W_Q(t; 1)$ for the price returns of three representative financial records: Dow Jones index (*circles*), IBM stock (*squares*), British Pound vs. US Dollar exchange rate (*diamonds*) for (**a**) $R_Q = 10$ and (**b**) $R_Q = 70$. The *dashed lines* shows the corresponding risk functions obtained from the MRC model obtained for a single representative configuration of size $L = 2^{14}$. In the *insets*, the corresponding pdfs of the return intervals are shown; the *dashed lines* show the numerical estimate for an average over 150 MRC records of length $L = 2^{21}$

extreme event $y_n > Q$. The strategies are mainly based on two approaches. In the first approach, one concentrates on the extreme events and their precursory patterns and determines the frequency of these patterns. In the second approach, one considers *all* patterns of k events $y_{n,k} : y_{n-k}, y_{n-k+1}, \ldots, y_{n-1}$ that precedes any event in the record and determines the probability that a given pattern is a precursor for an extreme event $y_n > Q$ [28, 29]. This pattern recognition technique (PRT) appears more profound, since it considers information about precursors of *all* events, thus providing additional information on the time series studied, as has been confirmed recently for short- and long-term correlated data [30, 31].

The PRT is illustrated in Fig. 13. To estimate the risk probability in this approach, one can either use the "learning" observational record itself or use a model representing the record (here the MRC model, where we based our estimations on 150 MRC records of length $L = 2^{21}$). First we choose a pattern length k and create a digital database of all possible patterns $y_{n,k}$ of length k. To this end, we divide the total range of the possible data y_i into l windows such that there is the same number of values in each window. Accordingly, there exist l^k different patterns. Next we determine how often each pattern is followed by an event above Q which after normalization yields the desired probability $P(y_n > Q|y_{n,k})$ that the following event y_n exceeds Q.

The major disadvantage of the PRT (compared with the RIA) is that it needs a considerable amount of fine-tuning for finding the optimum parameters l and k that yield the highest prediction efficiency. For transparency, we have kept the total number of patterns $l^k = const$ and concentrated on five pattern lengths $k = 2,3,4,5$ and 6. For the predictions in the MRC record, we have chosen $l^k = 10^6$ and obtained the

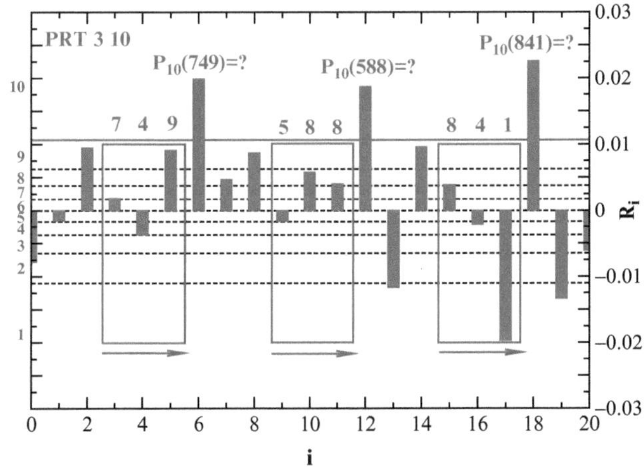

Fig. 13 Illustration of the precursory pattern recognition technique for a fragment of the daily returns of the Dow Jones index. For illustration, we have chosen for the pattern length $k = 3$ and for the magnitude resolution $l = 10$. The *dashed lines* divide the returns into 10 quantiles. The *horizontal solid line* shows the threshold Q corresponding to $R_Q = 10$. All patterns of three consecutive events are considered in a sliding window. The identificator (ID) of each pattern is a k-dimensional consisting of the quantile numbers. For each pattern, we determine the frequency of being a precursor of an event above Q from the learning cohort

best result for $k = 2$. Smaller and larger values of l^k did not improve the prediction efficiency. For predictions based on the observational records where the statistics is limited, it is usually not possible to exceed $l^k = 10^2$ for obtaining ROC curves (see below) that cover the whole area between zero and unit sensitivity. We obtained the best performance for $k = 1$ and 2.

7 Decision-making Algorithm and the ROC-Analysis

The common strategy for a decision-making algorithm is to seek for that pattern which has the highest probability to be followed by an extreme event and give an alarm when this pattern appears. In nonlinear complex records (e.g., in finance, geophysics, climate and physiology), this pattern may not be representative, since many other patterns may have comparable probabilities to be followed by an extreme event. In this case, a better approach is to give an alarm when the estimated probability for an extreme event to occur exceeds a certain threshold Q_P. The (arbitrary) selection of Q_P is usually optimized according to the minimal total cost of false predictions made, including false alarms and missed events, after a certain cost of a single false alarm and of a single missed event has been specified, see, e.g., [29].

In order to illustrate the decision-making algorithm, we show in Fig. 14a a representative fragment of the Dow Jones returns record (multiplied by (−1) such that large positive values now represent large losses), where we have indicated a

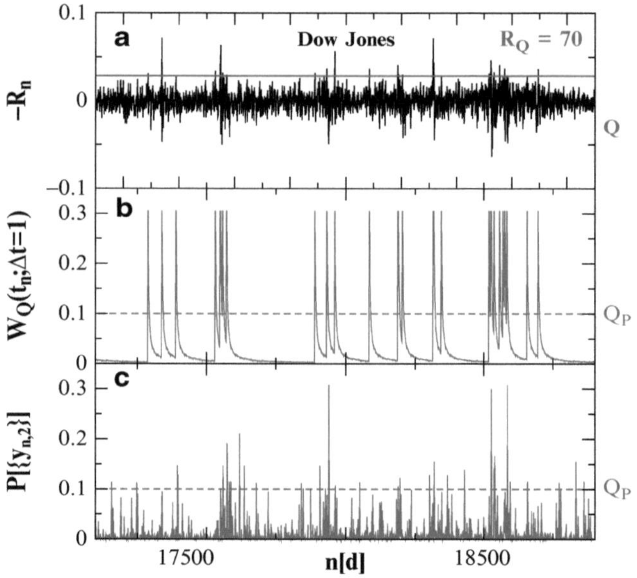

Fig. 14 (**a**) A representative fragment of the Dow Jones index returns record. (**b**) Risk function $W_Q(t;1)$ estimated by (9). (**c**) The same risk probabilities obtained by the precursory pattern recognition with $k = 2$ and $l = 10$. All quantities are given for $R_Q = 70$. An alarm is given, when the probabilities exceed a certain decision threshold Q_P

threshold Q corresponding to the mean return time $R_Q = 70$ between large negative returns. Figure 14b illustrates the estimated probabilities $W_Q(t;1)$ from (9) for the above record. Figure 14c illustrates the same probabilities estimated by the PRT with $k = 2$ and $l = 10$.

The decision threshold Q_P is shown as dashed lines in Figs. 14b,c. When the estimated occurrence probabilities exceed Q_P, an alarm is activated. For a certain Q_P value, the efficiency of the algorithm is generally quantified by the sensitivity *Sens*, which denotes the fraction of correctly predicted events, and the specificity *Spec*, which denotes the fraction of correctly predicted non-events. The larger *Sens* and *Spec* are, the better is the prediction provided by the algorithm. The overall quantification of the prediction efficiency is usually obtained from the "receiver operator characteristic" (ROC) analysis, where *Spec* is plotted versus *Sens* for all possible Q_P values. By definition, for $Q_P = 0$, $Sens = 1$ and $Spec = 0$, while for $Q_P = 1$, $Sens = 0$ and $Spec = 1$. For $0 < Q_P < 1$, the ROC curve connects the upper left corner of the panel with the lower right one. If there is no memory in the data, $Spec + Sens = 1$, and the ROC curve is a straight line between both corners (dashed lines in Fig. 15). The total measure of the predictive power *PP*, $0 < PP < 1$, is the integral over the ROC curve, which equals one for perfect prediction and equals one half for the random guess.

Figures 15a,b show the ROC-curves for a single MRC record of length $L = 2^{21}$ for $R_Q = 10$ and 70. The figure shows that in this case, where the statistics is

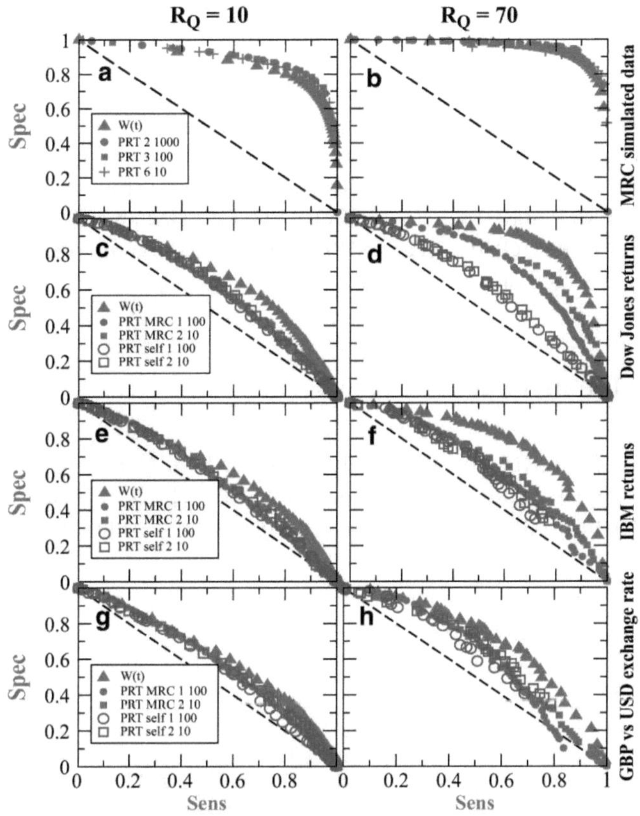

Fig. 15 ROC-curves quantifying the prediction efficiency obtained in the linearly uncorrelated MRC record, (**a**) for $R_Q = 10$ and (**b**) for $R_Q = 70$, based on the PRT for $k = 2$ (*filled circles*), 3 (*filled squares*) and 6 (*plus symbols*) as well as on RIA, (9) (*filled triangles*). Similar curves are presented in (**c**,**d**) for the Dow Jones returns, in (**e**,**f**) for the IBM stock and in (**g**,**h**) for the British Pound vs. US Dollar exchange rate. In (**c**–**h**) ROC-curves are shown for $k = 1$ (*filled circles*) and 2 (*filled squares*) for pattern database obtained from the MRC model and for $k = 1$ (*open circles*) and 2 (*open squares*) for the pattern database obtained directly from the observational record

excellent (compared with observational records), the prediction efficiency of both the RIA and the PRT approach, is quite high and comparable with each other. Figures 15c–h show the equivalent curves for the Dow Jones Index (c,d), the IBM Stock (e,f) and the exchange rate between the British Pound and the US Dollar (g,h). In these figures, we have also added the corresponding PRT results obtained from the observational records, where we "learned" from the precursors of large positive returns (gains) to predict large negative returns (losses). This is reasonable since the behavior of large positive returns is in quantitative agreement with that for negative returns [23], see also [32]. For both $R_Q = 10$ and 70, "learning" on the MRC model generally yields a higher prediction efficiency than "learning" on the observational records. The figure shows that for the three financial records, the ROC curves for the

RIA are systematically above the curves from the PRT, especially close to $Sens = 1$. Accordingly, for the same high sensitivities the RIA yields considerably less false alarms than the PRT. The superiority of the RIA approach increases with increasing return period R_Q when the finite size effects in the observational records become increasingly important. We like to note that we have obtained similar conclusions for all representative records analysed, consisting of four indices (DJ, FTSE, NASDAQ and S&P 500), four stocks (BOEING, GE, GM and IBM), four currency exchange rates (DKK, GBP, GM and SWF vs. USD) and four oil prices (Brent, WTI, Rotterdam and Singapore).

8 The Value-at-Risk

Finally, we use the RIA to estimate the Value-at-Risk (VaR), which is probably the best-known risk estimation technique in finances. The VaR is defined as the loss that, in a given time interval, can only be exceeded with a certain small probability q.

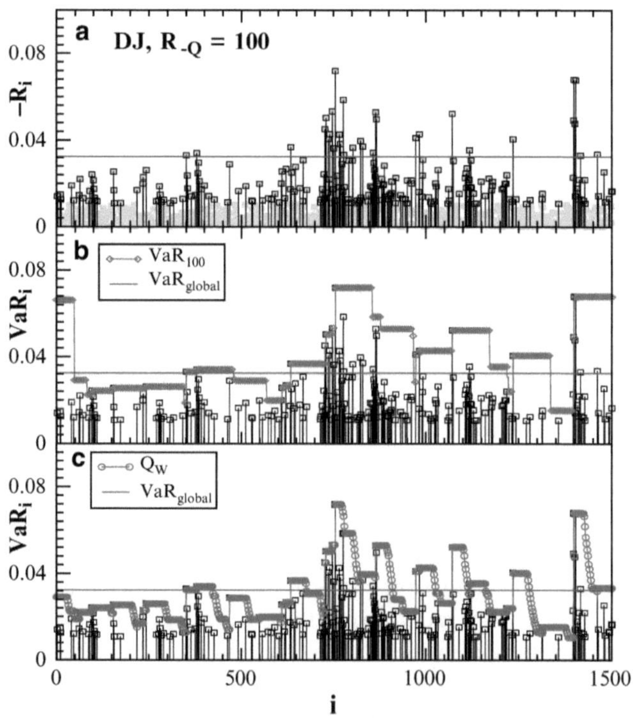

Fig. 16 Value-at-Risk estimates for the Dow Jones index. (**a**) A fragment of the DJ daily returns sequence from 09/10/1934 until 04/10/1940. (**b**) Value-at-Risk (VaR) estimates for the exceedance probability $q = 1/100$ for each day, obtained from (10) by using the global distribution (*straight line*) or the local distribution of the last 100 days (*diamonds*). (**c**) VaR estimates based on the return interval approach (*circles*)

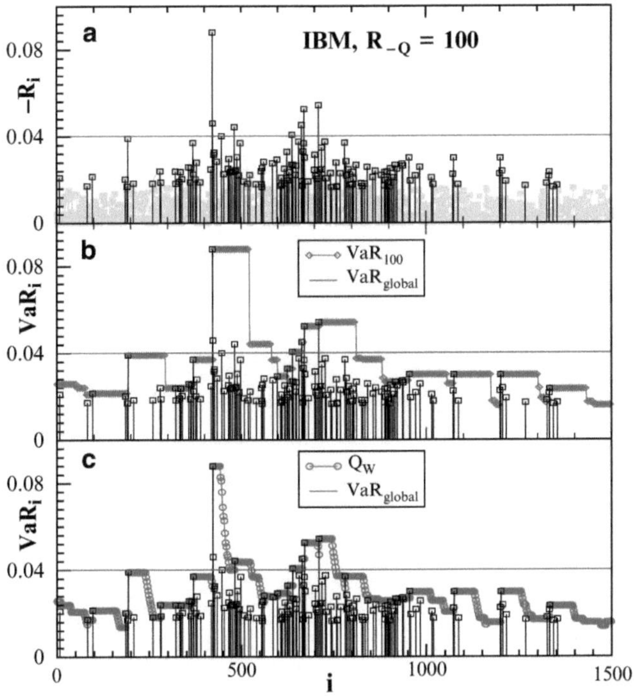

Fig. 17 Value-at-risk estimates for the IBM stock. (**a**) A fragment of the IBM daily returns sequence from 11/01/1972 until 19/12/1977. (**b**) Value-at-Risk (VaR) estimates for the exceedance probability $q = 1/100$ for each day, obtained from (10) by using the global distribution (*straight line*) or the local distribution of the last 100 days (*diamonds*). (**c**) VaR estimates based on the return interval approach (*circles*)

In the following, we consider as time interval 1 day. In a first order approximation, when memory effects are being neglected, the VaR can be simply determined from the (global) distribution $P(r)$ of the daily returns via

$$\int_{-Q}^{-\infty} P(r)dr = q. \tag{10}$$

By solving this equation, which is identical to $1/R_{-Q} = q$, one can obtain the return $-Q$ and the corresponding VaR. In order to take into account the fact that the fluctuations in the return vary in time, one often does not consider the global distribution to estimate $-Q$, but rather a local distribution of the returns based, e.g., on the last 100 days. This technique, which allows to distinguish between volatile and less volatile times, is usually a better estimator of the VaR. For a further improvement of the VaR, recently a method based on the return interval between the last two events below $-Q$ has been suggested [26].

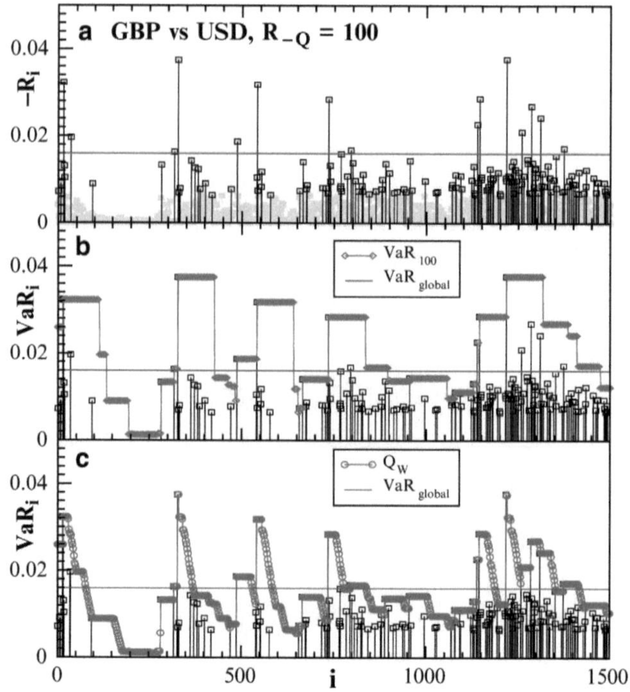

Fig. 18 Value-at-risk estimates for the GBP vs USD exchange rate. (**a**) A fragment of the GBP vs USD daily returns sequence from 09/12/1975 until 13/11/1981. (**b**) Value-at-Risk (VaR) estimates for the exceedance probability $q = 1/100$ for each day, obtained from (10) by using the global distribution (*straight line*) or the local distribution of the last 100 days (*diamonds*). (**c**) VaR estimates based on the return interval approach (*circles*)

Here we show how the RIA can be used to further improve the estimations of the VaR by using explicitly (9) for the risk function W_Q, which is a symmetric function of Q [33].

To obtain the VaR within the RIA, we proceed iteratively:

(1) In the first step, we choose q and determine from $1/R_{-Q} = q$ the corresponding loss $-Q$ in zero-th order approximation. Next we determine the time $t = t_{-Q}$ that has been elapsed after the last return below $-Q$ and use (9) to determine the new probability W_{-Q}.

(2) (a) If W_{-Q} is within a certain confidence interval Δq around q, the algorithm is stopped. (b) If W_{-Q} is above the confidence interval we multiply $-Q$ by $1 + \Delta Q$ and determine the new elapsed time $t_{-Q(1+\Delta Q)}$ after the last event below $-Q(1 + \Delta Q)$. If W is still above the confidence interval, we repeat this step until, in the n^* step, $W_{-Qn^*(1+\Delta Q)}$ is either within or below the confidence interval. Then we stop the algorithm and choose, as the estimate of the VaR, $-Q^* = -Qn^*(1 + \Delta Q)$. (c) If W_{-Q} is below the confidence interval, we proceed as in (b), but with a negative increment $\Delta Q < 0$ until we are within or above the confidence interval.

Figures 16–18 show the VaR for the three assets discussed in the preceding figures. Each figure consists of three panels. In panel (a), we show the negative returns $-R_i$. For transparency, we only highlight those negative returns with a return period above 10. We consider as extreme events returns with return period above 100 which is shown as straight line in the figures. In panel (b) we show an estimate of the VaR for each day i. We employ (4) with $q = 1/100$ and choose the (local) distribution of the returns from the last 100 days. In this case, the VaR for day i is identical to the maximum negative return between day $i - 100$ and $i - 1$. In panel (c), we show the estimate of the VaR by the RIA, obtained by he iterative procedure described above. The confidence interval was chosen between 0.0099 and 0.0101, and the size of the Q-increments was chosen as $|\Delta Q| = 0.025$.

Acknowledgements We like to thank our colleagues Jan F. Eichner, Jan W. Kantelhardt and Shlomo Havlin for valuable discussions.

References

1. Bunde A, Kropp J, Schellnhuber H-J (eds) (2002) The science of disasters – climate disruptions, heart attacks, and market crashes. Springer, Berlin
2. Pfister C (1998) Wetternachhersage. 500 Jahre Klimavariationen und Naturkatastrophen 1496–1995. Paul Haupt, Bern
3. Glaser R (2001) Klimageschichte Mitteleuropas. Wissenschaftliche Buchgesellschaft, Darmstadt
4. Yamasaki K, Muchnik L, Havlin S, Bunde A, Stanley HE (2005) PNAS 102:9424
5. Bunde A, Eichner JF, Kantelhardt JW, Havlin S (2005) Phys Rev Lett 94:48701
6. Altmann EG, Kantz H (2005) Phys Rev E 71:056106
7. Eichner JF, Kantelhardt JW, Bunde A, Havlin S (2007) Phys Rev E 75:011128
8. Bogachev MI, Eichner JF, Bunde A (2007) Phys Rev Lett 99:240601
9. Mudelsee M, Börngen M, Tetzlaff G, Grünewald U (2003) Nature London 425:166
10. Stanley HE, Amaral LAN, Goldberger AL, Havlin S, Ivanov PCh, Peng C-K (1999) Physica A 270:309
11. Ivanov PCh, Amaral LAN, Goldberger AL, Havlin S, Rosenblum MG, Stanley HE, Struzik ZR (2001) Chaos 11:641
12. Losa GA, Merlini D, Nonnenmacher TF, Weibel ER (eds) (2005) Fractals in biology and medicine. Birkhäuser, Basel
13. Lux T, Marchesi M (2000) Int J Theor Appl Finance 3:475
14. Hartmann P, Straetmans S, de Vries CG (2003) A global perspective on extreme currency linkages. In: Hunter W, Kaufman G, Pomerleano M (eds) Asset price bubbles: the implications for monetary, regulatory and international policies. MIT Press, Cambridge, MA
15. Hartmann P, Straetmans S, de Vries CG (2004) Rev Econ Stat 86:313
16. Riedi RH, Crouse MS, Ribeiro VJ, Baraniuk RG (1999) IEEE Trans Inf Theor 45:992
17. Helbing D, Farkas I, Viscek T (2002) In: Bunde A, Kropp J, Schellnhuber H-J (eds) The science of disasters – climate disruptions, heart attacks and market crashes. Springer, Berlin, p 331
18. Feder J (1989) Fractals. Plenum, New York
19. Peitgen H-O, Jürgens H, Saupe D (1992) Chaos and fractals: new frontiers of science. Springer, New York
20. Meneveau C, Sreenivasan KR (1987) Phys Rev Lett 59:1424
21. Greiner M, Eggers HC, Lipa P (1998) Phys Rev Lett 80:5333
22. Bacry E, Delour J, Muzy JF (2001) Phys Rev E 64:026103

23. Bogachev MI, Bunde A (2008) Phys Rev E 78:036114
24. Kantelhardt JW, Zschiegner SA, Koscielny-Bunde E, Havlin S, Bunde A, Stanley HE (2002) Physica A 316:87–114
25. Bendat JS, Piersol AG (1986) Random data: analysis and measurement procedures. Wiley, New York
26. Yamasaki K, Muchnik L, Havlin S, Bunde A, Stanley HE (2006) In: Takayasu H (ed) Practical fruits of econophysics. Springer, Tokyo, p 43
27. Bogachev MI, Eichner JF, Bunde A (2007) Phys Rev Lett 99:240601
28. Bishop CM (1995) Neural networks for pattern recognition. Oxford University Press, Oxford
29. Bernado JM, Smith AFM (1994) Bayessian theory. Wiley, New York
30. Hallerberg S, Altmann EG, Holstein D, Kantz H (2007) Phys Rev E 75:016706
31. Hallerberg S, Kantz H (2008) Phys Rev E 77:011108
32. Eisler Z, Perelló J, Masoliver J (2007) Phys Rev E 76:056105
33. Bogachev MI, Bunde A (2009) Phys Rev E 80:182908

Microstructure and Execution Strategies in the Global Spot FX Market*

Anatoly B. Schmidt

Abstract Modern global inter-bank spot foreign exchange is essentially a limit-order market. Execution strategies in such a market may differ from those in markets that permit market orders. Here we describe microstructure and dynamics of the EBS market (EBS being an ICAP company is the leading institutional spot FX electronic brokerage). In order to illustrate specifics of the limit-order market, we discuss two problems. First, we describe our simulations of maker loss in case when the EUR/USD maker order is pegged to the market best price. We show that the expected maker loss is lower than the typical bid/offer spread. Second, we discuss the problem of optimal slicing of large orders for minimizing execution costs. We start with analysis of the expected execution times for the EUR/USD orders submitted at varying market depth. Then we introduce a loss function that accounts for the market volatility risk and the order's P/L in respect to the market best price. This loss function can be optimized for given risk aversion. Finally, we apply this approach to slicing large limit orders.

1 Introduction

Modern inter-bank spot foreign exchange is conducted primarily via two electronic broking systems, EBS and Reuters. EBS being an ICAP business dominates the EUR/USD and USD/JPY exchanges. In January – September 2008, the average daily transacted volume in the EBS market has reached 225 billion USD. As a result, EUR/USD and USD/JPY rates posted on the EBS trading screens have become the reference prices quoted by dealers to their customers worldwide [1].

A.B. Schmidt (✉)
Business Development and Research, ICAP Electronic Broking LLC, One Upper Pond Road, Building F, Parsippany, NJ 07054, USA
e-mail: Alec.Schmidt@us.icap.com

*The information presented in this work is provided for educational purposes only and does not constitute investment advice. Any opinions expressed in this work are those of the author and do not necessarily represent the views of ICAP Electronic Broking LLC, its management, officers or employees.

M. Takayasu et al. (eds.), *Econophysics Approaches to Large-Scale Business Data and Financial Crisis*, DOI 10.1007/978-4-431-53853-0_3, © Springer 2010

Early empirical research of the high-frequency FX markets was overwhelmingly based on the Reuters indicative rates (see [2, 3] and references therein). The disadvantages of such rates, as compared to the "firm" rates at which the inter-bank currency transactions are done, are well documented [4]. In recent years, several studies of the high-frequency FX market based on the EBS data have been reported [1,5–10].

In the known publications on the FX trading strategies, it is usually assumed that when a trading model generates a signal to trade, the order is executed instantly at the current market price (see, e.g. [11–13] and references therein). While this assumption may be sufficient for analysis of low-frequency trading (e.g. for daily returns), it is not valid for the institutional high-frequency market where only limit orders are accepted and the bid-offer spread plays an important role in the realized P/L. While the trading strategies answer the question "*When* to trade", the execution strategies answer the question "*How* to trade".[1] Two typical problems to be solved prior to submitting an order are: how to split a large order into smaller pieces in order to prevent an adverse market impact, and where to place an order in respect to the current best price. Understanding of the order book dynamics is critical for implementing efficient execution strategies. Microstructure effects in equity markets have been discussed in [15–20] and references therein. In this work we review some findings on the EBS market microstructure and related execution strategies [6, 21–23].

This report is structured as follows. Description of the EBS market specifics is given in the next section. The work's relevant empirical findings are described in Sect. 3. In particular, we present the EUR/USD price distribution within the order book and the expected execution times for EUR / USD orders of various sizes and distances from the market best prices. We also describe the distributions of the EUR/USD order book volume at best price and its depletion rate. We conclude Sect. 3 with discussion of the memory processes in the FX order flows. In particular, we show that autocorrelations in FX order flows decay much faster than those in equity markets [24, 25].

In Sect. 4, we offer a theoretical framework for trading large amounts in a limit-order market by slicing them into smaller orders. Our approach is similar to that of suggested for slicing the market order in that it minimizes utility function comprising of volatility risk and execution cost [26,27]. The main specific of our approach is that the execution cost for market orders is determined primarily by their market impact while the cost of a limit order can be expressed in terms of the distance from its price to the market best price.

In Sect. 5, we describe our simulations of maker loss for limit orders submitted at current best price. According to our strategy, if price moves in the adverse direction before the order is executed, the order is canceled and resubmitted at a new best price [22, 23]. Our goal is to check whether such a strategy may lead to a loss exceeding

[1] With proliferation of alternative trading systems, dark liquidity pools, etc., another important problem for institutional trading has become answering the question "*Where* to trade" [14].

the taker loss (the bid/offer spread). We describe three models that differ in their calibration with the market data and show that all three models yield a maker loss below the bid/offer spread.

We offer concluding remarks in Sect. 6.

2 The Specifics of the EBS Market

The EBS system only accepts the limit orders. There are two types of orders: *quotes* and *hits*. Quotes (bid/offer orders) are submitted at a price specified by the trader. Quotes stay in the order book until they are filled or canceled by their owners. Hits (buy/sell orders) are submitted at the best price existing at the submission time: buy orders and sell orders are submitted at the current best offer and best bid prices, respectively. If the best price worsens by the time a hit arrives in the market, (e.g. the best offer price gets higher) or, in other words, the hit misses its counterparty, this hit is automatically canceled. A hit is always a taker order while a quote may be either a maker or a taker order. Namely, when one quote matches another quote on the opposite side of the market, the maker quote is the quote which arrived in the market earlier while the taker order is the quote that arrived later. A match between a maker order and a taker order is called a *deal*.

The EBS system has a distributed architecture with three matching engines (called *arbitrators*) in London, New York, and Tokyo that are responsible for regional trading in Europe, Americas, and in Asia, respectively. Orders arrive first at their local regional arbitrator. If a quote (or its part) is not immediately filled with the orders present in the order book, it is placed in all three regional order books according to its price/arrival time priority. Since the orders arrive at the remote regional arbitrators with some delays due to the finite light speed, the regional order books may be not fully identical.

An important feature of the EBS market is that trading can only be conducted between those counterparties that have bilateral credit. Every EBS customer establishes a credit limit with other EBS customers and can change it at any time. This implies that the *EBS best prices* (highest bid and lowest offer) may or may not be available to an EBS customer, depending on whether this customer has bilateral credit with the maker(s) of the best prices. As a result, a taker order may sometimes match a maker order that is not on the top of the order book. The credit constraints may be the cause of the power-law asymptote in the distribution of the quote life time on top of the order book [6].

Orders in the EBS market are submitted in units of millions (MM) of the base currency (the first currency in the name of the currency pair, e.g. USD in USD/JPY and EUR in EUR / USD). Exchange rates are usually quoted with five significant digits and two junior digits are named *pips*. For example, the difference in pips for EUR / USD rates 1.2345/1.2347 and for USDJPY rates 101.23/101.25 is the same: two pips.

3 The Microstructure and Dynamics of the Global FX Market

3.1 The Structure of the Order Book

Choosing an order price is always a compromise between the execution speed and cost. If a trader wants to quickly execute an order with amount higher than the amount available at the current best price (which is visible on the trader screen), he should place an order at a price better than the current best market price (e.g. a bid order at a price higher than the current best offer). On the other hand, a patient trader may submit an order at a price worse than the current best market price (e.g. a bid order at a price lower than the current best bid). Then in case of favorable market price move, execution cost will be lower. The very event of execution of such an order is not guaranteed, though.

Let P, BB, and BO be the order price, the best bid, and the best offer, respectively. We define the bid-side distance (in pips) from best price as $D = BB - P$ and the offer-side distance as $D = P - BO$. An example of the distribution of the number of orders for different distances D and for several quote sizes (V) is presented in Fig. 1. This distribution is clearly skewed to negative values of D as limit orders are more often placed inside the order book in expectation of a favorable price move.

3.2 Dynamics of the Top of the Order Book

Generally a quote submitted at best price on its side of the market (e.g. a bid order submitted at a current best bid price) must reach the top of the order book before it can be filled.[2] Important characteristics of the order book are the distributions of its size at best price and its depletion rate. Often these distributions can be described with the gamma distribution (see Fig. 2)

$$P(x) = x^{\alpha-1}\frac{\exp(-x/\beta)}{\Gamma(\alpha)\beta^{\alpha}}. \tag{1}$$

The gamma parameters $\{\alpha, \beta\}$ that describe the order book volume at best price and its depletion rate in Fig. 2 are equal $\{1.82, 5.96\}$ and $\{2.90, 0.76\}$, respectively. The expected values of the order book volume at best price and its depletion rate equal 11.3 and 2.2 MM EUR per second, respectively.

Note that the global FX market (at least its most liquid instruments, EUR / USD and USD/JPY) never sleeps and the probability for a zero-size or static order book is practically zero.

[2] Recall, however, that the credit constraints may violate this rule (see Sect. 2).

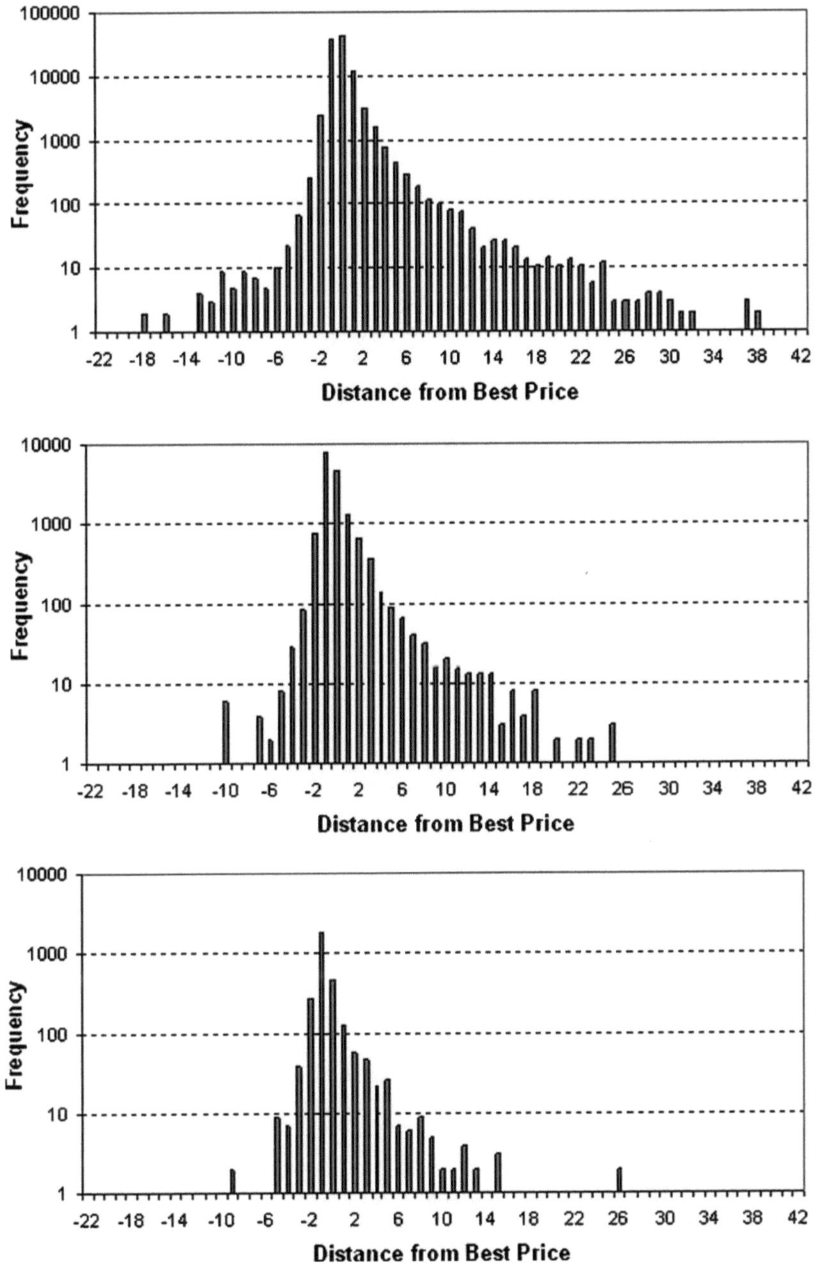

Fig. 1 Price distribution for EUR/USD orders. Data for 7:00–17:00 GMT, week ending 28 Apr 2006

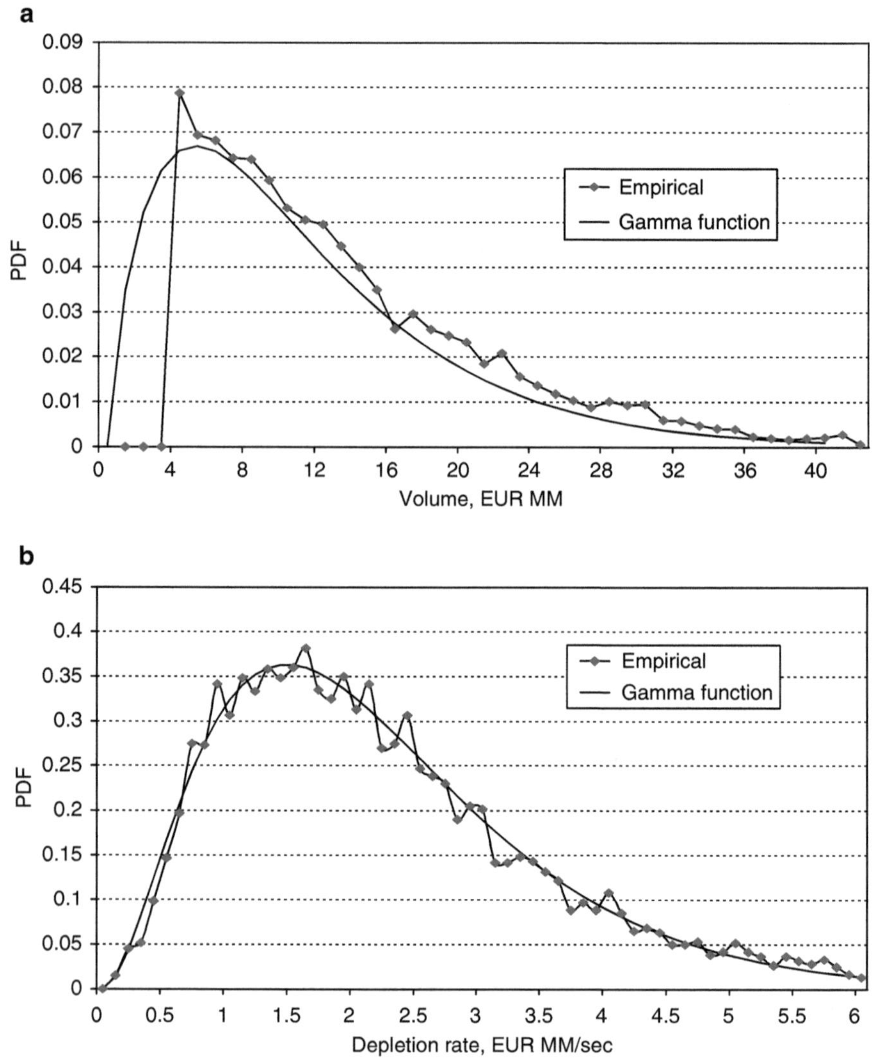

Fig. 2 Probability density functions for week ending 16 Nov 2007: (**a**) order book volume at best price; (**b**) depletion rate of the order book volume at best price

3.3 Order Execution Dynamics

Estimates of the expected quote execution time (T) are very important for implementing trading strategies. In general, T depends on the order size, its distance to the best price, and market volatility (σ). Estimating $T(V, D)$ for given time period (assuming $\sigma = \text{const}$) is complicated since limit orders are often cancelled before full or even partial filling. An example of order execution statistics is given in Table 1.

Table 1 Filling ratios of the EUR/USD quotes. Data for week ending 24 Apr 2006

Order size (MM EUR)	1	5	10	20
Partially filled (%)	–	6.5	3.0	41.3
Fully filled (%)	27.6	29.6	39.7	31.3

Table 2 Expected execution time, $T(V, D)$, (in seconds) for EUR/USD quotes. Standard error estimates are shown in parentheses. Data for the time window 7:00–17:00 GMT, 24 Apr 2006 to 2 Jun 2006

$T(V, D)$	Distance from best price, D (pips)				
Order size, V (MM EUR)	−2	−1	0	1	2
1	0.322 (0.005)	1.08 (0.02)	12.8 (0.1)	42.4 (0.6)	101 (2)
5	0.586 (0.036)	2.80 (0.11)	16.5 (0.5)	58.2 (2.7)	103 (7)
10	0.939 (0.080)	4.73 (0.47)	17.9 (0.9)	74.8 (9.5)	161 (26)
20	1.83 (0.25)	6.44 (0.71)	21.4 (2.7)	125 (50)	137 (30)

There may be different reasons for canceling unfilled or partially filled quotes. Order cancellations may be triggered by the algorithmic trading strategies reacting at changing market conditions. A trader may have firm belief that price will not revert within the accepted time horizon (e.g. due to some news), or may merely have lack of patience. Sometimes a trader may cancel a partially filled order if the trading terminal is set with a default order amount exceeding current trader's need. There have been attempts to censor cancelled orders using simple assumptions on the cancellation process [17, 20]. In this work, we have chosen to consider only those quotes that were fully executed. An example of estimates $T(V, D)$ for fully executed orders during the time window are given in Table 2.

It should be noted that the expected execution time is very sensitive to the return volatility, which for the entire period from Apr 2006 to 2 Jun 2006 on 1-s grid averaged to $\sim$0.5 pip. The quote execution time can be partitioned into the time for reaching the top of the order book (which does not depend on the order size), and the order filling time. The data in Fig. 3 demonstrate that the inverse time for reaching the top of the order book grows linearly with volatility.

3.4 Autocorrelations in FX Order Flows and Deal Flows

Long-range autocorrelations have been reported for signed order flows in equity markets in several studies (see [24, 25] for recent analysis and references therein). Note that the *signed* order flow is defined as the difference between the buy order volume and the sell order volume. In particular, autocorrelations of the signed order flow with values of approximately 0.05 or higher may last up to 100 five-minute intervals (see Fig. 4 in [24]). These autocorrelations may be explained with order slicing, which is widely used to prevent price impact of large market orders. In known research of FX markets, weak autocorrelations of the signed deal flow (decaying below 0.05 within about 5 min) were described for the interval of

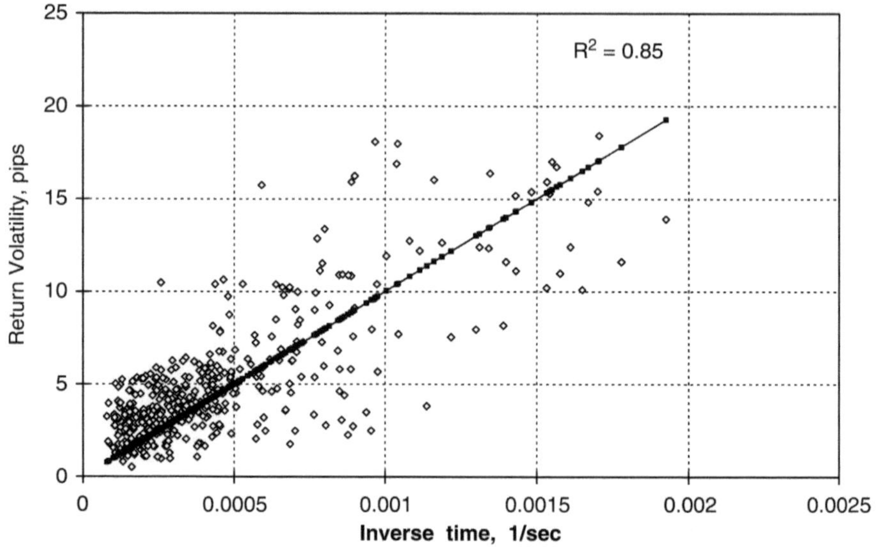

Fig. 3 Relation between the inverse time for reaching the top of the order book and return volatility for EUR/USD quotes submitted at the best price. The calculations were performed for the week ending 28 Apr 2006 on the 1-s grid for 5-min returns and volatility was calculated for every 30-min window

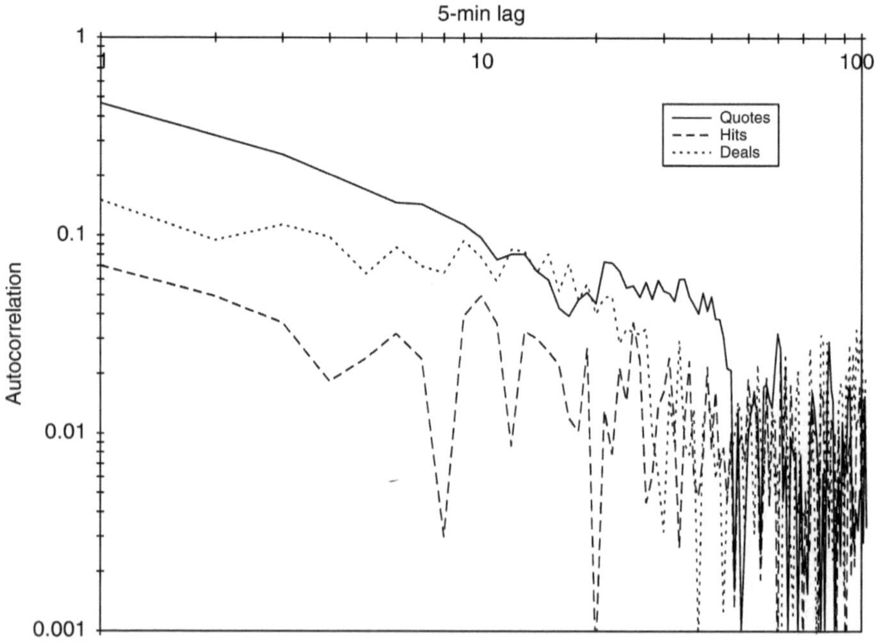

Fig. 4 Autocorrelations in the signed deal flow and the signed order flow. EUR/USD data for 24 Apr 2006 to 2 Jun 2006

1999–2003 [5]. Recall that a deal is a match between a maker order and a taker order. Our calculations of autocorrelations for EUR / USD in 2006 confirm fast decay of autocorrelations in the signed deal flow (see Fig. 4).

There is significant difference between autocorrelations in the hit flows and the quote flows. Namely, autocorrelations in the signed hit flows are practically nonexistent. However, the signed quote flow autocorrelations decay significantly slower than the signed deal autocorrelations: they remain above 0.1 for at least ten 5-min intervals. It would be interesting to derive a model in which the deal flow is treated as a product of two stochastic processes: a maker order flow with a long memory and a taker order flow with a short memory.

The FX quote autocorrelations are much shorter than the order autocorrelations in the equity markets. This may be caused by a less frequent usage of the order slicing in the EBS market due to the restriction on the minimal order size.

4 Algorithmic Trading of Large Limit Orders

In this section we show how the estimates of the order execution time can be used in algorithmic trading, namely in slicing large limit orders into small orders for minimizing trading loss. Optimal slicing of large equity orders is widely discussed in the literature (see [26–28] and references therein). In equity algorithmic trading, market orders are usually used, and the main focus is placed on analysis of the order impact on price. Indeed, a large market order can wipe out significant part of the order book and hence move price in the adverse direction (so-called *slippage*). However, a limit order can absorb only a small part of the order book. In this case, potential losses are determined primarily by the distance between the order price and the market best price.

Obviously, our estimates of the expected order execution time do not imply that every order *is* executed within the estimated time interval. A trader who locks his capital in a limit order is exposed to market risk as price may move in the adverse direction and may not revert within an accepted time horizon. To reduce market risk, one has to submit a more aggressive order than those already in the order book (e.g. a bid order at a price higher than the best offer). This, however, incurs immediate losses.

Here we follow the current literature on the optimal execution strategies [26–28] and introduce a "Risk + Cost" loss function $L_1(V, D)$ for an order of size V placed at distance D from the best price [21]:

$$L_1(V, D, \lambda) = 100V\left[\lambda\sigma\sqrt{T(V, D)} - D\right]. \tag{2}$$

The first term within the brackets of (2) is an estimate of potential losses due to the return volatility, σ (similar to value-at-risk estimates often used in risk management); $T(V, D)$ is the expected execution time and λ is the risk aversion coefficient.

Table 3 The loss function (2) for $\sigma = 0.47$ and data for $T(V, D)$ from Table 2

$L_1(V, D, \lambda)$	$\lambda = 1$					$\lambda = 2$				
	D					D				
V	−2	−1	0	1	2	−2	−1	**0**	**1**	**2**
1	2.31	1.53	1.72	2.12	2.89	2.62	2.05	3.45	5.24	7.78
5	2.36	1.79	1.92	2.61	2.81	2.72	2.58	3.83	6.21	7.61
10	2.46	2.02	2.02	3.12	4.03	2.92	3.03	4.04	7.25	10.06
20	2.64	2.19	2.18	4.25	3.50	3.27	3.39	4.35	9.51	9.00

The second term is the order P/L based on the order price distance to the current market best price. Hence in our approach, the cost of executing a bid (offer) order at current best bid (offer) price is assumed to be zero. While the bracketed value is measured in pips, V is given in millions of the base currency (i.e. EUR in case of EUR / USD) and the scaling factor outside the brackets, 100 V, presents the loss function in the local currency units (i.e. USD in the case of EUR / USD). Other utility functions used in optimal portfolio management, such as CARRA (see, e.g. [29]), can be also applied within this approach.

In Table 3, we provide an example of calculations of the loss function for $T(V, D)$ taken from Table 2 and $\sigma = 0.47$ (the EUR/USD volatility during the week ending 28 Apr 2006). This example shows that there may be an optimal value of D for given λ that minimizes the value of the loss function.

This approach can be applied to the algorithm of trading a large amount N by consecutively placed n small orders of amount $V (N = nV)$ so that every new small order is submitted right after the former one is filled. Within our approach, the nth order is locked during the time it takes to execute $(n - 1)$ previous orders as well as the nth order itself. Therefore the potential loss, L_n, for the nth order is

$$L_n(V, D, \lambda) = 100V \left[\lambda\sigma\sqrt{nT(V, D)} - D\right]. \tag{3}$$

As a result, the loss function for the total amount, N, is the sum of the individual loss functions L_1 through L_n:

$$L_{(N)}(V, D, \lambda) = 100V \left[\lambda\sigma \sum_{k=1}^{n} \sqrt{kT(V, D)} - nD\right]. \tag{4}$$

Expression (4) can answer the question whether for a chosen value of λ it is preferable to trade $N = 100$ with say $n = V = 10$ or with $n = 20$ and $V = 5$, etc. Our estimates with $T(V, D)$ taken from Table 2 show that if $\lambda = 1$, the loss function monotonically decreases with growing V (see Table 4). However for $\lambda = 2$ and $D = -2$, there is a marginal minimum at $V = 10$.

Table 4 The loss function $L_{(N)}$ for $N = 100$ calculated using values of $T(V, D)$ from Table 2

$L_{(N)}$	$\lambda = 1$		$\lambda = 2$	
V	$D = -1$	$D = 0$	$D = -1$	$D = -2$
1	46,932	120,754	83,864	63,649
5	35,416	61,611	60,832	43,012
10	33,843	47,341	57,685	41,461
20	30,847	38,037	51,693	42,225

5 Simulations of Maker Loss

In a limit-order market, a trader assuming a long position can submit a *maker order* at a price equal to (or lower than) the current best bid. Another option is to submit a *taker order* at a price equal to (or higher than) the current best offer price. A taker order has a much higher probability of fast filling yet has a loss in respect to a maker order. Here we describe simulations of a strategy in which a maker order is pegged to the best price until the order is executed [22, 23]. If price moves in the adverse direction before the maker order is executed, this order is canceled and resubmitted at a new best price. These simulations allow for estimating the expected maker loss in respect to the bid/offer spread.

We assumed that trading on the buy and sell sides of the market is symmetric and considered only the bid orders with the size of 1 MM. We used three models that differ in their calibration with the market data. The simulation process for *Model A* is shown in Fig. 5.

First, we simulated the order book volume at the best bid price and placed our order in the end of the order queue. The order book volume at best price and its depletion rate were simulated using the distributions shown in Fig. 2. Then we simulated the new best price. In Model A, we used a random trinomial tree function in which price may increase or decrease by one pip with probability $\gamma \leq 0.5$, or price may remain the same with probability $(1-2\gamma)$. The standard deviation for this price model equals $\sqrt{2\gamma}$. The standard deviation for our data set equals 0.37 pips; hence we used $\gamma = 0.068$. If the new price was lower than the current best bid, we assumed that entire order book at the best bid price including our order had been filled. If the new price was the same, we decreased the order queue by the simulated depletion rate. If the resulting order volume was lower than zero, we assumed that our order had been filled. Finally, if the new price was higher than the current best bid, we simulated a new order book volume at best price and its depletion rate, and resubmitted our order at the new price.

The simulations can be made more realistic if the model employs the real best prices and order book volumes at best price (which are observable to the EBS traders). Then only the order book depletion rate, R, (which is not observable) must be simulated. In *Model B*, empirical prices were used while both the order book volume at best price and its depletion rate were simulated. Unfortunately, the

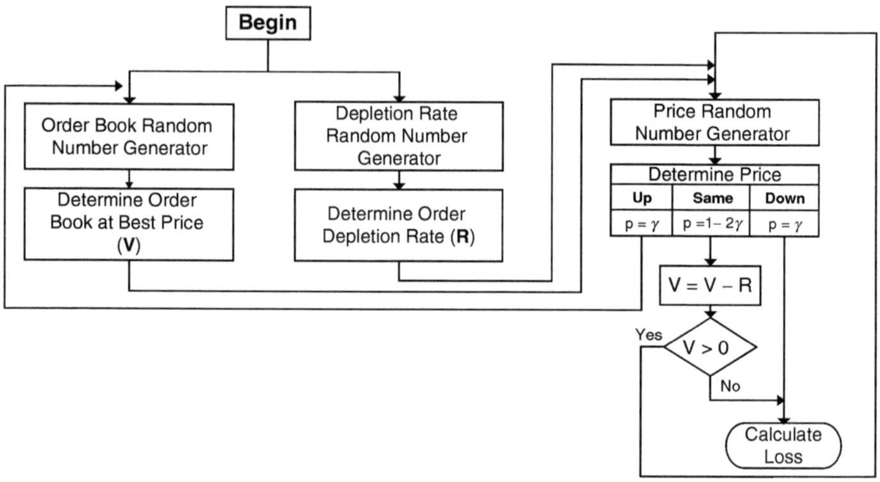

Fig. 5 Flow chart for simulations of maker loss

simulated and real order book volumes may become inconsistent. Let denote the empirical order book volume at best price in the beginning of the current simulation with $V_e(0)$. If price remains unchanged for n time steps, then the simulated order book volume equals $V_s(n) = V_e(0) - nR$ (note that $R = const$ within simulation of a given maker order). The case with $V_s(n) < V_e(n)$ can be explained with new orders arriving in the queue *after* our order. However the case with $V_s(n) > V_e(n)$ shows inconsistency between the simulated and empirical data. *Model C* differs from *Model B* in that the simulated order book volumes are defined as

$$V_s\,(0) = V_e(0), \quad V_s\,(n) = \min\left[V_s\,(n) - R, V_e\,(n)\right]. \tag{5}$$

The convergence of simulations for all three models is shown in Fig. 6. The expected loss simulated with all three models is listed in Table 5. These data show that all three models yield very close results of less than one pip (i.e. less than the bid/offer spread). This implies that the maker strategy may be statistically more advantageous than the taker strategy.

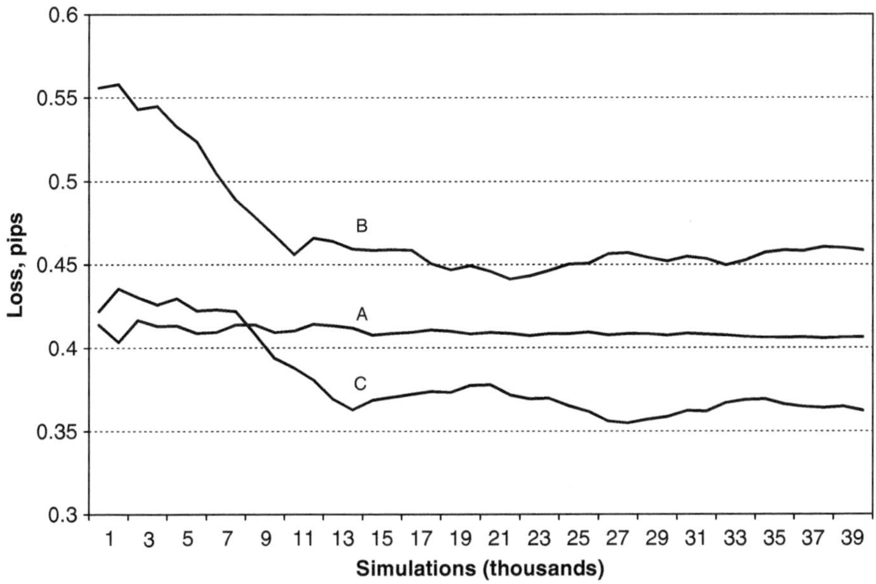

Fig. 6 Convergence of simulations of maker loss for three simulation models: A, B, and C (see model descriptions in the text). The market data used for model calibration are for EUR/USD, week ending 16 Nov 2007

Table 5 Expected maker loss for EUR/USD, week ending 16 Nov 2007

Model	A	B	C
Expected maker loss (pips)	0.41	0.46	0.36

6 Concluding Remarks

The results presented in Sects. 4 and 5 should be treated primarily as illustrations of concepts on how high-frequency market data can be used for back-testing the trading strategies in a limit-order market. Adequate calibration of trading model remains a challenge for every practitioner. In particular, there is a fine line in choosing a sample for the model calibration. On one hand, econometricians strive for long data samples to satisfy the law of large numbers. However, if a time series is not stationary (which may be the case for high-frequency market data sets), averaging upon a long sample may be not helpful for calibrating the model to specific market conditions (e.g. to given volatility). More generally, data snooping remains a real problem in analysis of performance of trading strategies [30]. Resampling with the bootstrap and Monte Carlo simulations has been a promising approach in this field [31, 32].

References

1. Chaboud AP, Chernenko SV, Howorka E, Krishnasami RS, Liu D, Wright JH (2004) The high-frequency effects of US macroeconomic data releases on prices and trading activity in the global interdealer foreign exchange market. International Finance Discussion Papers, N823
2. Dacorogna MM, Gencay R, Muller U, Olsen RB, Pictet OV (2001) An introduction to high-frequency finance. Academic, New York
3. Lyons RK (2001) The microstructure approach to exchange rates. MIT Press, Cambridge
4. Goodhart CAO, O'Hara M (1997) High frequency data in financial markets: issues and applications. J Empir Finan 4:73–114
5. Berger DW, Chaboud AP, Chernenko SV, Howorka E, Krishnasami RS, Liu D, Wright JH (2005) Order flow and exchange rate dynamics in electronic brokerage system data. International Finance Discussion Papers, N830
6. Howorka E, Schmidt AB (2006) Dynamics of the top of the order book in a global FX spot market. In: Computational finance and its applications. WIT Press, Southampton, pp 257–266
7. Ito T, Hashimoto Y (2006) Intra-day seasonality in activities of the foreign exchange markets: evidence from the electronic broking system. J Japanese Int Econ 20(4):637–664
8. Chaboud AP, Chiquoine B, Hjalmarsson E, Loretan M (2008) Frequency of observation and the estimation of integrated volatility in deep and liquid financial markets. BIS Working Paper 249
9. LeBaron B, Zhao Y (2008) Foreign exchange reversals in New York time. Working paper. Brandeis University
10. Hashimoto Y, Ito T, Ohnishi T, Takayasu M, Takayasu H, Watanabe T (2008) Random walk or a run: market microstructure analysis of the foreign exchange rate movements based on conditional probability. NBER Working Paper 14160
11. Dunis CL, Williams M (2003) Applications of advanced regression analysis for trading and investment. In: Applied quantitative methods for trading and investment. Wiley, New York, pp 1–40
12. Osler C (2003) Currency orders and exchange-rate dynamics: explaining the success of technical analysis. J Finance 58:1791–1819
13. James J (2005) FX trading models – how are they doing? Quant Finan 5(5):425–431
14. Smith C (2008) The rise of alternative trading venues. J Trading 3(1):56–58
15. Biais B, Hillion P, Spatt CS (1995) An empirical analysis of the limit order book and the order flow in the Paris bourse. J Finance 50(5):1655–1689
16. Harris L, Hasbrouck J (1996) Market versus limit orders: the superdot evidence on order submission strategy. J Finan Quant Anal 31:213–231
17. Lo AW, MacKinlay AC, Zhang J (2002). Econometric models of limit-order executions. J Fin Econ 65:31–71
18. Hollifield B, Miller RA, Sandas P, Slive J (2006). Estimating the gains from trade in limit-order markets. J Finance 61:2753–2804
19. Potters M, Bouchaud J-P (2003) More statistical properties of order book and price impact. Physica A 324:133–140
20. Eisler Z, Kertesz J, Lillo F, Mantegna RN (2009) Diffusive behavior and the modeling of characteristic times in limit order executions. Quant Finan 9:547–563
21. Howorka E, Nagirner E, Schmidt AB (2007). Analysis of order execution in a global FX spot market. In: 13th International conference on computing in economics and finance, Montreal
22. Howorka E, Nagirner E, Schmidt AB (2007) Maker or taker: simulations of trading loss in the EBS market. ICAP Memo
23. Schmidt AB (2008) Simulation of maker loss in the global inter-bank FX market. J Trading 3(4):66–70
24. Farmer JD, Gerig A, Lillo F, Mike S (2006) Market efficiency and the long-memory of supply and demand: is price impact variable and permanent or fixed and temporary? Quant Finan 6(2):107–112
25. Bouchaud J-P, Kockelkoren J, Potters M (2006) Random walks, liquidity molasses and critical response in financial markets. Quant Finan 6(2):115–123

26. Almgren R, Chriss N (2000) Optimal execution of portfolio transactions. Risk 3(2):5–39
27. Kissell R, Glantz M (2003) Optimal trading strategies. *AMACOM*
28. Kissell, R, Glantz M, Malamut R (2004) A practical framework for estimating transaction costs and developing optimal strategies to achieve best execution. Fin Res Lett 1(1):35–46
29. Fabozzi F, Kolm PN, Pachamanova D, Focardi SM (2007) Robust portfolio optimization and management. Wiley, New York
30. Sullivan R, Timmermann A, White H (1999) Data-snooping, technical trading rule performance, and the bootstrap. J Finance 54:1647–1691
31. Aronson RA (2006) Evidence-based technical analysis: applying the scientific method and statistical inference to trading signals. Wiley, New York
32. Fusai G, Ronoroni A (2008) Implementing models in quantitative finance: methods and cases. Springer, Berlin

Temporal Structure of Volatility Fluctuations

Fengzhong Wang, Kazuko Yamasaki, H. Eugene Stanley, and Shlomo Havlin

Abstract Volatility fluctuations are of great importance for the study of financial markets, and the temporal structure is an essential feature of fluctuations. To explore the temporal structure, we employ a new approach based on the return interval, which is defined as the time interval between two successive volatility values that are above a given threshold. We find that the distribution of the return intervals follows a scaling law over a wide range of thresholds, and over a broad range of sampling intervals. Moreover, this scaling law is universal for stocks of different countries, for commodities, for interest rates, and for currencies. However, further and more detailed analysis of the return intervals shows some systematic deviations from the scaling law. We also demonstrate a significant memory effect in the return intervals time organization. We find that the distribution of return intervals is strongly related to the correlations in the volatility.

1 Introduction

Understanding financial markets attracts many researchers in both economics and physics [1–10]. A key topic of economics and econophysics is fluctuations in price movement. Due to the recent development of electronic trading and data

F. Wang (✉) and H.E. Stanley
Center for Polymer Studies and Department of Physics, Boston University, 590 Commonwealth Avenue, Boston, MA 02215, USA
e-mail: fzwang@bu.edu; hes@bu.edu

K. Yamasaki
Department of Environmental Sciences, Tokyo University of Information Sciences, 4-1 Onaridai, Wakaba-ku, Chiba 265-8501, Japan
e-mail: yamasaki@edu.tuis.ac.jp

S. Havlin
Minerva Center and Department of Physics, Bar-Ilan University, Ramat-Gan 52900, Israel
e-mail: havlin@ophir.ph.biu.ac.il

M. Takayasu et al. (eds.), *Econophysics Approaches to Large-Scale Business Data and Financial Crisis*, DOI 10.1007/978-4-431-53853-0_4, © Springer 2010

storing, huge financial databases have become available, enabling the analysis of the dynamic properties of financial fluctuations. The fluctuations can be evaluated by the absolute value of logarithmic price change ("volatility"). One of the most fundamental problems is the temporal structure in the volatility time series. A new approach to analyze the temporal structure is to study statistics and organization of the time interval ("return interval") between two successive shocks larger than a threshold.

Recently Bunde and his colleagues [11–15] studied the return intervals for climate records and found scaling in the distribution and long-term memory. Analogous to the climate, one can use the *return intervals* approach to study financial fluctuations. The first effort was conducted by Yamasaki et al. who studied the daily data of currencies and US stocks and found scaling in the return intervals distribution and the long-term memory in the time series of the return intervals [16, 17]. Later, Wang et al. studied the intraday data of 30 stocks which constitute the Dow Jones Industrial Average (DJIA) index, Standard and Poor's 500 (S&P 500) index, currencies, interest rates, and commodities (oil and gold). They found similar behavior for all quantities studied [18, 19]. A similar analysis has been done for the Japanese [20] and Chinese [21,22] stock markets. To compare between models and empirical data, Vodenska-Chitkushev et al. examined return intervals from two well-known models, FIGARCH and fractional Brownian motion (fBm) and showed that both models capture the memory effects, but only fBm yields the scaling features [23]. Bogachev et al. related the nonlinear correlations in volatilities to the multiscaling behavior in the return intervals. They also showed that the return intervals distribution follows a power law for multifractal data sets [24]. Recently, Wang et al. systematically studied 500 components of the S&P 500 index, and demonstrated a systematic deviation from the scaling. They showed that this multiscaling behavior is related to the nonlinear correlations in the volatility sequence [25]. Further, Wang et al. analyzed the relation between multiscaling and several essential factors, such as capitalization and number of trades, and found a systematic dependence [26]. The multiscaling behavior was also found in the Chinese stock market [27]. A comprehensive review of the return intervals studies can be found in [28]. These studies help us to better understand the volatility and therefore may lead to better risk estimation and portfolio management [1–4, 29, 30]. Return intervals have also been studied in many other fields (see [31] and references therein). It is calculated in similar ways but with different names, such as waiting times, interoccurrence times or interspike intervals.

2 Data Analyzed

In this paper we analyze the volatility return intervals of the US stock market by analyzing the Trades And Quotes (TAQ) database from the New York Stock Exchange (NYSE). The period studied is from Jan 2, 2001 to Dec 31, 2002 (500 trading days). TAQ records every trade for all securities in the US market. To avoid many

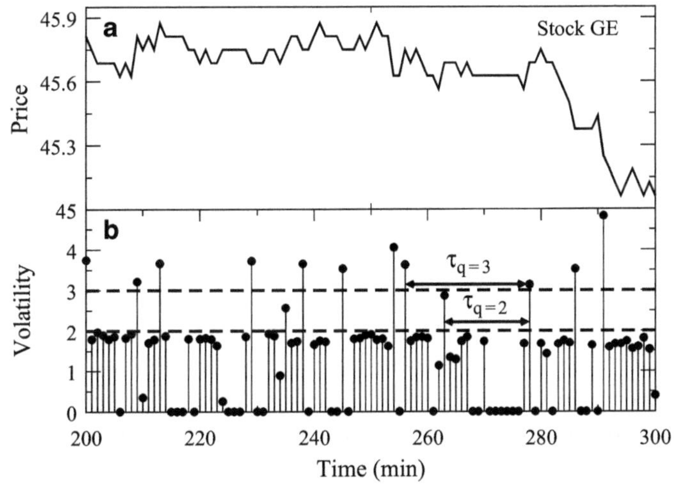

Fig. 1 Illustration of the return interval time series. Panel (**a**) is the stock price and (**b**) is the corresponding volatility for the stock GE on Jan 8, 2001. The two *dashed lines* in panel (**b**) represent two arbitrary thresholds $q = 2$ and $q = 3$. The time intervals $\tau_{q=2}$ and $\tau_{q=3}$ are between the volatility values that exceed these thresholds which form two time series. Note that the volatility is in units of its standard deviation

missing points in 1-min resolution, we choose to analyze only the 1,000 most-traded stocks. Their numbers of trades range from 600 to 60,000 per day. The volatility is defined the same as in [18]. First, we compute the absolute value of the logarithmic change of the 1-min price, then remove the intraday U-shape pattern [5–7], and finally normalize the series with its standard deviation. Therefore, the volatility is in units of standard deviations. With 1-min sampling interval, a trading day has 390 points (after removing the market closing hours), and each stock has about 195,000 records over the 500 trading days analyzed.

To illustrate the generation of the return interval time series, we plot the price and corresponding volatility of a typical stock, General Electric (GE), for a typical day, in Figs. 1a,b. We identify all volatility values above a certain threshold q, and we calculate the sequence of time intervals τ_q between them, as shown Fig. 1b. These time intervals form the return interval time series. The only free parameter in this procedure is the threshold q. The return interval time series reflects the temporal structure of the volatility time series at different size scales by choosing different thresholds q.

3 Scaling and Universality

Two important conceptual advances in statistical physics are *scaling* and *universality*. A system obeys a scaling law if its components can be characterized by the same power law form over a certain range of scales ("scale invariance"). A typical behavior for scaling is *data collapse*: all curves can be "collapsed" onto a single curve, after a certain scale transformation. In many systems, the *same* scaling function holds, suggesting universality laws.

We begin by examining the scaling and universality in the distribution of the return intervals. The distribution can be characterized by a probability density function (PDF) or cumulative distribution function (CDF). Previous studies [16–28] suggested that $P(\tau)$, the PDF for the return interval τ, can be well approximated by a scaling law if τ is scaled by its average $\langle\tau\rangle$,

$$P(\tau) = \frac{1}{\langle\tau\rangle} f\left(\frac{\tau}{\langle\tau\rangle}\right). \tag{1}$$

Here $\langle\cdots\rangle$ stands for the average over a data set. The scaling function f does not depend explicitly on the threshold q, but only through the mean interval $\langle\tau\rangle$. If $P(\tau)$ is known for one value of q, (1) can make predictions for other values of q – in particular for very large q (extreme events), which are difficult to study due to the lack of statistics.

Our first task is to examine the universality of this scaling behavior. There are several determining characteristics associated with the return interval. A unique return interval time series is determined by the threshold q. Previous studies shown that the scaling is valid for a broad range of thresholds [16–22]. As an example, we plot in Fig. 2a the return interval distributions of stock GE for thresholds $q = 2$ to 6. Clearly, the distributions show scaling and similar results are found for many other stocks [16–22].

The second characteristic is the sampling interval Δt. To test whether the system exhibits diverse behaviors at different time resolutions, we examine the distributions over various sampling intervals Δt, from $\Delta t = 1$ min to $\Delta t = 30$ min, as shown in Fig. 2b. Remarkably, all curves collapse onto a single curve, demonstrating the persistence of the scaling over a wide range of Δt. The scaling can even be extended to one trading day [19].

The next question is, how about different stocks? Is the same scaling still valid? To answer this question, we study the return interval distributions of 30 component stocks, which consist of the benchmark DJIA index, and plot all of them for $q = 2$ in Fig. 2c. Again we find that all curves collapse onto a single curve, suggesting that the same scaling holds for different stocks.

Until now all the results have been for equity markets. To test the scaling for different financial markets, we plot in Fig. 2d the distributions of the return interval for two typical commodities, one typical interest rate, two currencies, as well as a stock and a stock index (see [19] for a detailed description of the data sets). We can see that all symbols clearly fall onto a single curve, indicating the scaling of these distributions.

In conclusion, Fig. 2 strongly suggests the universality of the scaling in return interval distributions, which is valid over four characteristics: thresholds, sampling intervals, stocks, and markets.

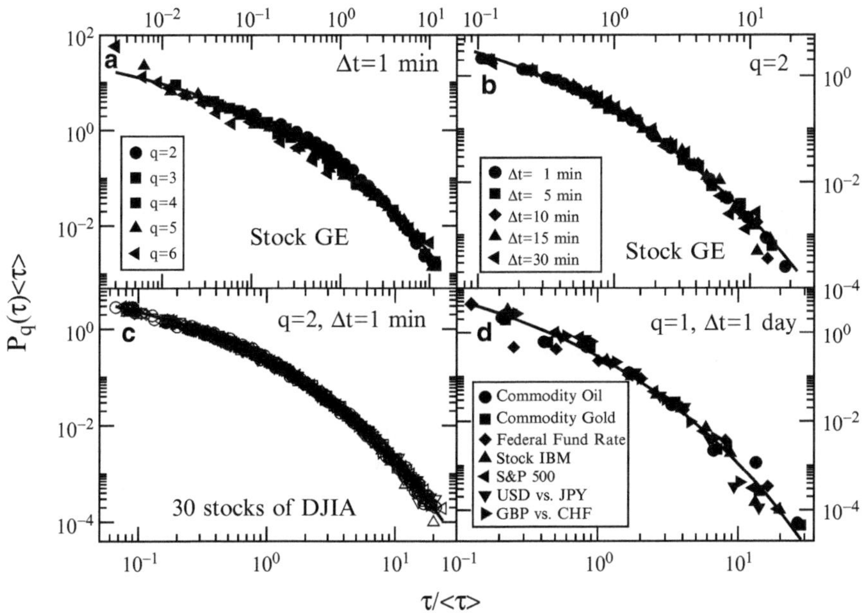

Fig. 2 Scaling in the return interval distributions for (**a**) stock GE for 500 trading days with thresholds $q = 2, 3, 4, 5$, and 6, (**b**) stock GE with sampling intervals $\Delta t = 1$ to 30 min, (**c**) 30 component stocks of the DJIA index, and (**d**) four different markets: stock, commodity, interest rate, and currency. Note that stock GE is only a representative example; other stocks have similar features. In all four panels we find good scaling, consistent with universality. The *solid lines* fit to the data are stretched exponential fits – i.e., the scaling function is consistent with a stretched exponential, $\exp(-x^{\gamma})$, with $\gamma \approx 0.3$

One question naturally arises, what is the form of the scaling function f? As shown by the solid lines in Fig. 2, the function was suggested to be in a good approximation to a stretched exponential [16–28],

$$f(x) \sim e^{-(x/x^*)^{\gamma}}. \tag{2}$$

Here x^* is the characteristic scale and γ is the shape parameter, which is related to the correlations in the volatility sequence and thus called "correlation exponent" [11–14]. For an uncorrelated time series, $\gamma = 1$ and f reduces to the exponential function. This is confirmed by shuffling the original volatility time series and examining the corresponding return interval [16–19]. Later we will further discuss the relation between the distribution of the return intervals and the correlations in the volatility.

From (2), the PDF function can be rewritten as

$$P(\tau) \sim e^{-(\tau/a)^{\gamma}}, \tag{3}$$

where a is the characteristic time scale. The two parameters in (3) are related, by the definition of a PDF and $\langle\tau\rangle$ [25, 31],

$$a = \frac{\langle\tau\rangle\Gamma(1/\gamma)}{\Gamma(2/\gamma)}. \tag{4}$$

Here $\Gamma(a) \equiv \int_0^\infty t^{a-1}e^{-t}dt$ is the Gamma function. To simplify the calculation and without loss of generality, we assume τ/a is continuous, then the corresponding CDF, $C(\tau)$, is the integral of the PDF,

$$C(\tau) \equiv \int_x^\infty P(\tau)d\tau \sim \Gamma(1/\gamma, (\tau/a)^\gamma), \tag{5}$$

where $\Gamma(a, x) \equiv \int_x^\infty t^{a-1}e^{-t}dt$ is the incomplete Gamma function.

Since the CDF accumulates the information of the time series and has better statistics than the PDF, we obtain the correlation exponent γ by fitting the CDF with (5).[1] In this way, we calculate γ values for the 1,000 most-traded stocks. In Fig. 3,

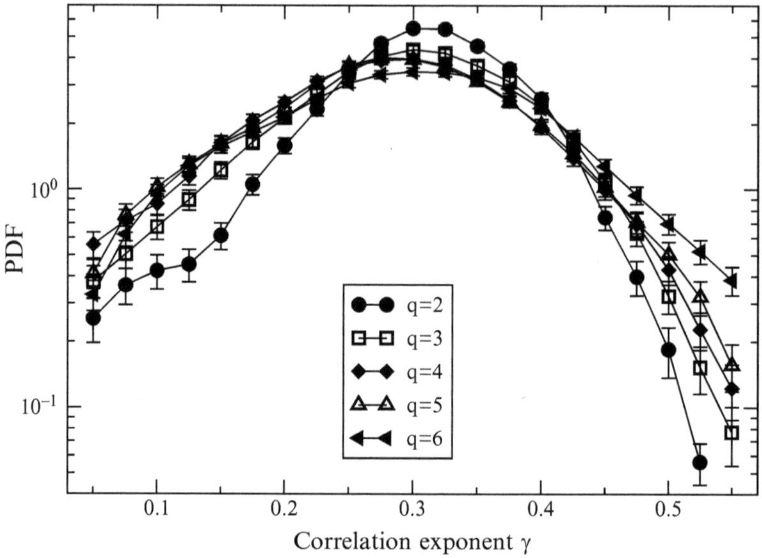

Fig. 3 PDF of the correlation exponent γ for the 1,000 most-traded US stocks (see Footnote 2). The results for five thresholds, $q = 2$ to 6, are displayed. Interestingly, all have similar distributions. Moreover, the PDF for $q = 3$, 4, and 5 collapse onto an approximate single curve, indicating a scaling over the entire equity market

[1] To avoid the discreteness for small τ (Eichner et al. [15] suggested a power law function for this range) and large fluctuations for very large τ, we choose the range of $0.01 \leq CDF \leq 0.50$ and also use a as a free parameter to perform the stretched exponential fit.

we plot the PDF of γ values for five thresholds, $q = 2$ to 6.[2] All five distributions have similar shapes, with the range from 0 to 0.6 and the peak around 0.3. Note, the differences between the curves of $q = 3$, $q = 4$, and $q = 5$ are in the range of their error bars. This significant similarity supports the validity of scaling for a wide range of thresholds for the entire equity market.

The distributions of returns and volatilities of many financial time series have power law tails [1–10]. As for the return interval, Yamasaki et al. suggested that the scaling function is also consistent with a power law tail for large intervals [16, 17]. Moreover, Bogachev and Bunde have shown that for multifractal time series the PDF of return intervals are governed by power laws [24]. Wang et al. comprehensively examined and compared the fittings of the stretched exponential PDF and power law tail for the 1,000 most-traded stocks [28]. They found that the power law tail can not be ruled out, though the stretched exponential function seem to better describes the PDF of return intervals.

Although the scaling is remarkably universal for the return interval distributions, we can see some systematic deviations from a single universal scaling law. As an example, the PDF for larger threshold has a lower probability in medium intervals, as shown in Fig. 2a. Also, the probability of larger γ values is increasing with the increasing of the threshold, as shown in Fig. 3. To explicitly examine the scaling features of the return intervals, Wang et al. studied the moments for 500 component stocks of the S&P 500 index. They found that the moments have certain tendencies with $\langle\tau\rangle$, indicating multiscaling in the return interval distribution [25]. Ren and Zhou found similar behavior for two Chinese indices [27].

To find the origin of the differences in γ values, and to further understand the complexity of the market, we study the dependence of the γ values on four characteristics of stocks, the market capitalization, the risk (measured by the standard deviation of logarithmic price changes), the number of trades, and the return (logarithmic price change). In Fig. 4, we plot these relations for four thresholds $q = 2$, 3, 4, and 5. We can see that the curves have a similar tendency and for each panel they are closer to each other for larger thresholds. For the capitalization (Fig. 4a), γ increases and then shows a slight decrease, suggesting that companies of a certain size have the smallest correlations. For the risk, there is no crossover but a negative dependence, as shown in Fig. 4b. Larger risk means that the probability of larger volatilities is higher and the corresponding correlations are stronger. Price movement is realized by trades and the temporal structure of the volatility probably relates to the size of the trades. Counterintuitively, γ is almost insensitive to market activity, as shown in Fig. 4c. A possible reason for this is that many investors do not change their strategies just because of a dramatic change of trading frequency. In Fig. 4d we show that the dependence on the return has similar shape as that of the

---

[2] To obtain the error bars for each point in the distribution, we produced 1,000 bootstrap resamples from the empirical data set, then compute the probability density of each of these resamples, and calculate the average and standard deviation (as shown as points and error bars in Figs. 3 and 6) for each point in the distribution. For the parameter C_1 and C_2 in (8), we use a similar method to obtain the error bars.

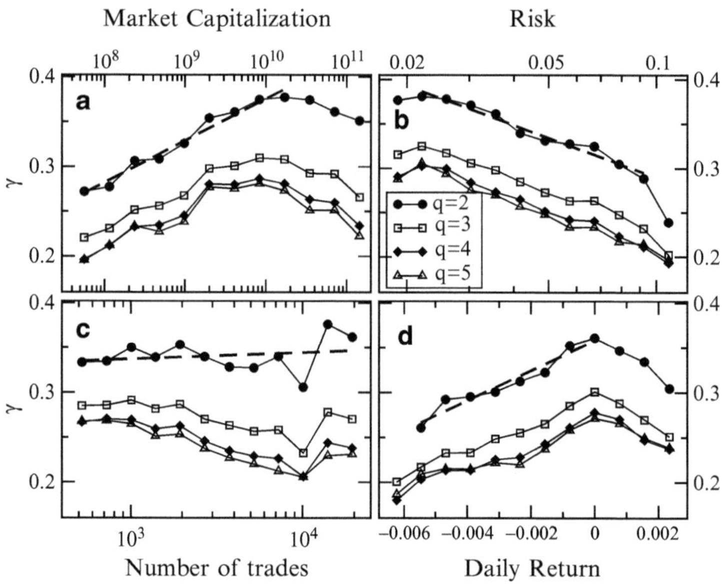

Fig. 4 Relation between the correlation exponent γ and four stock characteristics: (**a**) market capitalization, (**b**) risk, the standard deviation of daily return, (**c**) average daily number of trades, and (**d**) average daily return. Shown are curves of four thresholds $q = 2, 3, 4$, and 5. *Dashed lines* are logarithmic fits to the $q = 2$ curve (except in (**d**) where the fit is linear)

capitalization, but the peak is located around zero. This behavior suggests that the return is related to the size of the risk. For returns with large magnitude representing high volatilities, the corresponding risk is relatively high and therefore γ is small (see Fig. 4b). All these findings support the multiscaling nature of the return interval distributions, since except for Fig. 4c that the data are not horizontal.

4 Long-Range Correlations

Many financial time series are *auto-correlated*, where a value in the sequence "remembers" the previous values therefore the time series has a memory. Previous studies have shown that the return does not exhibit any linear correlations extending over more than a few minutes, but the volatility exhibits long-term correlations (see [3] and [5–7]). This leads to long periods of high volatility as well as other periods where the volatility is low ("volatility clustering"). The return intervals, generated from the volatility time series, consist of a time series. Thus, the temporal structure in the return interval time series itself is also of interest. To test the memory effect in the return interval time series, first we employ a direct method, the mean interval conditional on the immediate earlier value of the return intervals $\langle\tau|\tau_0\rangle$. Here we use a subset of values τ_0 instead of a single τ_0 value, since the statistics for the latter

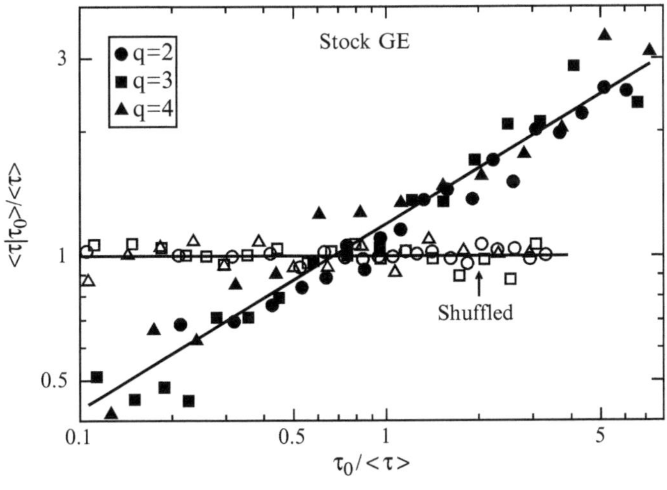

Fig. 5 Mean conditional return interval $\langle\tau|\tau_0\rangle/\langle\tau\rangle$ vs. $\tau_0/\langle\tau\rangle$ for the GE stock. Symbols are for three different thresholds $q = 2$, 3 and 4. To compare with the original data results (*filled symbols*), we also plot the corresponding results for shuffled records (*open symbols*). The distinct difference between the two records implies the memory effect in the original return interval

is poor. If there is no correlation among return intervals, $\langle\tau|\tau_0\rangle$ should be independent on τ_0 and identical to $\langle\tau\rangle$, or the quantity $\langle\tau|\tau_0\rangle/\langle\tau\rangle$ should be independent on $\tau_0/\langle\tau\rangle$ (to compare return intervals for different thresholds, we normalize τ_0 with $\langle\tau\rangle$). Interestingly, we find a power law relation between $\langle\tau|\tau_0\rangle/\langle\tau\rangle$ and $\tau_0/\langle\tau\rangle$ with exponent 0.4 ± 0.1, as shown in Fig. 5. Note that large (small) τ tend to follow large (small) τ_0. To find the origin of this dependence, we shuffle the original volatility sequence, then generate the return interval time series. We find that the corresponding mean conditional return intervals are almost constant, as plotted in Fig. 5 (open symbols). This behavior suggests that the memory effect in the return intervals is due to the long-range power law correlations in the volatility time series.

To further study the correlations in the return interval, we employ the detrended fluctuation analysis (DFA) method [32–39]. After removing trends, DFA computes the root-mean-square fluctuation $F(\ell)$ of a time series within windows of ℓ points, and determines the Hurst exponent α from the scaling function,

$$F(\ell) \sim \ell^{\alpha}. \tag{6}$$

The correlation is characterized by the Hurst exponent $\alpha \in (0, 1)$. If $\alpha > 0.5$, the records have positive long-range correlations. if $\alpha = 0.5$, no correlation (white noise), and if $\alpha < 0.5$, it has long-range anti-correlations.

Similar to the volatility [5–7], there is also a crossover in the DFA curves of the return intervals. Thus, the entire regime can be split into two regimes, $\ell < \ell^*$ and $\ell > \ell^*$, where ℓ^* is approximately 390 min or one trading day [18]. Without loss of generality, we investigate the return intervals of the 1,000 most-traded stocks

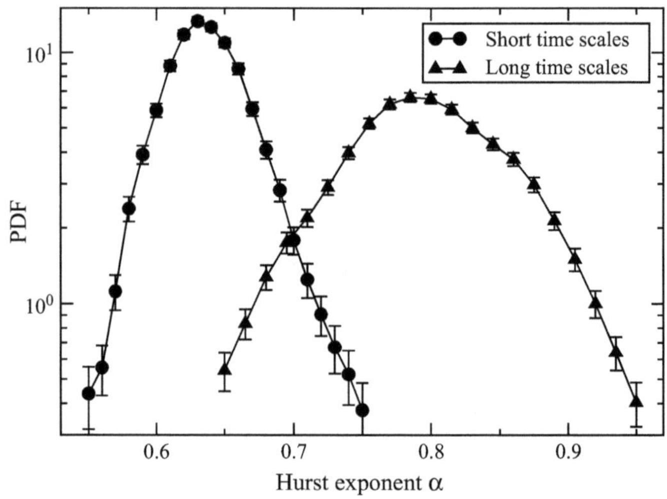

Fig. 6 Distribution of the Hurst exponent α for the return intervals of the 1,000 most-traded US stocks. The α values are obtained from two time scales: (**a**) short time scales (<1 day) and (**b**) long time scales (>1 day), split at the window size of one trading day (see Footnote 2). The two distributions are significantly separated

for a threshold $q = 2$. The distributions of the Hurst exponent α for these two regimes are displayed in Fig. 6. There is a significant difference between them, for the short time scales the α values are centered around 0.6, and for the long time scales around 0.8. Remarkably, this behavior is also consistent with that for the volatility, suggesting a common origin for the strong persistence of correlations in both volatility and return interval records. In fact the clustering in return intervals is related to the known effect of volatility clustering [8–10]. However, the correlations in the return intervals suggest also that long return intervals tend to follow long return intervals which is not trivially revealed by the correlations in the volatility.

We already know that the distribution of the return interval depends on the correlations in the volatility. What is the exact relation between the Hurst exponent α and the correlation exponent γ? Based on studying artificial long-term linear correlated sequences, Bunde et al. suggested it to be [11–14]

$$\alpha = 1 - \frac{\gamma}{2}. \tag{7}$$

To test this relation in real data, we plot in Fig. 7 the α values for the volatility vs. the γ values for the return intervals. Here the α values are obtained for the long time scale. There is an apparent dependence between the two exponents, which can be expressed as

$$\alpha = C_1 - C_2\,\gamma. \tag{8}$$

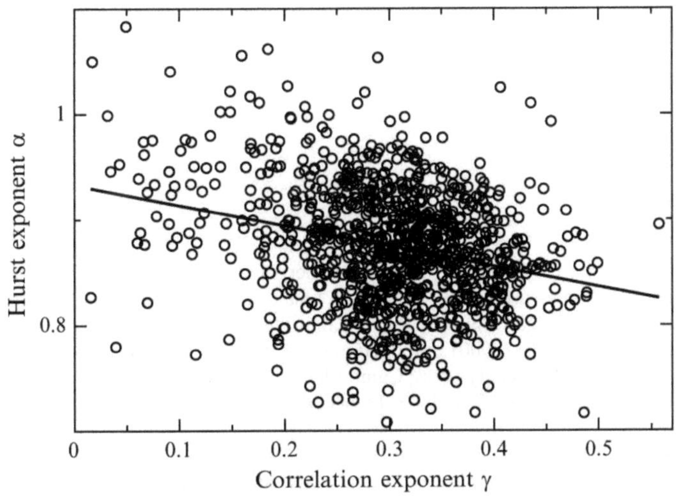

Fig. 7 Scatter plot for the 1,000 most-traded US stocks showing the dependence of the Hurst exponent α on the correlation exponent γ. The exponent α is obtained from the long time scales for the volatility, and γ is obtained from the stretched exponential fit of the CDF of the return intervals. Without loss of generality, the return intervals are for the threshold $q = 2$. Clearly, there is a significant dependence between the two exponents, suggesting that the distribution of return intervals is strongly related to the correlations in volatility. The regression line is the function $\alpha = 0.93 - 0.19\gamma$ (cf. (8))

We find that $C_1 = 0.93 \pm 0.01$ and $C_2 = 0.19 \pm 0.02$ (see Footnote 2), which are not the same as that in (7). As we know, both the volatility correlations and the return interval PDF are complicated and non-linear process exist. Nevertheless, Fig. 7 shows a relation between α and γ. This relation of (8) may help us understand the non-linearities in the volatility, which will be the subject of a future study.

5 Summary

In summary, to study the temporal structure in the volatility, we analyzed the properties of the volatility return interval for the 1,000 most-traded stocks in the US equity market. We found that there is good scaling in the distribution, which can be approximated by a stretched exponential (although the power law tail can not be ruled out). Importantly, the scaling is persistence for a wide range of thresholds, a broad scale of sampling intervals, many stocks, and diverse financial markets. Further analysis showed that the return interval distribution has a systematic tendency away from a single scaling law, and the correlation exponent γ has a weak dependence on the company size, risk and return. We also showed a significant memory effect in the return intervals time series. Furthermore we showed that the correlation exponent γ significantly depends on the Hurst exponent α, indicating that the distribution of the return intervals is strongly related to the correlations in the volatility.

Acknowledgements We thank A. Bunde, L. Muchnik, P. Weber, W.-S. Jung and I. Vodenska-Chitkushev for collaboration on many aspects of this research, and the NSF and Merck Foundation for financial support.

References

1. Kondor I, Kertész J (1999) Econophysics: an emerging science. Kluwer, Dordrecht
2. Bouchaud J-P, Potters M (2000) Theory of financial risk: from statistical physics to risk management. Cambridge University Press, Cambridge
3. Mantegna R, Stanley HE (2000) Introduction to econophysics: correlations and complexity in finance. Cambridge University Press, Cambridge
4. Johnson NF, Jefferies P, Hui PM (2003) Financial market complexity. Oxford University Press, New York
5. Liu Y, Gopikrishnan P, Cizeau P, Meyer M, Peng C-K, Stanley HE (1999) Phys Rev E 60:1390
6. Plerou V, Gopikrishnan P, Gabaix X, Amaral LAN, Stanley HE (2001) Quant Finance 1:262
7. Plerou V, Gopikrishnan P, Stanley HE (2005) Phys Rev E 71:046131. For application to heartbeat intervals, see Ashkenazy Y, Ivanov PCh, Havlin S, Peng C-K, Goldberger AL, Stanley HE (2001) Phys Rev Lett 86:1900
8. Lux T, Marchesi M (2000) Int J Theor Appl Finance 3:675
9. Giardina I, Bouchaud J-P (2001) Physica A 299:28
10. Lux T, Ausloos M (2002) In: Bunde A, Kropp J, Schellnhuber HJ (eds) The science of disasters: climate disruptions, heart attacks, and market crashes. Springer, Berlin, p 373
11. Bunde A, Eichner JF, Havlin S, Kantelhardt JW (2004) Physica A 342:308
12. Bunde A, Eichner JF, Kantelhardt JW, Havlin S (2005) Phys Rev Lett 94:048701
13. Livina VN, Havlin S, Bunde A (2005) Phys Rev Lett 95:208501
14. Eichner JF, Kantelhardt JW, Bunde A, Havlin S (2006) Phys Rev E 73:016130
15. Eichner JF, Kantelhardt JW, Bunde A, Havlin S (2007) Phys Rev E 75:011128
16. Yamasaki K, Muchnik L, Havlin S, Bunde A, Stanley HE (2005) Proc Natl Acad Sci USA 102:9424
17. Yamasaki K, Muchnik L, Havlin S, Bunde A, Stanley HE (2005) In: Takayasu H (ed) Proceedings of the third Nikkei econophysics research workshop and symposium, the fruits of econophysics, Tokyo, November 2004. Springer, Berlin, p 43
18. Wang F, Yamasaki K, Havlin S, Stanley HE (2006) Phys Rev E 73:026117
19. Wang F, Weber P, Yamasaki K, Havlin S, Stanley HE (2007) Eur Phys J B 55:123
20. Jung W-S, Wang FZ, Havlin S, Kaizoji T, Moon H-T, Stanley HE (2008) Eur Phys J B 62:113
21. Qiu T, Guo L, Chen G (2008) Physica A 387:6812
22. Ren F, Guo L, Zhou W-X (2008) Physica A 388:881
23. Vodenska-Chitkushev I, Wang FZ, Weber P, Yamasaki K, Havlin S, Stanley HE (2008) Eur Phys J B 61:217
24. Bogachev MI, Eichner JF, Bunde A (2007) Phys Rev Lett 99:240601; Bogachev MI, Bunde A (2008) Phys Rev E 78:036114
25. Wang F, Yamasaki K, Havlin S, Stanley HE (2008) Phys Rev E 77:016109
26. Wang F, Yamasaki K, Havlin S, Stanley HE (2009) Phys Rev E 79:016103
27. Ren F, Zhou W-X (2008) Europhys Lett 84:68001
28. Wang F, Yamasaki K, Havlin S, Stanley HE (2009) In: Zhou J (ed) Complex 2009, Part I. Lecture Notes of ICST, vol 4. Springer, Shanghai, p 3
29. Black F, Scholes M (1973) J Polit Econ 81:637
30. Cox JC, Ross SA (1976) J Financ Econ 3:145; Cox JC, Ross SA, Rubinstein M (1979) J Financ Econ 7:229
31. Altmann EG, Kantz H (2005) Phys Rev E 71:056106
32. Peng C-K, Buldyrev SV, Havlin S, Simons M, Stanley HE, Goldberger AL (1994) Phys Rev E 49:1685

33. Peng C-K, Havlin S, Stanley HE, Goldberger AL (1995) Chaos 5:82
34. Bunde A, Havlin S, Kantelhardt JW, Penzel T, Peter J-H, Voigt K (2000) Phys Rev Lett 85:3736
35. Hu K, Ivanov PCh, Chen Z, Carpena P, Stanley HE (2001) Phys Rev E 64:011114
36. Chen Z, Ivanov PCh, Hu K, Stanley HE (2002) Phys Rev E 65:041107
37. Xu L, Ivanov PCh, Hu K, Chen Z, Carbone A, Stanley HE (2005) Phys Rev E 71:051101
38. Chen Z, Hu K, Carpena P, Bernaola-Galvan P, Stanley HE, Ivanov PCh (2005) Phys Rev E 71:011104
39. Kantelhardt JW, Zschiegner S, Koscielny-Bunde E, Havlin S, Bunde A, Stanley HE (2002) Physica A 316:87

Theoretical Base of the PUCK-Model with Application to Foreign Exchange Markets

Misako Takayasu, Kota Watanabe, Takayuki Mizuno, and Hideki Takayasu

Abstract We analyze statistical properties of a random walker in a randomly changing potential function called the PUCK model both theoretically and numerically. In this model the center of the potential function moves with the moving average of the random walker's trace, and the potential function is given by a quadratic function with its curvature slowly changing around zero. By tuning several parameters the basic statistical properties fit nicely with those of real financial market prices, such as power law price change distribution, very short decay of autocorrelation of price changes, long tails in autocorrelation of the square of price changes and abnormal diffusion in short time scale.

1 Introduction

The first random walk model was introduced by Bachelier in 1900 as a model of market price fluctuations [1]. Independently Einstein proposed a random walk model for the motion of colloid particles [2]. In the case of Einstein's theory the driving force of random walker is random collision of molecules in thermal equilibrium, while in the case of Bachelier's theory the driving force is mass psychology of fickle traders who are predicting the future price individually. Mathematical

M. Takayasu (✉) and K. Watanabe
Department of Computational Intelligence and Systems Science, Interdisciplinary Graduate School of Science and Engineering, Tokyo Institute of Technology, 4259-G3-52 Nagatsuta-cho, Midori-ku, Yokohama 226-8502, Japan
e-mail: takayasu@dis.titech.ac.jp; watanabe@smp.dis.titech.ac.jp

T. Mizuno
The Institute of Economic Research, Hitotsubashi University, 2-1 Naka, Kunitachi, Tokyo 186-8603, Japan
e-mail: mizuno@ier.hit-u.ac.jp

H. Takayasu
Fundamental Research Group, Sony Computer Science Laboratories, 3-14-13 Higashigotanda, Shinagawa-ku, Tokyo 141-0022, Japan
e-mail: takayasu@csl.sony.co.jp

M. Takayasu et al. (eds.), *Econophysics Approaches to Large-Scale Business Data and Financial Crisis*, DOI 10.1007/978-4-431-53853-0_5, © Springer 2010

formulation of Bachelier's theory was developed into highly sophisticated manner after his death in the field of financial technology and it is now widely accepted in the real world finance.

Earnest scientific verification of random walk assumptions for market price fluctuations started in the middle of 1990s with the advent of the new field, Econophysics, until that time high frequency tick-by-tick market data was not accessible for scientists. Roughly speaking the amount of tick-by-tick data is about ten thousand times denser than traditional daily market data, and intensive analysis of high quality financial market data from physicists' view has clarified that market price fluctuations are not simple random walks [3]. Empirically stylized facts which clearly deviate from a naive random walk model can be summarized by the following four characteristics:

1. The distribution of price change in a unit time has nearly symmetric long tails approximated roughly by power laws. This property was firstly pointed out by Mandelbrot [4]. A typical value of the power law exponent is 3, however, it seems not to be a universal constant but depends on market conditions [5].
2. The autocorrelation of price change decays quickly to zero often accompanied by a negative correlation for very short time [3].
3. The magnitude of price changes called the volatility, defined by the square of price changes, is known to have a long correlation often approximated by a power law [3].
4. For large time scale the diffusion property of market price generally follows the normal diffusion in which the variance is proportional to time, however, for short time scale abnormal diffusion is observed, i.e., the variance is approximated by a fractional power of time. The estimated exponent of the power is not universal, but it seems to depend on the market condition and it is closely related to the Hurst exponent of market price fluctuations [6].

There are many variants of random walk models of market prices, however, it is not easy to reproduce all of these characteristics. For example, the Nobel prize laurelled ARCH model [7] roughly satisfies characteristics 1, 2 and 3, however, it misses the abnormal diffusion characteristics 4.

The present authors already proposed a new type of random walk model in which random walker moves in a quadratic potential function whose center is given by the moving average of the random walker's trace [8]. This model is named as PUCK model from the abbreviation of Potentials of Unbalanced Complex Kinetics. As the potential function moves according to the motion of the random walker the statistical properties are very different from the case of a fixed potential function, i.e., the Ornstein–Uhlenbeck process. We show that the characteristic 4 becomes accessible by the effect of such moving potential function. Also considering the case that the coefficient of the potential function is changing randomly, the price change dynamics is approximated by a random multiplicative process and the distribution of price changes follows a power law satisfying the characteristic 1. Long correlation of volatility is also satisfied by taking into account the effect that the coefficient of the potential function fluctuates with a long correlation.

In the next section we introduce the PUCK model and analyze its basic properties with a fixed potential coefficient in Sect. 2.1. In Sect. 2.2 we consider the case that the coefficient fluctuates randomly and show that the above four characteristics are roughly satisfied in some parameter ranges. In Sect. 3 we analyze real market data of Dollar–Yen exchange rates and confirm the above basic characteristics comparing with the PUCK model's results. The final section is devoted for discussion.

2 The PUCK Model

We consider the following form of random walk in a moving potential function $U_M(x;t)$:

$$x(t+1) - x(t) = -\frac{d}{dx} U_M(x;t)|_{x=x(t)-x_M(t)} + f(t), \tag{1}$$

$$U_M(x;t) \equiv \frac{b(t)}{M-1}\frac{x^2}{2}, \tag{2}$$

$$x_M(t) \equiv \frac{1}{M}\sum_{k=0}^{M-1} x(t-k), \tag{3}$$

where $f(t)$ is a random external noise, $b(t)$ is the coefficient of the quadratic potential, M is the size of moving average to define the center of potential function, $x_M(t)$. This model has been used as a new type of time series data analysis characterizing time-dependent stability of markets. Here, we analyze its basic properties both numerically and theoretically.

2.1 *The Case of Constant b*

Let us first consider the simplest case of a constant $b(t) = b$. The case of $b = 0$ corresponds to the simplest random walk. For $b > 0$ the random walker is attracted to the moving average of its own traces, so that the diffusion becomes slower than the case of $b = 0$. On the other hand when $b < 0$ the random walker is pushed away from the moving average of its traces and the walker diffuses faster than normal random walk. This property can be confirmed in Fig. 1a in which traces for different values of b are plotted for the same random number seed. For larger value of M the behaviors of $x(t)$ are smoother as shown in Fig. 1b. There is a sharp transition in diffusion property at $b = -2$. For $0 > b > -2$ the diffusion is faster than the normal case of $b = 0$, however, its long time behavior follows the normal diffusion law, that is, the variance is proportional to the time as shown later. When $b \leq -2$ the repulsive force from the center of the potential function is larger than the effect of additive random force, $f(t)$, and the motion of $x(t)$ is approximated by an exponential growth as shown in Fig. 1c. These cases are considered to be related to crashes or bubbles in markets. In such a case the direction of growth, either going up

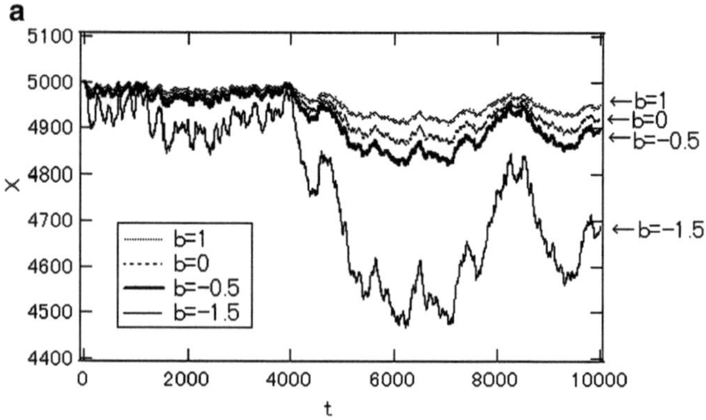

Fig. 1a Examples of $x(t)$ for $b = -1.5, -0.5, 0, 1.0$ with $M = 10$ in (3). $f(t)$ is a Gaussian random number with the mean value 0 and the standard deviation unity

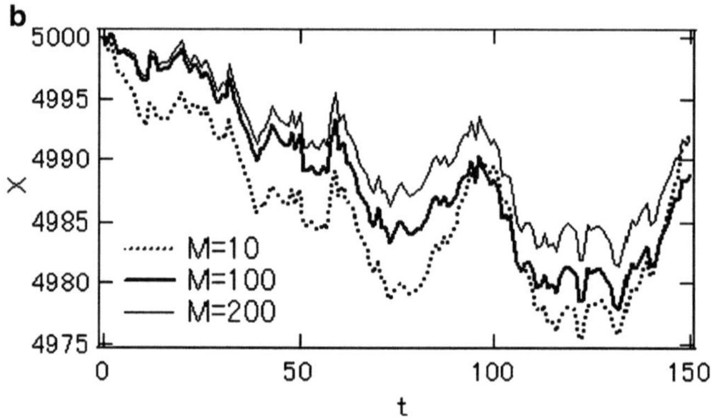

Fig. 1b Examples of $x(t)$ in the case of $b = -1.0$ for $M = 10, 100$ and 200

Fig. 1c An example of b-dependence around $b = -1.9$. For $b = -2.0$ and -2.1 the value of $x(t)$ diverges. Here, the curves look smooth as the scale of the vertical axis is about 20 times larger

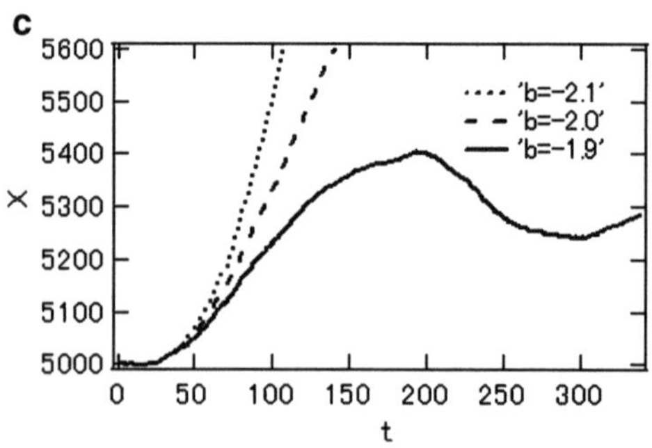

or down, is determined by the initial value condition or by the external noise, $f(t)$, as long as the potential function is symmetric. On the other hand for positive large number of b the potential force is so strong that the motion becomes a diverging oscillation as shown in Fig. 1d.

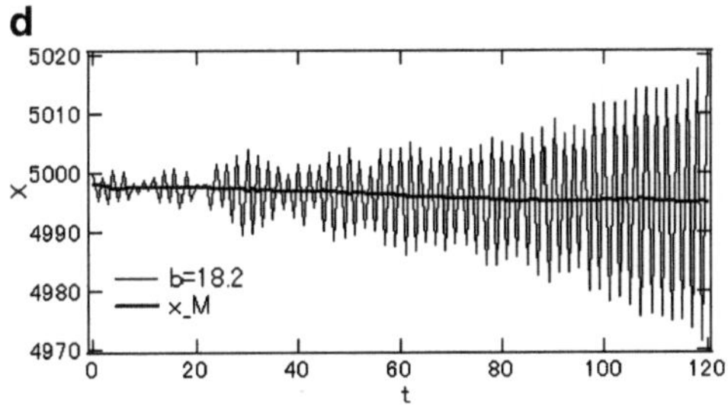

Fig. 1d An example of oscillating divergence observed in the case of $b = 18.2$ with $M = 10$

The threshold value, $b = -2$, of this exponential divergence can be analyzed theoretically in the following way by representing the basic equation (1)–(3) in terms of the velocity defined by the price difference $v(t) \equiv x(t) - x(t-1)$;

$$v(t+1) = -\frac{b}{2}\sum_{k=1}^{M-1} \omega_k v(t-k+1) + f(t), \tag{4}$$

where the weight function ω_k is given by

$$\omega_k \equiv \frac{2(M-k)}{M(M-1)}, \quad \sum_{k=1}^{M-1} \omega_k = 1. \tag{5}$$

As (4) can be viewed as an Auto-Regressive process, the condition for realization of statistically steady state can be determined by the condition that all solutions of z of the following equation is within a radius 1 in the complex plain:

$$z^{M-1} = -\frac{b}{2}\sum_{k=1}^{M-1} \omega_k z^{M-k-1}. \tag{6}$$

From this analysis it is shown that the stochastic process governed by (4) is non-stationary when $b \le -2$. At the boundary case of $b = -2$ with $M = 2$ it is confirmed that the velocity satisfies the basic random walk instead of $x(t)$,

$$v(t+1) = v(t) + f(t). \tag{7}$$

Therefore, in this case that the velocity is a non-stationary variable and the diffusion of $x(t)$ is much faster than that of normal diffusion.

Next we observe the basic statistical properties of PUCK-model with a constant b when the external noise follows a white Gaussian noise. It is easy to show that the first moment, $< v(t) >$, is always zero from (4). For the second order moments such as the variance, $< v(t)^2 >_c$, we have the following Yule–Walker equation:

$$< v(t+T)v(t) > = \sum_{k=1}^{M-1}\left(-\frac{b}{2}\right)\omega_k < v(t+T-k)v(t) > + F\delta_T, \tag{8}$$

where F is the variance of the white noise, $< f(t) f(t') > = F\delta_{t-t'}$, and δ_T is the Kronecker delta which is 1 when $T = 0$ and is 0 otherwise. For given b and M the variance, $< v(t)^2 >$, is obtained for $T = 0$ in (8). In the special case of $M = 2$, the solution is given as follows:

$$< v(t)^2 > = \frac{F}{1 - \left(\frac{b}{2}\right)^2}. \tag{9}$$

This representation is valid for $-2 < b < 2$. In this special case it is easy to prove that the distribution of $v(t)$ follows a Gaussian with the variance given by (9). Figure 2a shows a typical example of numerical result of distribution of price difference $\Delta x(T;t) \equiv x(t+T) - x(t)$ for a general case of b and M, in each case the distribution is well approximated by a Gaussian distribution.

The autocorrelation of $v(t)$, $C_v(T) \equiv < v(t+T)v(t) >_c / < v(t)^2 >_c$, is obtained directly from (8). In the case of $M = 2$ the solution is given as,

$$C_v(T) \equiv \left(-\frac{b}{2}\right)^T. \tag{10}$$

For $0 > b > -2$ this is an exponential damping, and it is an exponential damping with oscillation for $2 > b > 0$. In Fig. 2b the autocorrelations for some combinations of b and M are plotted. As confirmed from this figure, the autocorrelation is

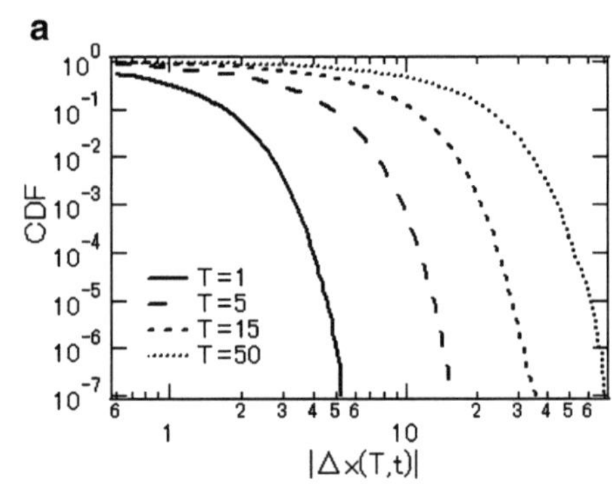

Fig. 2a Log-log plot of the cumulative distribution of price difference. $M = 10$, $b = 1.5$ with $T = 1, 5, 15, 50$

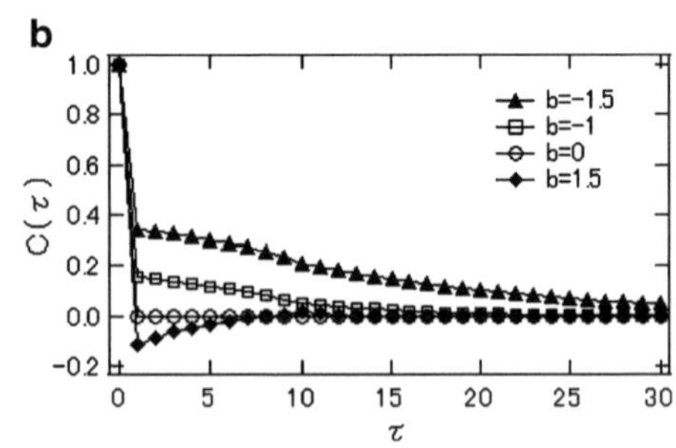

Fig. 2b The autocorrelation function for $v(t)$ with $M = 10$, $b = -1.5, -0.5, 0.0, 1.5$

always positive and decay exponentially for any negative b-value. On the other hand for a positive b-value we can find an oscillatory behavior in general.

The volatility time series is defined by $\{v(t)^2\}$ and its autocorrelation is also analyzed theoretically using (8). Here, we show analytical result for $M = 2$. In this case we have the following equation by taking a square of (8), and multiplying $v(t-T+1)^2$, then taking average over $\{f(t)\}$:

$$< v(t+1)^2 v(t+1-T)^2 > = \frac{b^2}{4} < v(t)^2 v(t+1-T)^2 > + < f(t)^2 >< v(t+1-T)^2 > . \quad (11)$$

From this equation we have the following solution for the volatility autocorrelation:

$$C_{v^2}(T) \equiv < v^2(t+T)v^2(t) >_c / \{< v(t)^2 >\}^2 = \left(\frac{b}{2}\right)^{2T} . \quad (12)$$

Examples of general cases are numerically estimated as shown in Fig. 2c. As known from this result the autocorrelation of volatility always decays exponentially like the case of theoretical solution of $M = 2$, and no long-correlation can be observed as far as the value of b is a constant.

The diffusion property can be characterized by observing the time evolution of variance $\sigma^2(t) \equiv < \{x(t) - x(0)\}^2 >$ as numerically obtained in Fig. 2d. For $b > 0$

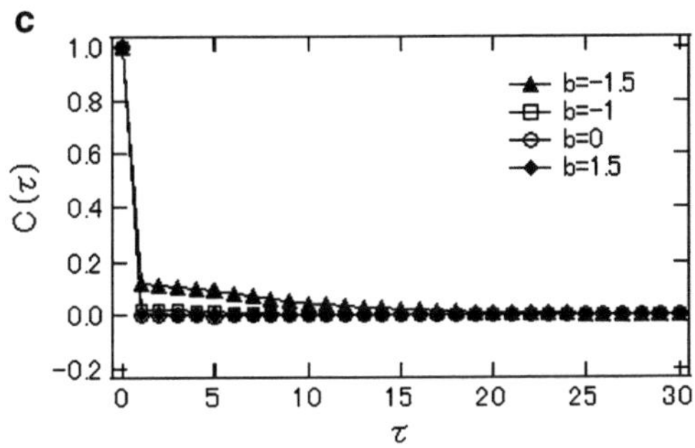

Fig. 2c The autocorrelation of $v(t)^2$ with $M = 10$, $b = 1.5, 0.5, 0.0, 1.5$

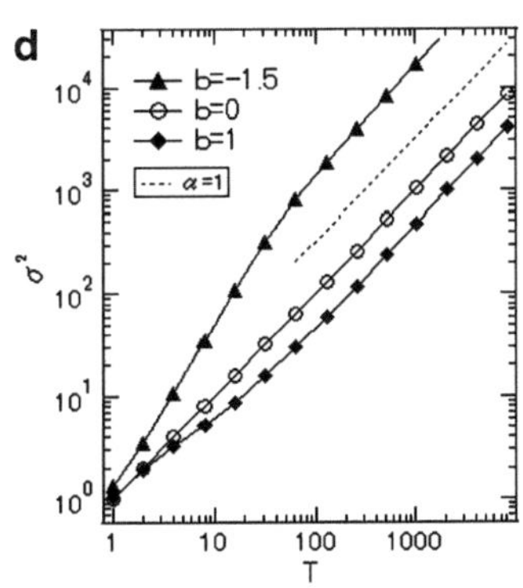

Fig. 2d Log-log plot of the variance of price diffusion for $M = 10$, $b = 1.5, 0.5$

the diffusion is slower than the case of $b = 0$ for short time scale, and for large time scale the slope becomes 1 which is equivalent to the normal diffusion. In the case of $M = 2$ we can obtain an exact solution also for this quantity. By solving (4) with $M = 2$, we have the following exact representation:

$$v(t) = \left(-\frac{b}{2}\right)^t v(0) + \sum_{s=1}^{t} \left(-\frac{b}{2}\right)^{s-1} f(t-s), \tag{13}$$

$$x(t) - x(0) = \frac{1-\left(-\frac{b}{2}\right)^t}{1-\left(-\frac{b}{2}\right)} v(0) + \sum_{s=0}^{t-1} \frac{1-\left(-\frac{b}{2}\right)^{t-s-1}}{1-\left(-\frac{b}{2}\right)} f(s). \tag{14}$$

Then, taking average over the square of (14), we have the solution for $t \geq 1$:

$$< \{x(t) - x(0)\}^2 > = \frac{\left(1-\left(-\frac{b}{2}\right)^t\right)^2}{\left(1+\frac{b}{2}\right)^2} < \{v(0)\}^2 >$$
$$+ \frac{F}{\left(1+\frac{b}{2}\right)^2} \left\{ t + b\frac{1-\left(-\frac{b}{2}\right)^{t-1}}{1+\frac{b}{2}} + \frac{b^2}{4}\frac{1-\left(\frac{b^2}{4}\right)^{t-1}}{1-\frac{b^2}{4}} \right\}. \tag{15}$$

We have the diffusion constant for large t as follows [9]:

$$D_x = \frac{4F}{(2+b)^2}. \tag{16}$$

Abnormal diffusion at small t can be approximated by assuming the following fractional power law,

$$< \{x(t) - x(0)\}^2 > \propto t^{\alpha}, \tag{17}$$

where α can be determined approximately from small t. As an extreme case we can evaluate α from $t = 0$, 1 and 2:

$$\alpha = \frac{\log\{< x(2)^2 > / < x(1)^2 >\}}{\log 2} = 1 + \frac{\log\left(1 - \frac{b}{2} + \frac{b^2}{8}\right)}{\log 2}. \tag{18}$$

For b close to 0 we have the abnormal diffusion exponent as

$$\alpha \approx 1 - 0.72b. \tag{19}$$

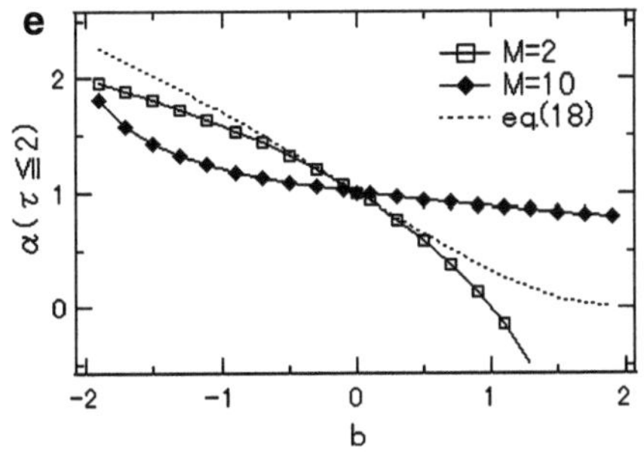

Fig. 2e Numerically estimated value of α for small t. The *dotted line* shows the theoretical value given by (18)

This approximation holds for the range of $-0.5 < b < 0.5$ as shown in Fig. 2e. For larger value of M the behavior of estimated α deviates from (18), however, qualitative behaviors are the same, i.e., slower abnormal diffusion for $b > 0$ and faster abnormal diffusion for $b < 0$.

2.2 *The Case of Random $b(t)$*

Next we consider the case that the potential coefficient value changes randomly with time. As empirically estimated value of $b(t)$ is known to be fluctuating around 0 [10], we assume that $b(t)$ follows a random walk in a fixed potential function, i.e., an Ornstein–Uhlenbeck process as follows:

$$b(t+1) - b(t) = -c_0 b(t) + g(t), \tag{20}$$

where c_0 is a positive constant in the range of [0,1] and $g(t)$ is a normal Gaussian noise with zero mean and the variance G. By this effect the value of $b(t)$ fluctuates spontaneously and the statistics of $x(t)$ changes accordingly. Examples of the set of time evolutions of $b(t)$, $x(t)$ and the volatility $|v(t)|$ are shown in Fig. 3a–c for typical values of G and c_0. The parameters G and c_0 plays the central role for the statistical properties of $x(t)$ and $v(t)$.

In the case that the value of c_0 is relatively large compared with the variance of the noise term G as typically shown in Fig. 3a, fluctuation of the value of $b(t)$ is concentrated around 0, then the behavior of $x(t)$ looks similar to the simple normal random walk (Fig. 3a). In this case the statistical properties of prices are confirmed to be nearly equivalent to the case of normal random walk, i.e., a normal distribution of the price difference, quick decay of the autocorrelations for both the market price and the volatility, and no abnormal diffusion of prices.

As shown in Fig. 3b when the value of c_0 is intermediate, it is confirmed that the volatility clustering caused by the fluctuation of $b(t)$ can be observed. When the value of c_0 is close to 0 like the case of Fig. 3c the fluctuation amplitude of $b(t)$ becomes very large, and there is a finite possibility that the value of $b(t)$ falls into the parameter range of non-stationary condition $b(t) \leq -2$. In such a case the behavior

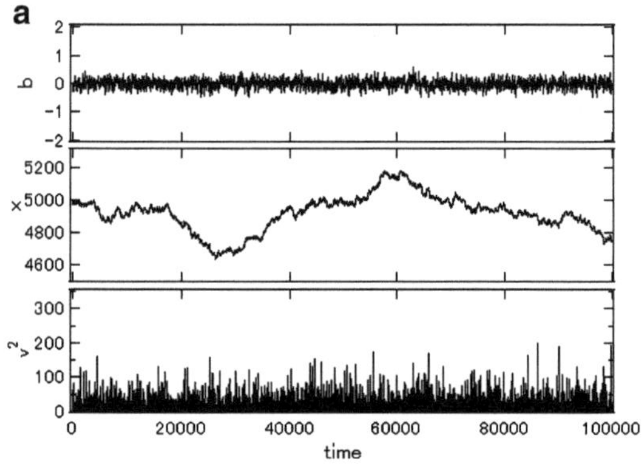

Fig. 3a Simulation results based on (20) with parameters $c_0 = 0.02$ and $G = 0.000784$. Value of $b(t)$ (*top*), market prices (*middle*), volatility (*bottom*)

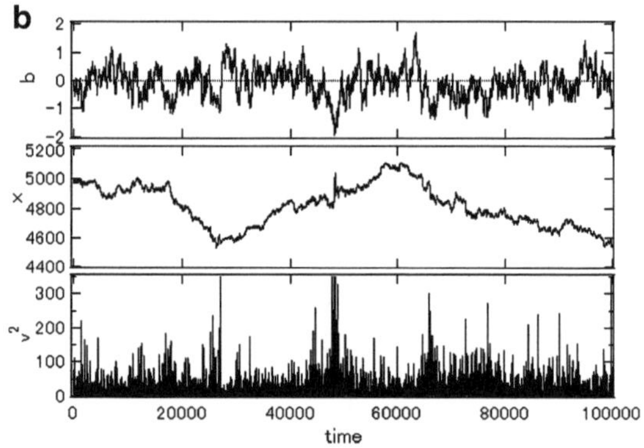

Fig. 3b Simulation results based on (20) with parameters $c_0 = 0.0015$ and $G = 0.000784$. Value of $b(t)$ (*top*), market prices (*middle*), volatility (*bottom*)

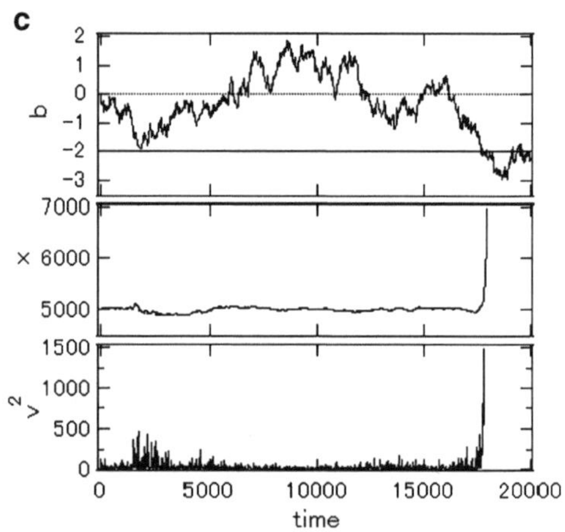

Fig. 3c Simulation results based on (20) with parameters $c_0 = 0.0001$ and $G = 0.000784$. Value of $b(t)$ (*top*), market prices (*bottom*). The plot of volatility is skipped as the price changes in the encircled period in which $b(t) < -2$ are too large compared with the other periods

of $x(t)$ switches between random walk phase and exponential growth phase in a random manner and the whole process becomes quite unstable and non-stationary.

The steady distribution of $b(t)$ is solved analytically just like the case of $v(t)$ with $M = 2$ and it is given by the following normal distribution when the random

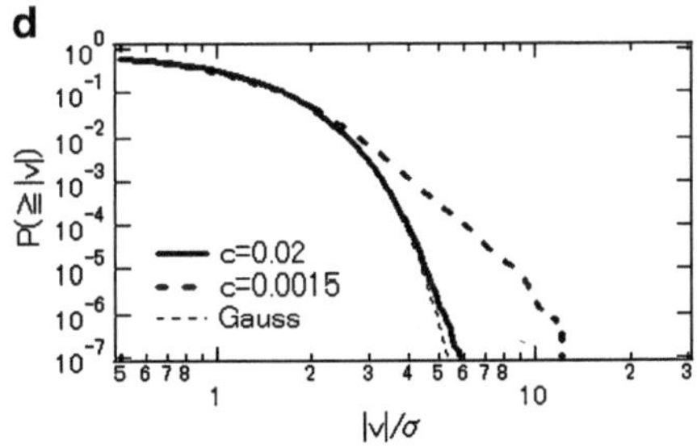

Fig. 3d Log-log plot of the cumulative distribution of price difference. The parameters are $c_0 = 0.02, 0.0015, 0.0001$ with $G = 0.000784$

noise $g(t)$ is an independent Gaussian distribution with variance G:

$$p(b(t)) = \frac{1}{\sqrt{2\pi}\sigma_b(c_0)} e^{-\frac{b(t)^2}{2\sigma_b(c_0)^2}}, \quad \sigma_b(c_0)^2 = \frac{G}{2c_0 - c_0^2}. \tag{21}$$

As known from this solution the probability of occurrence of $b(t) \leq -2$ always takes a finite value, therefore, theoretically there is a finite possibility that the market price moves nearly monotonically for a finite period. However, in the following discussion we consider the case that c_0 is not so small compared with G that the probability of realization of such non-stationary behavior is negligibly small.

In Fig. 3d a typical distributions of $v(t)$ is plotted for different values of c_0. Contrary to the case of constant b it is confirmed that the distributions are well-approximated by power laws in any case. Such power law like behaviors can be understood by considering the case of $M = 2$. As mentioned above the distribution of $v(t)$ in the case of fixed $b(t) = b$ is given by a normal distribution with the variance given by $F/(1 - b^2/4)$. Assuming that the change of $b(t)$ is slow enough and we can evaluate the distribution of $v(t)$ by superposition of such normal distributions with the weight of the distribution of $b(t)$:

$$p(v) \approx \int_{-2}^{2} \frac{\sqrt{1 - b^2/4}}{\sqrt{2\pi F}} e^{-\frac{1-b^2/4}{2F}v^2} \frac{\sqrt{2c_0 - c_0{}^2}}{\sqrt{2\pi G}} e^{-\frac{2c_0 - c_0{}^2}{2G}b^2} db. \tag{22}$$

By introducing a new variable $B = 1 - b^2/4$, we have the following form,

$$p(v) \approx \frac{\sqrt{2c_0 - c_0{}^2}}{2\pi\sqrt{FG}} e^{-2\frac{2c_0 - c_0{}^2}{G}} \int_0^1 \frac{\sqrt{B}}{\sqrt{1-B}} e^{-\left(\frac{v^2}{2F} - \frac{2c_0 - c_0{}^2}{2G}\right)B} dB. \tag{23}$$

Evaluating the integral with respect to B, the asymptotic behavior for large $|v|$ is estimated as

$$p(v) \approx \frac{\sqrt{2c_0 - c_0{}^2}}{2\pi\sqrt{FG}} e^{-2\frac{2c_0 - c_0{}^2}{G}} \left\{ 0.88 \left(\frac{v^2}{2F}\right)^{-\frac{3}{2}} + 0.66 \left(\frac{v^2}{2F}\right)^{-\frac{5}{2}} + \cdots \right\}. \tag{24}$$

We have symmetric power law tails in the distribution of $v(t)$. In this case the cumulative distribution is approximated by the following power law form with $\beta = 2$.

$$P(\geq |v|) \propto |v|^{-\beta}. \tag{25}$$

Another limit case is solved theoretically when the change of $v(t)$ is very fast in the case of $M = 2$. Applying the formula of random multiplicative noise [11] for (4), the power law exponent β of the steady state cumulative distribution of $v(t)$ is given by solving the following equation for β:

$$1 = < \left|\frac{b}{2}\right|^{\beta} > = \frac{1}{\sqrt{2^{\beta}\pi}}\Gamma\left(\frac{\beta+1}{2}\right)\left(\frac{G}{2c_0 - c_0{}^2}\right)^{\beta}, \tag{26}$$

where $\Gamma\left(\frac{\beta+1}{2}\right)$ is the gamma function. In view of this limit the power law exponent is given by the parameters G and c_0 characterizing the distribution of $b(t)$. Empirical values of β are known to lie mainly in the range $2 < \beta < 4$, which is realizable by (26) by tuning the value of G and c_0.

The autocorrelations of $v(t)$ in the case of random $b(t)$ are plotted in Fig. 3e. It is confirmed that the autocorrelations decay always rapidly to zero. This vanish of autocorrelation is not trivial as the autocorrelation for a fixed $b(t)$ is not zero as mentioned above. Accumulation of various values of $b(t)$ with 0 mean causes this phenomenon.

The volatility autocorrelation is plotted in Fig. 3f. We can find that the tail part becomes longer for smaller value of c_0. In the case that non-stationary price motion is included, the autocorrelation tends to converge to a non-zero value for large time difference.

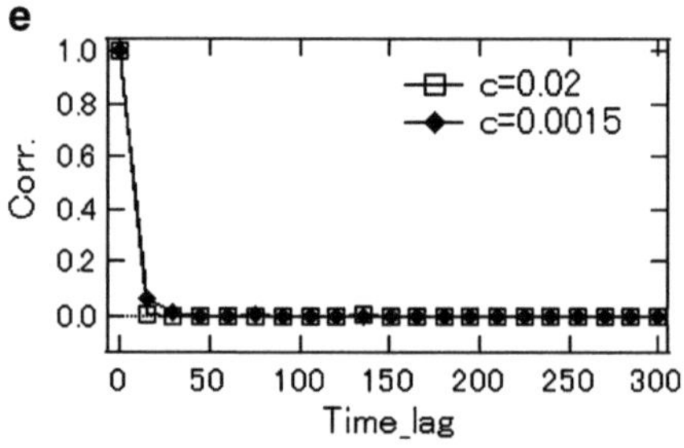

Fig. 3e The autocorrelation function for $v(t)$. The parameters are $c_0 = 0.02, 0.0015, 0.0001$ with $G = 0.000784$

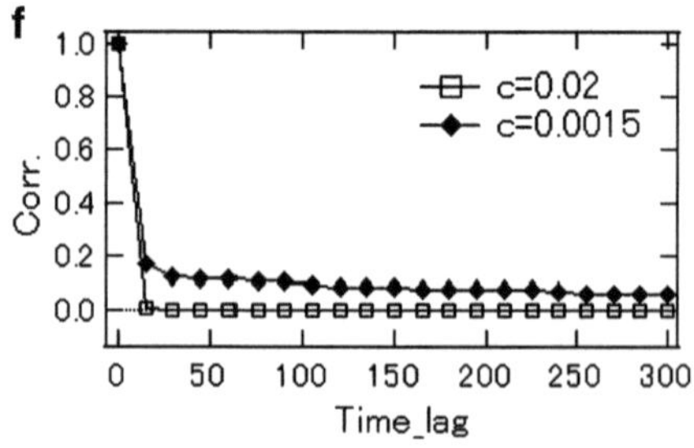

Fig. 3f The autocorrelation of $v(t)^2$. The parameters are $c_0 = 0.02, 0.0015, 0.0001$ with $G = 0.000784$

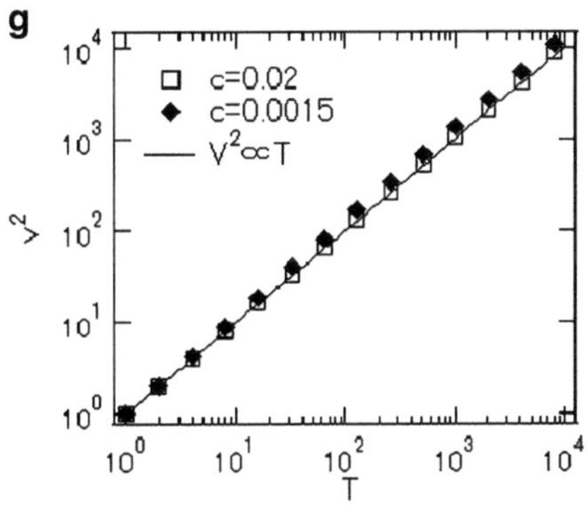

Fig. 3g Log-log plot of the variance of price diffusion. The parameters are $c_0 = 0.02, 0.0015, 0.0001$ with $G = 0.000784$

The diffusion properties are analyzed in Fig. 3g. We can find abnormal diffusion for small time scales, however, for large time scales the normal diffusion property is retained in any case.

3 Statistical Properties of a Real Financial Market

In this section we compare the model's properties with the real market data of US-dollar Japanese-yen exchange rates. This is the price of 1 dollar paid by Yen in the electronic broker system provided by ICAP. In this broker system major international banks are dealing continuously 24 h excepts weekends with the minimum unit of a deal 1 million dollar. The observation period is 2006 with time stamp in second. The number of total data points is about 3 million ticks. In the data there are four kinds of prices, deal-ask, deal-bid, best-ask and best-bid. Here, we apply deal prices which are actually transacted prices taken on the bid-side and ask-side, namely, offered orders to buy and sell.

For given time series of market prices, the PUCK analysis is done in the following procedures [8]: Firstly, we calculate the autocorrelation of price changes from the raw time series, and we apply the noise separation process based on Yule–Walker method to derive the optimal moving average [12]. After this procedure the smoothed market price time series $\{x(t)\}$ is used to define the super-moving average $x_M(t)$ by (3) with a fixed value of moving average size M. Then, we plot $x(t+1) - x(t)$ vs. $\{x(t) - x_M(t)\}/(M-1)$ for N data points using $\{x(t-N-M+2), x(t-N-M+1), \cdots, x(t+1)\}$, where the data number N is typically 500. The slope of the best-fit line for these scattered points gives the value of $-b(t)$, the curvature of the potential force.

By numerical tests using the artificial random data produced by the PUCK model with a constant value of $b(t)$, the magnitude of estimation error of $b(t)$ is estimated to be less than 0.3 and the occurrence probability of $b(t) < -1$ or $b(t) > 1$ are less than 0.1% in the statistical test [10].

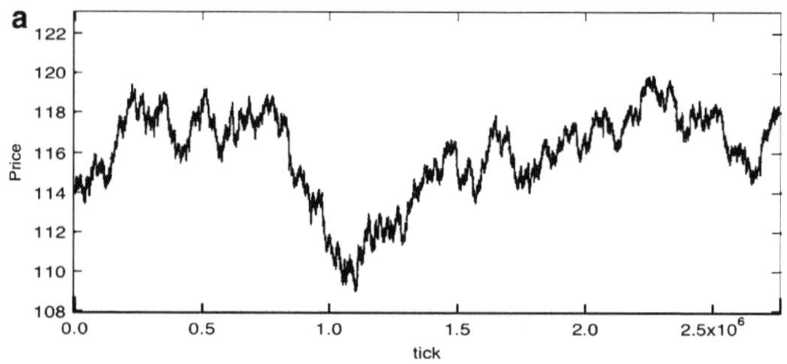

Fig. 4a An example of Dollar–Yen exchange rate time series in 2006

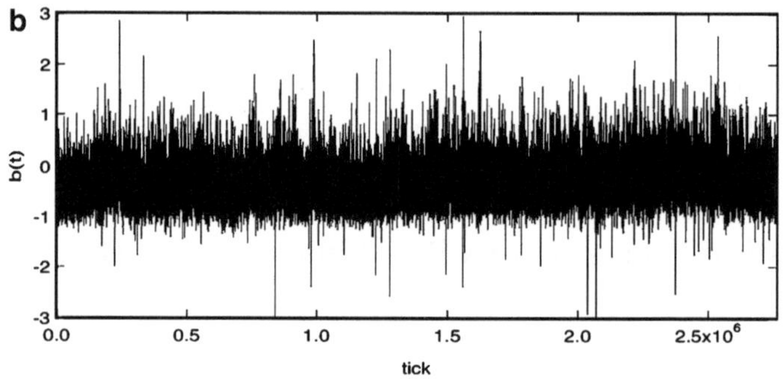

Fig. 4b The estimated values of $b(t)$ for Fig. 4a

Figure 4a shows an example of Dollar–Yen exchange rate time series, and Fig. 4b gives an estimated time series of $b(t)$ for the time series of Fig. 4a. As known from these figures, it is found that the value of $b(t)$ is always fluctuating in various scales in the market.

Here, we apply (20) as a numerical model of time evolution of $b(t)$. For the time sequence shown in Fig. 4b, we can approximate the dynamics by (20). Using such empirically estimated potential function of $\phi(b)$, the time evolution of $b(t)$ can be simulated by applying a Gaussian white noise for $g(t)$ [13]. Figure 5a is a simulated variant for real data version Fig. 4b. Figure 5b gives a simulated Dollar–Yen exchange rate fluctuation corresponding to the real data of Fig. 4a.

In Figs. 6a–d the basic statistics of this empirical model are compared with the real data [13]: The distribution of price difference, the autocorrelation of price change, the autocorrelation of volatility and the diffusion properties. It is confirmed that basic properties are roughly satisfied in all cases.

As demonstrated in these figures our model can reproduce most of the basic market properties. However, it should be noted that the basic statistics of markets are not universal in quantitative sense. For example, the slopes of the distribution or the functional forms for different observation periods may be slightly different. In the case of material systems non-stationary properties of this type rarely appear

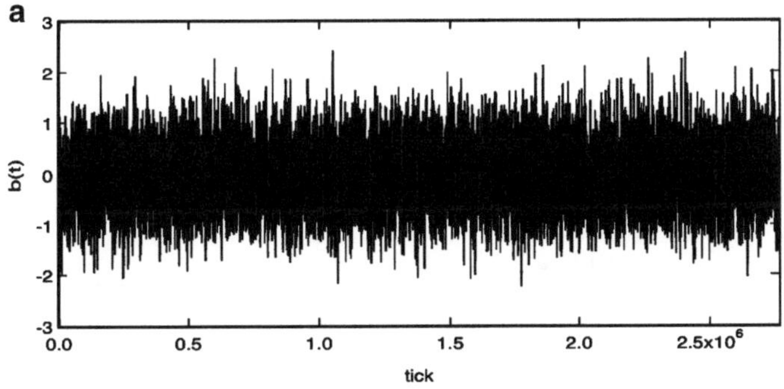

Fig. 5a An example of simulated time series of $b(t)$

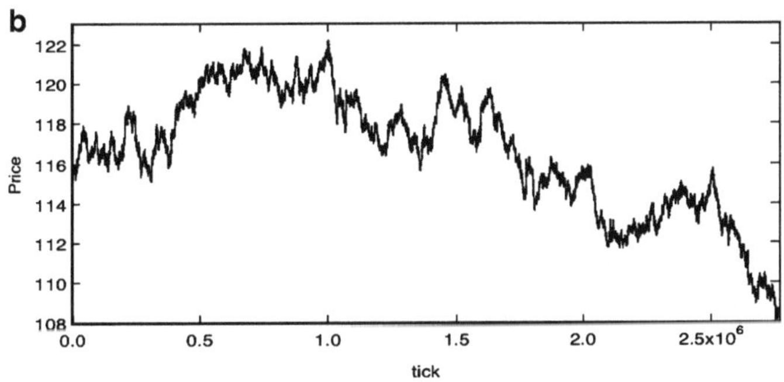

Fig. 5b An example of time series of market price assuming the fluctuation of potential coefficient $b(t)$ shown in Fig. 5a

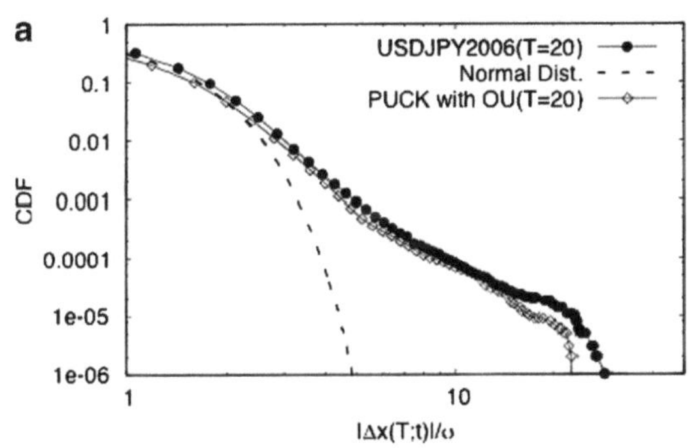

Fig. 6a Log-log plot of the cumulative distribution of Dollar–Yen rate changes, $|\Delta x(20; t)|/\sigma$, in 2006. The *line with black circle* shows the real data and the *line with diamond* shows the model

in general and we can expect permanent properties, however, for modeling human systems like the financial markets it is important to prepare a model with a wide variety and flexibility to cope with such non-stationary phenomena. In this sense the PUCK model and its generalized variants are flexible enough for description of

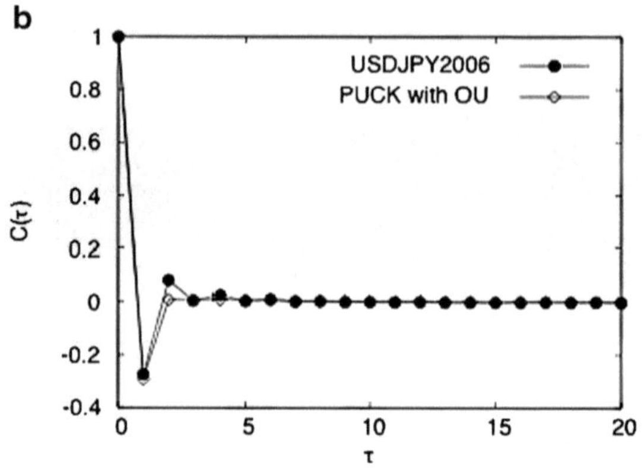

Fig. 6b The autocorrelation function for Dollar–Yen rate changes

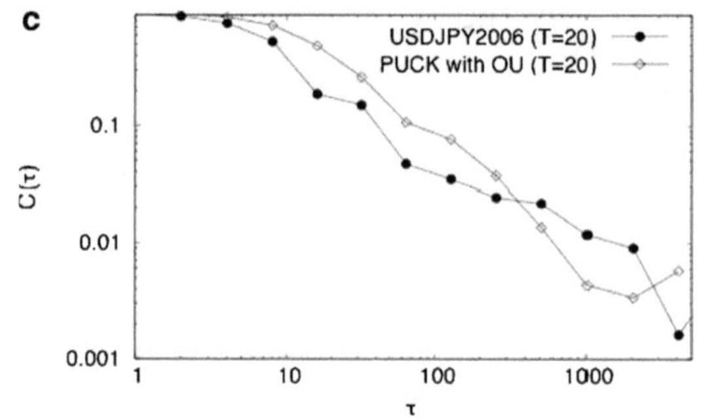

Fig. 6c The autocorrelation of volatility, $\Delta x(20;t)$^2, of Dollar–Yen exchange rates. The *line with black circle* shows the real data and the *line with diamond* shows the model

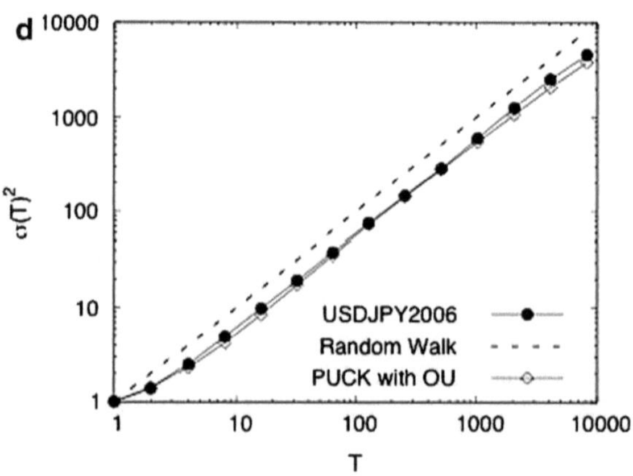

Fig. 6d Log-log plot of the variance of exchange rate diffusion. The *line with black circle* shows the real data, the *line with diamond* shows the model and the dotted line shows the case of normal diffusion

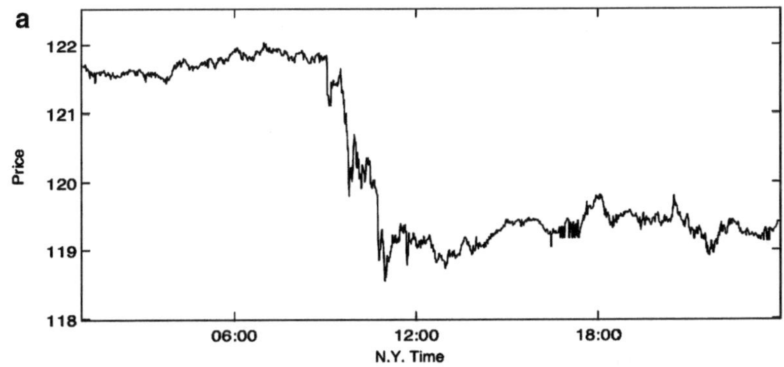

Fig. 7a Dollar–Yen exchange rates on September 11, 2001

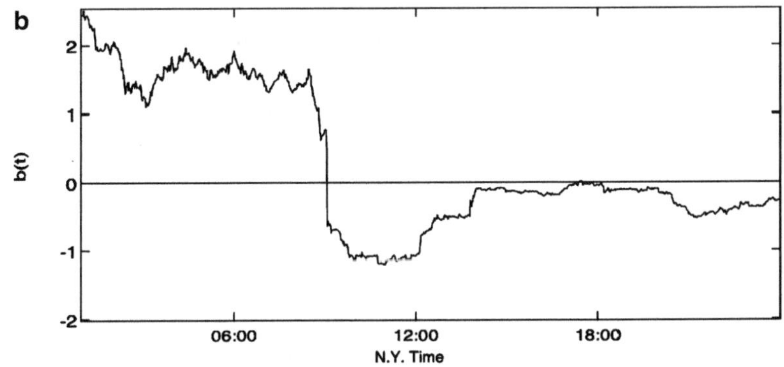

Fig. 7b Estimated value of $b(t)$ for Fig. 7a

markets as it can represent normal random walk, slower diffusion, faster diffusion and even an exponential growth by tuning the parameter, $b(t)$, and its statistics.

Figure 7a gives the special example of Dollar–Yen rate in which the market condition changed suddenly by an external news, the 9/11 terrorism in 2001, and Fig. 7b shows the corresponding value of $b(t)$. Before the terrorism Dollar–Yen market was calm and the value of $b(t)$ was larger than 1, the attractive stable state. A little before 9 a.m. in New York time, the first airplane hit the World Trade Center. The market price did not respond to this event clearly, however, the market became a neutral state, namely, the estimated value of $b(t)$ became close to 0. Right after the second airplane crashed the World Trade Center, the Dollar rate sharply dropped and accordingly the $b(t)$ value went into a negative region as clearly seen from Fig. 7b. The market kept a highly turbulent state for several hours.

As known from this example the response of $b(t)$ value is rather quick, about a few minutes delay. This quick response is non-trivial as the value of $b(t)$ is calculated using several hundreds data points which is about a few hours in real time. The reason for this sharp response is that when the market price changed suddenly the super moving average shifts certain amount and this shift causes change of the plot of $x(t+1)-x(t)$ vs. $\{x(t)-x_M(t)\}/(M-1)$. This example clearly shows that the market is reflecting the external news sensitively. Generally speaking any market model neglecting the effect of real time external news has only limited ability of description of real market.

4 Summary and Discussions

In this paper we theoretically analyzed the detail statistical properties of the new type of random walk model with a moving potential force, the PUCK model. As the center of the potential force is given by the moving average, the model with a constant potential curvature shows peculiar characteristics. When the curvature $b(t)$ is positive the random walk is slower than the case of simple random walk and when $-2 < b(t) < 0$ the random walk shows abnormally fast diffusion in short time scale, however, the large scale diffusion properties are the same as the case of simple random walk. When $b(t) \leq -2$ the system no longer shows a random walk behavior, rather it follows a dynamical exponential movement. Considering the generalized case with random autoregressive motion of $b(t)$ the basic statistical properties of this stochastic system becomes similar to that of real market price fluctuations as demonstrated in this paper.

From empirical point of view this PUCK model reproduces the major four statistical properties of market price fluctuations. Power law of price difference is explained by the effect of temporal change of $b(t)$. The quick decay of autocorrelation of price changes is realized by the statistical property that the average of the curvature is nearly zero, $< b(t) > \approx 0$. The long time autocorrelation of volatility is caused by the long time correlation in the changes of $b(t)$. The abnormal diffusion in small time scale is a direct reflection that the value of $b(t)$ takes non-zero values. The normal diffusion property in large time scale is due to the effect that most of the values of $b(t)$ satisfies the stationary condition, $|b(t)| < 2$.

From a theoretical view point the appearance of market's potential force has already been analyzed by our group using the theoretical dealer model [14]. It is shown that when dealers are all trend-followers who predict the near future price by a linear trend of latest market price changes, then, there appear a negative curvature in the potential function, $b(t) < 0$, in the market price fluctuations. On the contrary when the dealers are contrarians who predict that the market trend will turn in the near future, then, there appears a positive potential force in the market. Namely, the moving potential reflects the averaged response of the dealers to the market price changes. The value of $b(t)$ also depends both on the number of dealers and on other characteristics, so that the market potential function slowly changes in the real market spontaneously and sometimes very quickly by news like the case of 9/11 terrorism.

The PUCK model can be generalized in various ways. One direction of generalization is taking into account higher order terms in the potential function, $U_M(x;t)$ in (2). In the case that this potential function includes odd order terms of x such as x^3, it implies that the potential force works either up or down in uneven way causing a directional motion of market price [15]. It is already shown that in the special case that an asymmetric potential function changes its sign randomly at every time step, the PUCK model is reduced to the ARCH model in financial technology [10].

Another direction of generalization is the continuum limit and macroscopic limit of the PUCK model [16]. In this paper we consider discrete tick-by-tick time as a standard of time, however, we can consider a continuous time version of the model. In the continuum limit it is shown that the PUCK model is described by a Langevin type stochastic equation with time dependent mass and viscosity. In the case of stable potential function the corresponding mass is negative and the viscosity is positive. In the special case $b(t) = 0$ the corresponding mass is 0 and the price follows an ordinary diffusion equation. In the case $-2 < b(t) < 0$ both the mass and viscosity are positive, a situation similar to colloid particles in water. For $b(t) < -2$ the mass is positive and the viscosity takes a negative value, hence any small fluctuation is magnified indefinitely.

Macroscopic limit can be considered by applying renormalization to the PUCK model [16]. It is shown that the PUCK model becomes a macroscopic inflation equation by a renormalization limit. Applying such renormalization technique to the PUCK model we may expect to bridge the microscopic market phenomena and macroscopic social behaviors.

The potential forces in the market are expected to be strongly related to the distribution of demand and supply in the market called the order book. We hope to apply our model to market data with full order book information.

Interaction with other market prices is also an interesting open problem. It is well accepted that each market superficially look changing independently, however, in the case of foreign exchange market any three combination of currencies are interacting through so called the triangular arbitrage. Triangular arbitrage is the chance of getting more money by simply circulating currencies, for example, buy US dollar with Yen, buy Euro with US dollar, and buy Yen with Euro [17]. If each combination of currency exchange is done independently there occurs situation that such circulation of money causes an increase of dealer's asset. Then, those arbitrage transactions will change the market prices so that the arbitrage opportunity vanishes quickly. Also, many market prices are intuitively interacting with many others, however, mathematical description of such interaction is yet to be done.

Finally, as known from the mathematical formulation the applicability of the present model is not limited to market data only, but also to any time sequential data for finding hidden potential dynamics by analyzing the motion of moving averages in various scales. We hope that this method will contribute to analyze complicated dynamics from given data in wide field of science.

Acknowledgement The authors acknowledge Professors Takatoshi Ito and Tsutomu Watanabe for helpful discussions.

References

1. Bachelier L (1900) Théorie de la Spéculation, Doctoral disseration. Annales Scientifiques de l'Ecole Normale Superieure. Translation: Cootner PH (ed) (1964) The Random Character of Stock Market Prices. MIT Press, Cambridge, MA, pp 21–86
2. Einstein A (1905) Analen der Physik 17:549–560
3. Mantegna TN, Stanley HE (2000) An introduction to econophysics: correlation and complexity in finance. Cambridge University Press, Cambridge
4. Mandelbrot BB (1963) J Business 36:394–419
5. Gopikrishnan P, Meyer M, Amaral LAN, Stanley HE (1998) Eur Phys J B 3:139–140
6. Vandewalle N, Ausloos M (1998) Int J Mod Phys C 9:711–719
7. Engle R (1982) Econometrica 50:987–1008
8. Takayasu M, Mizuno T, Ohnishi T, Takayasu H (2005) In: Takayasu H (ed) Proceedings of practical fruits of econophysics. Springer, Tokyo, pp 29–32
9. Takayasu M, Mizuno T, Takayasu H (2006) Physica A 370:96–97
10. Takayasu M, Mizuno T, Takayasu H (2007) Physica A 383:115–119
11. Takayasu H, Sato A-H, Takayasu M (1997) Phys Rev Lett 79:966–969
12. Ohnishi T, Mizuno T, Aihara K, Takayasu M, Takayasu H (2004) Physica A 344:207–210
13. Takayasu M, Watanabe K, Takayasu H (2010) Phys Rev Lett (submitted)
14. Yamada K, Takayasu H, Ito T, Takayasu M (2009) Phys Rev E 79:051120
15. Watanabe K, Takayasu H, Takayasu M (2010) Phys Rev E 80:056110
16. Takayasu M, Takayasu H (2009) Prog Theor Phys 179(suppl):1–7
17. Aiba Y, Hatano N, Takayasu H, Marumo K, Shimizu T (2003) Physica A 324:253–257

Part 2
Financial Crisis and Macroeconomics

Financial Bubbles, Real Estate Bubbles, Derivative Bubbles, and the Financial and Economic Crisis

Didier Sornette and Ryan Woodard

Abstract The financial crisis of 2008, which started with an initially well-defined epicenter focused on mortgage backed securities (MBS), has been cascading into a global economic recession, whose increasing severity and uncertain duration has led and is continuing to lead to massive losses and damage for billions of people. Heavy central bank interventions and government spending programs have been launched worldwide and especially in the USA and Europe, with the hope to unfreeze credit and bolster consumption. Here, we present evidence and articulate a general framework that allows one to diagnose the fundamental cause of the unfolding financial and economic crisis: the accumulation of several bubbles and their interplay and mutual reinforcement have led to an illusion of a "perpetual money machine" allowing financial institutions to extract wealth from an unsustainable artificial process. Taking stock of this diagnostic, we conclude that many of the interventions to address the so-called liquidity crisis and to encourage more consumption are ill-advised and even dangerous, given that precautionary reserves were not accumulated in the "good times" but that huge liabilities were. The most "interesting" present times constitute unique opportunities but also great challenges, for which we offer a few recommendations.

D. Sornette (✉)
Department of Management, Technology and Economics, ETH Zurich, Kreuzplatz 5, 8032 Zurich, Switzerland
and
Swiss Finance Institute, c/o University of Geneva, 40 Blvd Du Pont d'Arve, 1211 Geneva 4, Switzerland
e-mail: dsornette@ethz.ch

R. Woodard
Department of Management, Technology and Economics, ETH Zurich, Kreuzplatz 5, 8032 Zurich, Switzerland
e-mail: rwoodard@ethz.ch

M. Takayasu et al. (eds.), *Econophysics Approaches to Large-Scale Business Data and Financial Crisis*, DOI 10.1007/978-4-431-53853-0_6, © Springer 2010

1 Diagnostics, Proximate and Systemic Origins of the Financial Crisis

At the time of writing (first half of April 2009), the World is suffering from a major financial crisis that has transformed into the worst economic recession since the Great Depression, perhaps on its way to surpass it. The purpose of the present paper is to relate these developments to the piling up of five major bubbles:

1. The "new economy" ICT bubble starting in the mid-1990s and ending with the crash of 2000.
2. The real-estate bubble launched in large part by easy access to a large amount of liquidity as a result of the active monetary policy of the US Federal Reserve lowering the Fed rate from 6.5% in 2000 to 1% in 2003 and 2004 in a successful attempt to alleviate the consequence of the 2000 crash.
3. The innovations in financial engineering with the CDOs (collateralized Debt Obligations) and other derivatives of debts and loan instruments issued by banks and eagerly bought by the market, accompanying and fueling the real-estate bubble.
4. The commodity bubble(s) on food, metals and energy.
5. The stock market bubble peaking in October 2007.

Since mid-2007, the media have been replete with news of large losses by major institutions and by operational and regulatory mishaps. One big question is: how deep will be the losses? Another one is: how severe could be the ensuing recession(s)?

These questions are stupendous because financial markets have transformed over the past decades from thermometers and liquidity providers of the real economy (tail moving with the dog) into "the tail wagging the dog," that is, financial markets now seem to drive the economy. To mention just one example, there are numerous indications that the corporate strategy of a given firm is significantly influenced by the value of its stock quoted in the capital markets. This is due to many different factors, including incentives (stock options held by CEOs and other top managers), and the financing channels for firm growth offered by higher market valuation, such as during mergers and acquisition operations [9].

Our starting point is that financial markets play an essential role in fostering the growth of economies in developed as well as in emergent countries. This impact of financial markets has been growing so much that it is not any longer an exaggeration to suggest that the economy has become in part controlled by a kind of "beauty contest," to paraphrase John Maynard Keynes, where one of the rules of the game for a firm is to appear "beautiful" to the financial analysts' eyes and to the investors, by meeting or even beating analysts' earning expectations. In this context, bubbles and crashes exemplify the resulting anomalies.

1.1 Nature of the Financial and Economic Crisis

Better than one thousand words, Fig. 1 compares the estimated losses for three asset classes [5]:

- Losses of US subprime loans and securities, estimated as of October 2007, at about $250 billion
- Expected cumulative loss in World output associated with the crisis, based on forecasts as of November 2008, estimated at $4,700 billion, that is, about 20 times the initial subprime loss
- Decrease in the value of stock markets, measured as the sum, over all markets, of the decrease in stock market capitalization from July 2007 to November 2008, estimated at about $26,400 billion, that is, 100 times the initial subprime loss!

While emphasizing dramatically the cascade from a relatively limited and localized event (the subprime loan crisis in the United States) to the World economy and the World stock markets, this starting point is deceptive in many ways, as will become clear below. The main misconception from our viewpoint is reducing the discussion to just the last few years. The present essay builds an argument that the present turmoil has its roots going back about 15 years in the past.

Figures 2 and 3 provide additional insights on the extraordinary character of the developments of the present crisis. First, Fig. 2 shows the total amount of

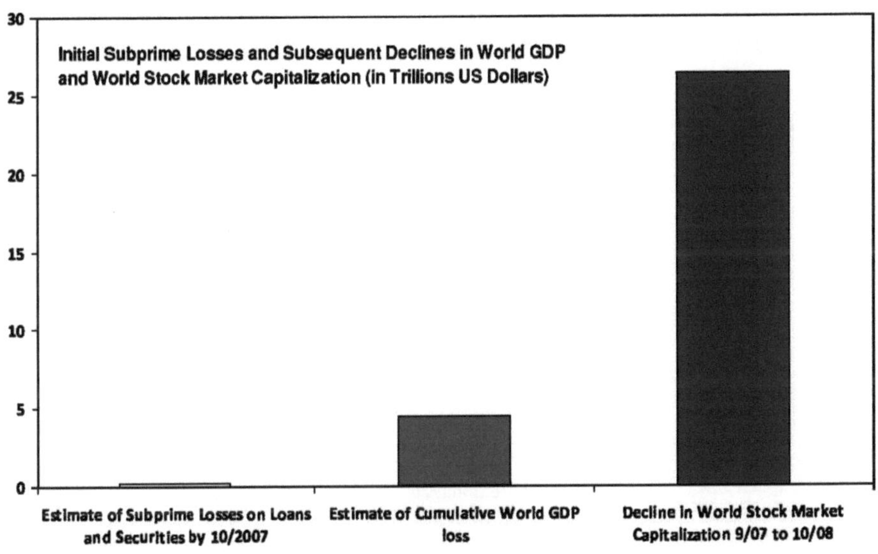

Fig. 1 Initial subprime losses (almost invisible on the left of the figure) and subsequent declines up to November 2008 in World GDP and World stock market capitalization (in trillions US dollars). Source: IMF Global Financial Stability Report; World Economic Outlook November update and estimates; World Federation of Exchanges. Reproduced from Blanchard [5]

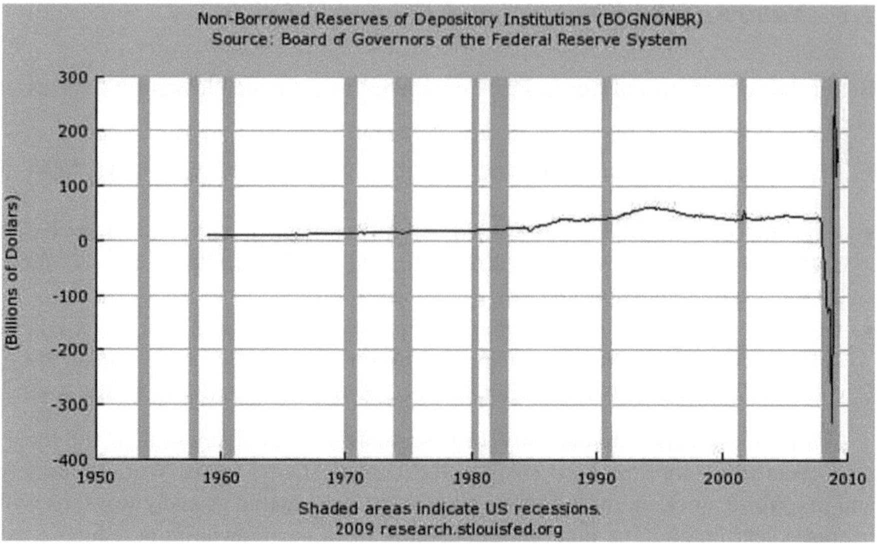

Fig. 2 Non-borrowed reserves of depository institutions from the late 1950s to April 2009. The *vertical scale* is expressed in billions of dollars. Source: Board of governors of the Federal Reserve system (2008 Federal Reserve Bank of St. Louis, http://research.stlouisfed.org)

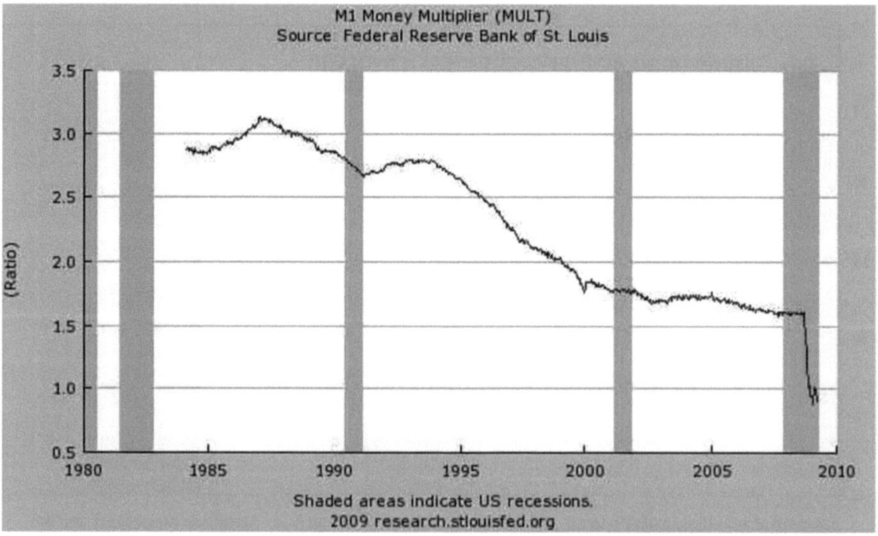

Fig. 3 Money multiplier M1 defined as the ratio of M1 to the St. Louis adjusted monetary base (http://research.stlouisfed.org/publications/mt/) from 1983 to March 2009. Source: Board of governors of the Federal Reserve system (Federal Reserve Bank of St. Louis, http://research.stlouisfed.org)

non-borrowed reserves of depository institutions (savings banks which are regulated by the Federal Deposit Insurance Corporation [FDIC]) from the late 1950s to April 2009. Notice the almost vertical drop from a level of slightly above +40 billion dollars to almost (minus!) −350 billion dollars that occurred in the last quarter of 2008, followed by a dramatic rebound to +300 billion dollars. In the last quarter of 2008, under-capitalized banks continued to hemorrhage money via losses and write-downs of over-valued assets. These banks had to borrow money from the Federal Reserve to maintain their reserves and their viability. What is striking in Fig. 2 is the exceptional amplitudes of the drop and rebound, which represent variations completely beyond anything that could have been foreseen on the basis of the previous 60 years of statistical data. Elsewhere, we refer to events such as those shown in Figs. 2 and 3, which blow up the previous statistics, as "outliers" [33, 36] or "kings" [44].

Figure 3 shows the time evolution of the M1 multiplier, defined as the ratio of M1 to the Adjusted Monetary Base estimated by the Federal Reserve Bank of St. Louis. Recall that M1 is defined as the total amount of money[1] in a given country (here the data is for the USA). Figure 3 again exhibits an extraordinary behavior, with an almost vertical fall to a level below 1! This reveals clearly the complete freezing of lending by financial institutions. Normally, the M1 multiplier is larger than 1 since money put on a checking account is used at least in part by banks to provide loans. The M1 money multiplier has recently slipped below 1. So each $1 increase in reserves (monetary base) results in the money supply increasing by $0.95. This expresses the fact that banks have substantially increased their holding of excess reserves while the M1 money supply has not changed by much. This recent development in the M1 multiplier is another illustration of the extraordinary occurrence that is presently unfolding.

This concept of "outliers" or "kings" is important in so far as it stresses the appearance of transient amplification mechanisms. As we will argue below, the occurrence of the crisis and its magnitude was predictable and was actually predicted by some serious independent economists and scholars. They were not taken seriously at a time when everything seems rosy, leading to what we refer to as an illusion of the "perpetual money machine." Of course, we are not claiming deterministic predictability for the specific unfolding scenario of the crisis, only that it was clear that the last 15 years of excesses have led to an unsustainable regime that could only blow up. In a series of papers to be reviewed below, our group has repeatedly warned about the succession of bubbles and their unsustainable trajectories [34, 64, 65, 71–73].

[1] Currency in circulation + checkable deposits (checking deposits, officially called demand deposits, and other deposits that work like checking deposits) + traveler's checks, that is, all assets that strictly conform to the definition of money and can be used to pay for a good or service or to repay debt.

1.2 Standard Explanations for the Financial Crisis

Before we construct our arguments and present the evidence, let us review briefly the standard proximal explanations that have been proposed in the literature. They all share a part of the truth and combine to explain in part the severity of the crisis. But, the full extent of the problem can only be understood from the perspective offered in Sects. 2 and 3.

1.2.1 Falling Real Estate Values

It has been argued that the immediate cause for the financial crisis is the bursting of the house price bubble principally in the USA and the UK and a few other countries, leading to an acceleration of defaults on loans, translated immediately into a depreciation of the value of mortgage-backed security (MBS) [18]. After a peak in mid-2006 (see Sect. 3.3), the real-estate market in many states plateaued and then started to decrease. A number of studies have shown indeed a strong link between house price depreciation and defaults on residential mortgages (see [16] and references therein). In particular, Demyanyk and van Hemert [17] explain that all along since 2001 subprime mortgages have been very risky, but their true riskiness was hidden by rapid house price appreciation, allowing mortgage termination by refinancing/prepayment to take place. Only when prepayment became very costly (with zero or negative equity in the house increasing the closing costs of a refinancing), did defaults took place and the unusually high default rates of 2006 and 2007 vintage loans occurred.

The explanation of the crisis based on falling real estate prices is both right and wrong: right mechanically as understood from the previous paragraph; wrong because it takes as exogenous the fall in house prices, which would suggest that it comes as a surprise. In contrast, Sect. 3 will argue that the fall in real estate value occurred as part of a larger scheme of events, all linked together.

1.2.2 Real-Estate Loans and MBS as a Growing Asset Class Held by Financial Institutions

A mortgage-backed security (MBS) is a pool of home mortgages that creates a stream of payments over time paid to its owner. The payments are taken from those produced by borrowers who have to service the interests on their debts. Figure 4 summarizes the network of agents interacting to give life to the MBS.

Developing along with the real-estate bubble, the explosive exponential growth of the nominal market value of all MBS issued from 2002 to 2007, together with its subsequent collapse, justifies referring to it as a “bubble.” According to the Securities Industry and Financial Markets Association, aggregate global CDO (collateralized Debt Obligations) issuance grew from US $150 billion in 2004, to close to US $500 billion in 2006, and to $2 trillion by the end of 2007. From 0.6

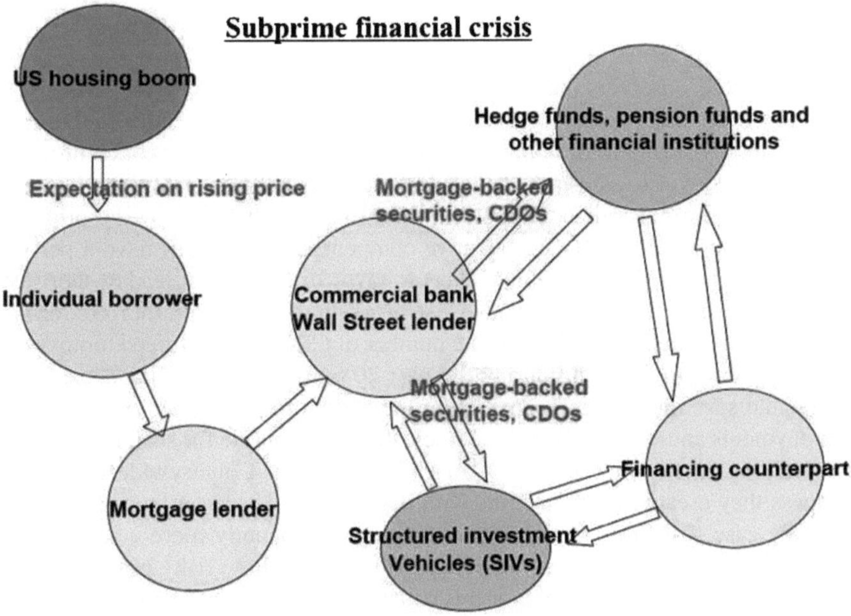

Fig. 4 Securitization, a form of structured finance, involves the pooling of financial assets, especially those for which there is no ready secondary market, such as mortgages, credit card receivables, student loans. The pooled assets are transferred to a special purpose entity and serve as collateral for new financial assets issued by the entity. The diagram shows the many involved parties

trillion dollars, the cumulative notional value of CDOs grew to 26 trillion dollars at the end of 2006. This bubble was fueled firstly by the thirst for larger returns for investors in the USA and in the rest of the World. It was made possible by a wave of financial innovations leading to the illusion that the default risks held by lenders, principally banks, could be diversified away. These innovations in financial engineering include the CDOs and other derivatives of debts and loan instruments eagerly bought by insurance companies, mutual fund companies, unit trusts, investment trusts, commercial banks, investment banks, pension fund managers, private banking organizations and so on. Since 2007, large losses by major institutions and often related operational and regulatory mishaps have been reported.

The sheer size of the nominal value of MBS held in the books of banks, insurance companies and many other institutions explains in part the amplitude of the crisis: when the deflation of the real-estate bubble started, the rate of defaults skyrocketed and the holders of MBS started to suffer heavy losses. As a consequence, many financial institutions have found themselves with insufficient equity and capital, leading to bankruptcies, fire sale acquisitions or bailouts by governments.

While compelling, this explanation is incomplete because it does not address the question of why did the MBS bubble develop. The underlying mechanisms for bubble formation are addressed in Sect. 2. This will help us understand the kind of inevitability associated with the current crisis.

1.2.3 Managers' Greed and Poor Corporate Governance Problem

It is clear to all observers that banks have acted incompetently in the recent MBS bubble by accepting package risks, by violating their fiduciary duties to the stockholders, and by letting the compensation/incentive schemes run out of control.

From executives to salesmen and trading floor operators, incentive mechanisms have promoted a generalized climate of moral hazard. Justified by the principles of good corporate governance, executive compensation packages have a perverse dark side of encouraging decision makers to favor strategies that lead to short-term irreversible profits for them at the expense of medium and long-term risks for their firm and their shareholders. Even if the number of CEOs facing forced turnover has increased three- to four-fold during the past 20 years while, simultaneously, most contractual severance agreements require the forfeiture of invested options, lump-sum payments and waiving forfeiture rules often compensate for such losses. There is something amiss when the CEOs of Citibank and of Countrywide walk out of the mess they created for their firms with nine figure compensation packages. It is often the case that firms finally turn out losing significantly more when the risks unravel than their previous cumulative gains based on these risky positions, while the decision makers responsible for this situation keep their fat bonuses. As long as the risks are borne by the firm and not equally by the decision makers, the ensuing moral hazard will not disappear. It is rational for selfish utility maximizers and it will therefore remain a major root of future financial crises.

Herding effects amplify the moral hazard factor just discussed. Indeed, performance is commonly assessed on the basis of comparisons with the average industry performance. Therefore, each manager cannot afford to neglect any high yield investment opportunity that other competitors seem to embrace, even if she believes that, on the long run, it could turn out badly. In addition, herding is often rationalized by the introduction of new concepts, e.g. "the new economy" and new "real option" valuation during the Internet bubble. And, herding provides a sense of safety in the numbers: how could everybody be so wrong? Evolutionary psychology and neuro-economics inform us that herding is one of the unavoidable consequences of our strongest cognitive ability, that is, imitation. In a particularly interesting study using functional magnetic resonance imaging on consumption decisions performed by teenagers, Berns et al. [4] have recently shown that the anxiety generated by the mismatch between one's own preferences and others' motivates people to switch their choices in the direction of the consensus, suggesting that this is a major force behind conformity.

Greed, anxiety, moral hazard and psychological traits favoring risk taking in finance were prevalent in the past and are bound to remain with us for the foreseeable future. Therefore, the question whether greed and poor governance was at the origin of the crisis should be transformed into the question of timing, that is, why these traits were let loose to foster the development of anomalous excesses in the last few years.

1.2.4 Poor Lending Standards and Deteriorating Regulations and Supervision

Philippon and Reshef provide an informative view on the question posed by the title of this section, based on detailed information about wages, education and occupations to shed light on the evolution of the US financial sector from 1906 to 2006 [51]. They find that financial jobs were relatively skill-intensive, complex, and highly paid until the 1930s and after the 1980s, but not in the interim period. They find that the determinants of this evolution are that financial deregulation and corporate activities linked to IPOs and credit risk increase the demand for skills in financial jobs, while computers and information technology play a more limited role. Philippon and Reshef's analysis shows that wages in finance were excessively high around 1930 and from the mid 1990s until 2006 [51]. It is particularly interesting to note that these two periods have been characterized by considerable excesses in the form of many bubbles and crashes. The last period is particularly relevant to our arguments presented in Sect. 3 over which a succession of five bubbles developed.

Evidence of deteriorating regulations abounds. Keys et al. [40] found that (observed) lending standards in the subprime mortgage market did deteriorate; and the main driving force of the deterioration was the securitization of those loans. Poser [52] provides important clues on the failures of the US Securities and Exchange Commission. Its most visible fault was its inability or reluctance to detect the alleged Madoff Ponzi scheme. But Poser [52] points out that the decline in SEC's regulatory and enforcement effectiveness began three decades ago. While in part explained by insufficient resources and inadequate staff training, the main cause of the SEC decline can probably be attributed to the growing prevalence of the ethos of deregulation that pervaded the US government [52].

This ethos is well exemplified by the failure to pass any legislation on financial derivatives. Going back to the 1990s, Alan Greenspan, supported successively by then Treasury Secretaries Robert Rubin and Laurence Summers, convinced the US Congress to make the fateful decision not to pass any legislation that would have supervised the development and use of financial derivatives, notwithstanding various attempts by legislators and the call from expert financiers of the caliber of Warren Buffet and Georges Soros who warned years before the present crisis about these "weapons of financial mass destruction". After being one of the most vocal supporters of the self-regulation efficiency of financial markets, Alan Greenspan is now writing in his memoirs that the villains were the bankers whose self-interest he had once bet upon for self-regulation.

The story would remain incomplete without distinguishing between the banking system which is highly regulated and the parallel or shadow banking system which is much less so [42]. In a speech in June 2008, T.F. Geithner (US Treasury secretary since January 26, 2009) said:

> The structure of the financial system changed fundamentally during the boom, with dramatic growth in the share of assets outside the traditional banking system. This non-bank financial system grew to be very large, particularly in money and funding markets. In early 2007, asset-backed commercial paper conduits, in structured investment vehicles, in

auction-rate preferred securities, tender option bonds and variable rate demand notes, had a combined asset size of roughly $2.2 trillion. Assets financed overnight in triparty repo grew to $2.5 trillion. Assets held in hedge funds grew to roughly $1.8 trillion. The combined balance sheets of the then five major investment banks totaled $4 trillion.

Given the coexisting two banking systems, the regular system being explicitly guaranteed with strict capital requirements and the shadow system being implicitly guaranteed with looser capital requirements, wealth utility maximizing bankers and investors have been naturally attracted to the second, which provided new ways to get higher yield [42]. Here, the implicit guarantee is that Bear Stearns, AIG and Merrill Lynch, while not protected by the FDIC, were protected – as the facts showed – by the belief that some firms are too big to fail.

1.2.5 Did the Fed Cause the Housing Bubble?

As a logical corollary of the previous subsection, several notable economists have blamed the Federal Reserve and the US government for failing to recognize that the shadow banking system, because it was serving the same role as banks, should have been regulated [42]. Stanford economist J.B. Taylor goes further by pointing out the errors that the Federal Reserve made in creating and fueling the crisis [67, 68], starting with the incredible monetary expansion of 2002–2003 (described more in Sect. 3.2), followed by the excesses of the expansion of government-sponsored Fannie Mae and Freddie Mac who were encouraged to buy MBS. These errors continued with the misguided diagnostic that the crisis was a liquidity problem rather than one fundamentally due to counter-party risks.

Actually, A. Greenspan, the former Chairman of the Federal Reserve stated on October 23, 2008 in a testimony to the US Congress, in reply to questions by Congressman H.A. Waxman: "I made a mistake in presuming that the self-interests of organizations, specifically banks and others, were such as that they were best capable of protecting their own shareholders and their equity in the firms." Referring to his free-market ideology, Mr. Greenspan added: "I have found a flaw. I don't know how significant or permanent it is. But I have been very distressed by that fact." Mr. Waxman pressed the former Fed chair to clarify his words. "In other words, you found that your view of the world, your ideology, was not right, it was not working," Mr. Waxman said. "Absolutely, precisely," Mr. Greenspan replied. "You know, that's precisely the reason I was shocked, because I have been going for 40 years or more with very considerable evidence that it was working exceptionally well." Greenspan also said he was "partially" wrong in the case of credit default swaps, complex trading instruments meant to act as insurance against default for bond buyers, by believing that the market could handle regulation of derivatives without government intervention.

However, in an article in the Wall Street Journal of March 11, 2009, A. Greenspan responded to J.B. Taylor by defending his policy on two arguments: (1) the Fed controls overnight interest rates, but not "long-term interest rates and the home-mortgage rates driven by them"; and (2) a global excess of savings was "the presumptive cause of the world-wide decline in long-term rates." Neither argument remains solid under scrutiny. First, the post-2002 period was characterized by one-year adjustable-rate mortgages (ARMs), teaser rates that reset in, say, two or three years. Five-year ARMs became "long-term" money. The overnight federal-funds rate that the Fed controls substantially influences the rates on such mortgages. Second, Greenspan offers conjecture, not evidence, for his claim of a global savings excess. Taylor has cited evidence from the International Monetary Fund (IMF) to the contrary, however. Global savings and investment as a share of world GDP have been declining since the 1970s, as shown by the data in Taylor's book [68].

1.2.6 Bad Quantitative Risk Models in Banks (Basel II)

Since mid-2007, an increasing number of economists, policy-makers and market operators have blamed the Basel II framework for banks' capital adequacy to be a major cause for the subprime financial crisis.

Basel II is the second of the Basel Accords, which provide recommendations on banking laws and regulations issued by the Basel Committee on Banking Supervision. Basel II was initially published in June 2004, with the purpose of creating an international standard that banking regulators can use when creating regulations on how much capital banks need to put aside to guard against the types of financial and operational risks banks face. The specific goals of Basel II are to ensure that capital allocation is more risk sensitive, to separate operational risk from credit risk, to quantify both types of risks, and to synchronize economic and regulatory capital.

First, one should point out that the implementation of Basel II was delayed by different revisions announced on September 30, 2005 by the four US Federal banking agencies (the Office of the Comptroller of the Currency, the Board of Governors of the Federal Reserve System, the Federal Deposit Insurance Corporation, and the Office of Thrift Supervision) [3]. Second, describing the actual role played by the new prudential regulation in the crisis and discussing the main arguments raised in the current debate, Cannata and Quagliariello [12] discriminate between more constructive criticisms and weaker accusations and conclude that there are no sound reasons for abandoning the philosophy underlying the Basel II framework.

The dotcom and housing bubbles as well as the development of an inflated financial sphere were actually apparent to many people. While imperfect, the so-called failure of models has played a relatively limited role in the unraveling of the crisis. More important is the desire of economists to think "things are different this time." This is reminiscent of the "new economy" mantra of the 1920s preceding the crash of Oct. 1929, the "new economy" claim of 1962 during the tronic boom preceding a severe downturn of the stock market and the "new economy" sentiment of the 1990s during the ICT bubble. Things change, but some things remain the same, such as greed and the belief that something fundamentally new is happening that calls for a downward revision of risk assessment. Herding is further amplified by the political difficulties in acknowledging independently what data tells us. B. DeLong, P. Krugman and N. Roubini are among those prominent vocal economists who have been worried about the development of the economy and the unsustainable succession of bubbles over the last decade, but they did not have the influence to make a significant impact on the US Congress or on Main Street (not to speak of Wall Street). Unfortunately, few see any pressing need to ask hard questions about the sources of profits when things are doing well. And even fewer will accept the "pessimistic" evidence that the "dancing" is going to stop, when all (superficial) evidence points to the contrary. Furthermore, one little discussed reason for the present crisis was the lack of adequate education of top managers on risks in all its dimensions and implications. How does one expect a CEO without risk culture to act on the face of the contradictory evidence of, on the one hand, a negative recommendation of the director of its risk management department and, on the other hand, great short-term potential gains in a global exuberant market? These factors, more than the "bad" models, were probably the problem with the use of quantitative models.

1.2.7 Rating Agency Failures

Credit rating agencies have been implicated as principal contributors to the credit crunch and financial crisis. They were supposed to create transparency by rating accurately the riskiness of the financial products generated by banks and financial actors. Their rating should have provided the basis for sound risk-management by mortgage lenders and by creators of structured financial products. The problem is that the so-called AAA tranches of MBS have themselves exhibited a rate of default many times higher than expected and their traded prices are now just a fraction of their face values.

To provide the rating of a given CDO or MBS, the principal rating agencies – Moody's, Fitch and Standard & Poor's – used quantitative statistical models based on Monte Carlo simulations to predict the likely probability of default for the mortgages underlying the derivatives. One problem is that the default probabilities fed into the calculations were in part based on historical default rates derived from the years 1990–2000, a period when mortgage default rates were low and home prices were rising. In doing so, the models could not factor in correctly the possibility of a general housing bust in which many mortgages are more likely to go into default. The models completely missed the possibility of a global meltdown of the real estate markets and the subsequent strong correlation of defaults. The complexity of the packaging of the new financial instruments added to the problem, since rating agencies had no historical return data for these instruments on which to base their risk assessments. In addition, rating agencies may have felt compelled to deliberately inflate their ratings, either to maximise their consulting fees or because the issuer could be shopping for the highest rating.

Recently, Skreta and Veldkamp [69] showed that all these issues were amplified by one single factor, the complexity of the new CDO and MBS. The sheer complexity makes very difficult the calibration of the risks from past data and from imperfect models that had not yet stood the test of time. In addition, the greater the complexity, the larger the variability in risk estimations and, thus, of ratings obtained from different models based on slightly different assumptions. In other words, greater complexity introduces a large sensitivity to model errors, analogous to the greater sensitivity to initial conditions in chaotic systems. If the announced rating is the maximum of all realised ratings, it will be a biased signal of the asset's true quality. The more ratings differ, the stronger are issuers' incentives to selectively disclose (shop for) ratings. Skreta and Veldkamp think that the incentives for biased reporting of the true risks have been latent for a long time and only emerged when assets were sufficiently complex that regulation was no longer detailed enough to keep them in check. Note that the abilities of ratings manipulation and shopping to affect asset prices only exist when the buyers of assets are unaware of the games being played by the issuer and rating agency. This was probably true until 2007, when the crisis exploded.

While these elements are important to understand the financial crisis, they treat the occurrence of the triggering real estate meltdown as exogenous. In addition, the extension of the leveraging on the new MBS and CDO derivatives is not explained. Overall, we need much more to fully grasp the full underpinning factors of the financial crisis.

1.2.8 Underestimating Aggregate Risks

As explained above, the wave of financial innovations has led to the illusion that the default risks held by lenders, principally banks, could be diversified away. This expectation reflects a widely spread misconception that forgets about the effects of stronger inter-dependencies associated with tighter firm networks.

Recent multidisciplinary research on self-organizing networks [6, 29, 55, 59] has shown unambiguously that loss of variety, lack of redundancy, removal of compartments, and stronger ties are all recipes for disaster. This is all the more so because the medium-sized risks are decreased, giving a false impression of safety based on the illusion that diversification works. And there is the emergence of an extremely dangerous collective belief that risks have disappeared. This led to the so-called "great moderation" in the fluctuations of GDP growths of developed economies and to absurd low risk pricing in financial markets in the last decade.

Due to globalization and the intricate networks of bank interdependencies (thousands of banks borrow and lend to each other every day in a complex ballet) [7, 23], the explosively growing losses on their MBS books and the realization that other banks were in the same situation have led to a flight for safety. As a consequence, banks have basically stopped inter-bank lending for fear of defaults of their financial counterparties. Correlatively, banks have made more rigid their previously lax lending practices into ridiculously stringent procedures offered to firms and private customers, basically threatening to freeze the real economy, which is becoming strangled by cash flow problems.

1.3 The Illusion of the "Perpetual Money Machine"

The different elements described above are only pieces of a greater process that can be aptly summarized as the illusion of the "perpetual money machine." This term refers to the fantasy developed over the last 15 years that financial innovations and the concept that "this time, it is different" could provide an accelerated wealth increase. In the same way that the perpetual motion machine is an impossible dream violating the fundamental laws of physics, it is impossible for an economy which expands at a real growth rate of 2–3% per year to provide a universal profit of 10–15% per year, as many investors have dreamed of (and obtained on mostly unrealized market gains in the last decade). The overall wealth growth rate has to equate to the growth rate of the economy. Of course, some sectors can exhibit transient accelerated growth due to innovations and discoveries. But it is a simple mathematical identity that global wealth appreciation has to equal GDP growth.

However, in the last decade and a half, this identity has been violated by an extraordinary expansion of the financial sphere. Consider first the evidence given in Fig. 5, which shows the total household net worth in the USA. expressed as a fraction of GDP from 1952 to March 2009. This ratio was relatively stable between 300% and 350% for more than 40 years. Since 1995, two major peaks towering above 450% can be observed to be followed by their collapse. The last rightmost arrow points to the peak attained in the third quarter of 2007, which is followed by a drastic drop. The figure suggests that the drop may have to continue for another 50–100% of GDP to come back to historical values. This could occur via a combination of continuing house value depreciation and stock market losses.

The second peak to the left coincides with the top of the dotcom bubble in 2000 that was followed by more than two years of strong bearish stock markets. The two other arrows to the left, one in 1962 and the other one in 1987 also coincide remarkably with two other bubbles previously documented in the literature: in 1962, the tronic "new economy" bubble collapsed with a cumulative loss of about 35% in

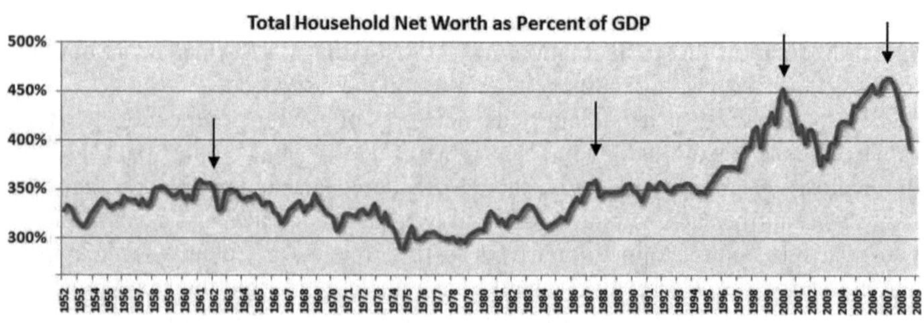

Fig. 5 Household Net Worth as a percent of GDP from 1952 to March 2009. This includes real estate and financial assets (stocks, bonds, pension reserves, deposits, etc.) net of liabilities (mostly mortgages). The data is from http://www.federalreserve.gov/releases/z1/Current/z1r-5.pdf (11 Dec. 2008). Adapted from http://www.calculatedriskblog.com

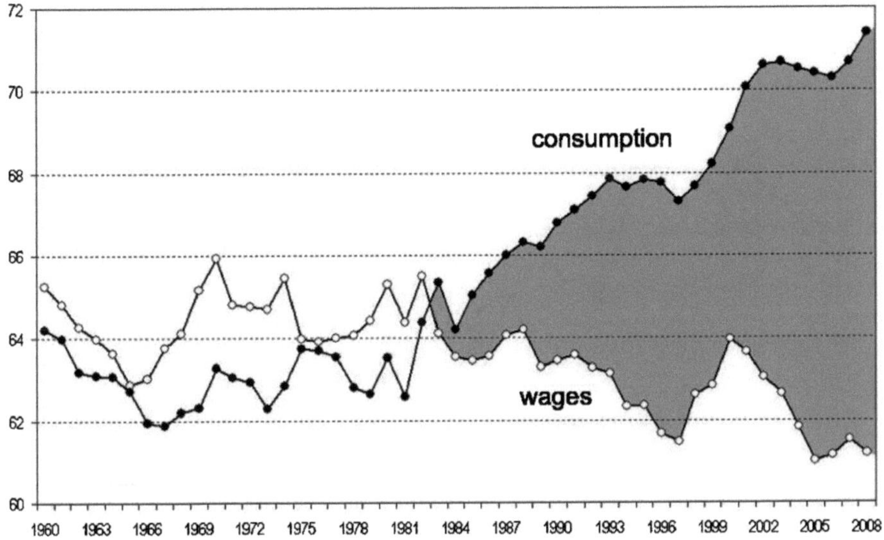

Fig. 6 Share of wages and of private consumption in Gross Domestic Product (GDP) for the United States + European Union + Japan. Source of data and graphics: Michel Husson (http://hussonet.free.fr/toxicap.xls)

3 months; on 19 October 1987, the famous Black Monday crash occurred that ended a strong spell of stock market appreciation over the previous few years.

The two Figs. 6 and 7 provide another vantage to appreciate fully the impact of the past financial sphere expansion on the global USA, European Union and Japan economies. First, Fig. 6 compares the time evolution of private consumption in the USA, European Union and Japan expressed in percentage of the GDP to the total wages. One can see that, until 1981, wages funded consumption. After 1984, the gap between consumption and wages has been growing dramatically. This means of course that consumption had to be funded by other sources of income than just wages. Figure 7 suggests that this other source of income is nothing but the increasing profits from investments, while the diminishing level of savings only partially covered the increased consumption propensity. The gap widens between profit and accumulation (gray zones) shown in Fig. 7, so as to compensate for the difference between the share of wages and the share of consumption (gray zones) shown in Fig. 6. In a nutshell, these two figures tell us that households in the USA, European Union and Japan have increased their overall level of consumption from about 64% of GDP to almost 72% of GDP by extracting wealth from financial profits. Figures for the USA alone confirm and amplify this conclusion. The big question is whether the financial profits were translated into real productivity gains and, therefore, were sustainable. It seems obvious today to everybody that financial innovations and their profits, which do not provide productivity gains in the real economy, cannot constitute a source of income on the long-term. This evidence was, however, lost as several exuberant bubbles developed during the last 15 years.

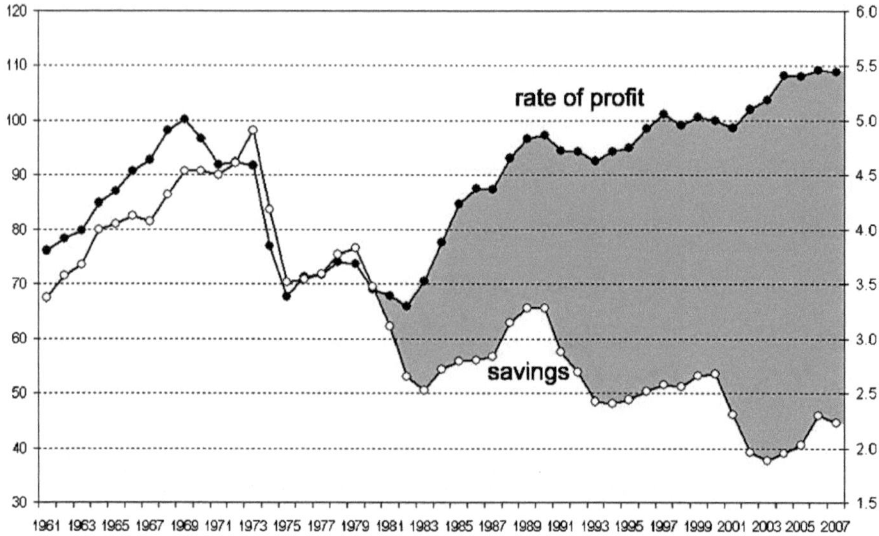

Fig. 7 Rate of profit (*left scale*) and rate of accumulation or savings (*right scale*) for the United States + European Union + Japan. The rate of accumulation is defined as the rate of growth rate of the net volume of capital × rate of profit = profit/capital (base: 100 in 2000), Source of data and graphics: Michel Husson (http://hussonet.free.fr/toxicap.xls)

The impact of financial profits on the wealth of households is well-illustrated by Fig. 8. This graph demonstrates the very strong correlation between US household wealth and the level of the stock market proxied by the Dow Jones Industrial Average. This supports the concept that financial profits have played a crucial role in the increase of household consumption discussed above. The component of wealth due to real estate appreciation during the housing bubble may have actually played an even bigger role, as it is well documented that the so-called wealth effect of house value is about twice that of the financial markets [11].

As long as the incomes drawn from financial assets are re-invested, the fortunes increase independently of any material link with the real sphere and the variation can potentially increase without serious impediment. But, financial assets represent the right to a share of the surplus value that is produced. As long as this right is not exercised, it remains virtual. But as soon as anyone exercises it, they discover that it is subject to the law of value, which means one cannot distribute more real wealth than is produced. The discrepancy between the exuberant inflation of the financial sphere and the more moderate growth of the real economy is the crux of the problem.

The lack of recognition of the fundamental cause of the financial crisis as stemming from the illusion of the "perpetual money machine" is symptomatic of the spirit of the time. The corollary is that the losses are not just the downturn phase of a business or financial cycle. They express a simple truth that is too painful to accept for most, that previous gains were not real, but just artificially inflated values

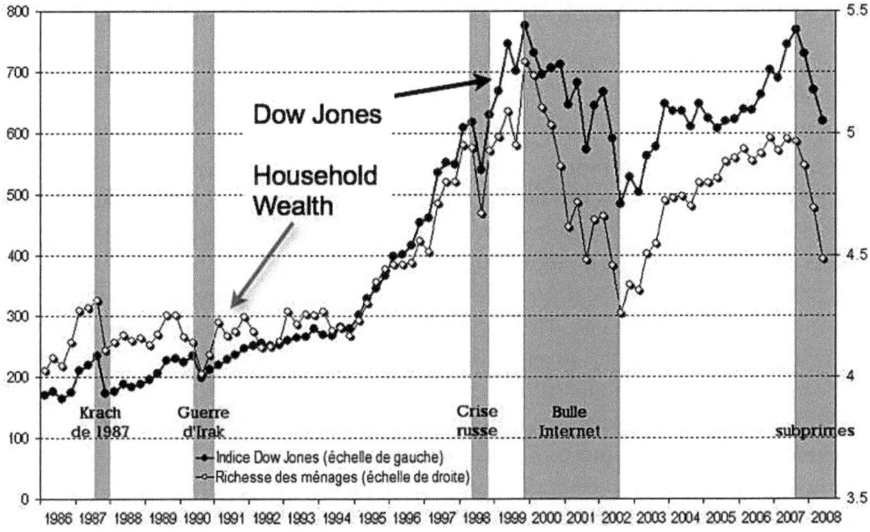

Fig. 8 The stock market level (*left scale*) and household wealth in the United States (*right scale*). The Dow Jones Industrial Average is shown with base 100 in 1960. The net wealth of households is given as a multiple of their current income. The five *vertical grey zones* outline five significant events, which are *from left to right*: the crash of 1987, the Iraq war of 1991, the Russian crisis of 1998, the crash and aftermath of the Internet bubble and the final subprime episode. Source of data and graphics: Michel Husson (http://hussonet.free.fr/toxicap.xls)

that have bubbled in the financial sphere, without anchor and justification in the real economy. In the last decade, banks, insurance companies, Wall Street as well as Main Street and many of us have lured ourselves into believing that we were richer. But this wealth was just the result of a series of self-fulfilling bubbles. As explained in more details below, in the USA and in Europe, we had the Internet bubble (1996–2000), the real-estate bubble (2002–2006), the MBS bubble (2002–2007), an equity bubble (2003–2007), and a commodity bubble (2004–2008), each bubble alleviating the pain of the previous bubble or supporting and justifying the next bubble.

The painful consequence of this brutal truth is that trying to support the level of valuation based on these bubbles is like putting gas in the "perpetual money machine." Worse, it misuses scarce taxpayer resources, increasing long-term debts and liabilities, which are already at dangerous levels in many countries.

A vivid example is provided by the market valuation of funds investing in brick-and-mortar companies often observed to be much higher at times of bubbles than the sum of the value of their components. Objective measures and indicators can be developed to quantify the ratio of wealth resulting from finance compared with the total economy. For instance, when it is measured that, on average, 40% of the income of major US firms result from financial investments, this is clearly a sign that the US economy is "building castles in the air" [47].

2 General Framework for Bubbles and Crashes in Finance

2.1 Introduction

Before reviewing the unfolding of the five bubbles over the 15 years that led to the mother of all crises, we review our approach to the diagnostic of bubbles and the explanation of crashes. A general review on models of financial bubbles encompassing much of the literature can be found in [39].

Consider the seven price trajectories shown in Fig. 9. They are seven bubbles that ended in very severe crashes. This figure illustrates the common future that crashes occur after a spell of very strong value appreciation, following a similar pattern. This suggests a common underlying mechanism.

According to the consecrated academic view that markets are efficient, only the revelation of a dramatic piece of information can cause a crash, yet in reality even the most thorough post-mortem analyses are typically inconclusive as to what this piece of information might have been. This is certainly true for the seven cases shown in Fig. 9 (see [59] for a detailed discussion).

Most approaches to explaining crashes search for possible mechanisms or effects that operate at very short time scales (hours, days, or weeks at most). Here, we

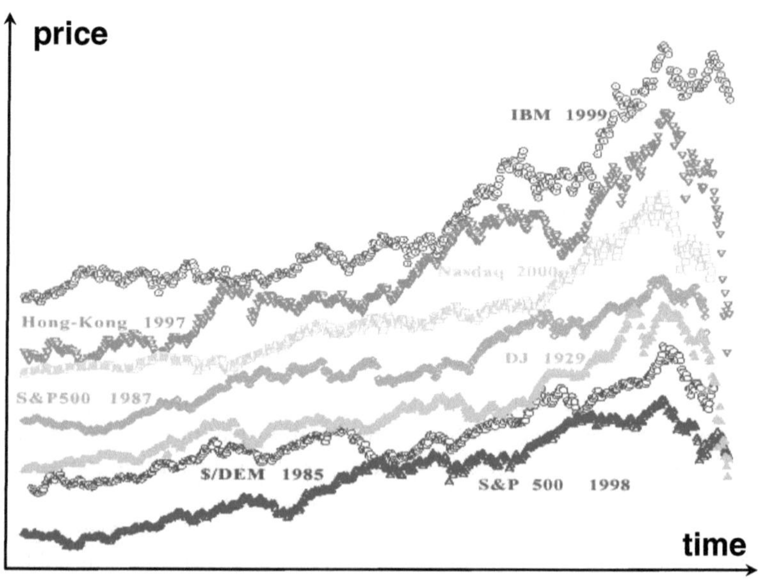

Fig. 9 Seven bubbles that ended in severe crashes. The bubble examples include stock market indices, individual companies, currencies, and for different epochs in the twentieth century. Each bubble has been rescaled vertically and translated to end at the time of the crash on the *right* of the graph. The *horizontal axis* covers approximately 2.5 years of data. The legend for each of the seven bubbles indicates the name of the asset supporting the bubble and the year when the crash occurred

build on the radically different hypothesis [59] that the underlying cause of the crash should be found in the preceding months and years, in the progressively increasing build-up of market cooperativity, or effective interactions between investors, often translated into accelerating ascent of the market price (the bubble). According to this "critical" point of view, the specific manner by which prices collapsed is not the most important problem: a crash occurs because the market has entered an unstable phase and any small disturbance or process may reveal the existence of the instability. Think of a ruler held up vertically on your finger: this very unstable position will lead eventually to its collapse, as a result of a small (or an absence of adequate) motion of your hand or due to any tiny whiff of air. The collapse is fundamentally due to the unstable position; the instantaneous cause of the collapse is secondary. In the same vein, the growth of the sensitivity and the growing instability of the market close to such a critical point might explain why attempts to unravel the proximal origin of the crash have been so diverse. Essentially, anything would work once the system is ripe.

What is the origin of the maturing instability? A follow-up hypothesis underlying this paper is that, in some regimes, there are significant behavioral effects underlying price formation leading to the concept of "bubble risks." This idea is probably best exemplified in the context of financial bubbles, such as the recent Internet example culminating in 2000 or the real-estate bubble in the USA culminating in 2006. Many studies have suggested that bubbles result from the over-optimistic expectation of future earnings (see, for instance, [56]), and many works have argued contrarily for rational explanations (for example, [24]). History provides a significant number of examples of bubbles driven by unrealistic expectations of future earnings followed by crashes. The same basic ingredients have been documented to occur repeatedly [59]. According to this view, fuelled by initially well-founded economic fundamentals, investors develop a self-fulfilling enthusiasm by an imitative process or crowd behavior that leads to the building of castles in the air, to paraphrase Malkiel [47]. Our previous research suggests that the ideal economic view, that stock markets are both efficient and unpredictable, may be not fully correct. We propose that, to understand stock markets, one needs to consider the impact of positive feedbacks via possible technical as well as behavioral mechanisms, such as imitation and herding, leading to self-organized cooperativity and the development of possible endogenous instabilities. We thus propose to explore the consequences of the concept that most of the crashes have fundamentally an endogenous, or internal, origin and that exogenous, or external, shocks only serve as triggering factors. As a consequence, the origin of crashes is probably much more subtle than often thought, as it is constructed progressively by the market as a whole, as a self-organizing process. In this sense, the true cause of a crash could be termed a systemic instability.

By studying many empirical historical examples, C. Kindleberger has identified the universal scenario associated with the development of bubbles [41] as follows (see also [59]):

$$\text{displacement} \rightarrow \text{credit creation} \rightarrow \text{euphoria} \rightarrow \text{critical financial distress} \rightarrow \text{revulsion}. \quad (1)$$

The upswing usually starts with an opportunity ("displacement") – new markets, new technologies or some dramatic political change – and investors looking for good returns. The scenario proceeds through the euphoria of rising prices, particularly of assets, while an expansion of credit inflates the bubble. In the manic euphoric phase, investors scramble to get out of money and into illiquid things such as stocks, commodities, real estate or tulip bulbs: a larger and larger group of people seeks to become rich without a real understanding of the processes involved. Ultimately, the markets stop rising and people who have borrowed heavily find themselves overstretched. This is distress, which generates unexpected failures, followed by revulsion or discredit. The final phase is a self-feeding panic, where the bubble bursts. People of wealth and credit scramble to unload whatever they have bought at greater and greater losses, and cash becomes king. The sudden fall, first in the price of the primary object of speculation, then in most or all assets, is associated with a reverse rush for liquidity. Bankruptcies increase. Liquidation speeds up, sometimes degenerating into panic. The value of collateral (credit and money) sharply contracts. Then, debt deflation ends as productive assets move from financially weak owners (often speculators or the original entrepreneurs) to financially strong owners (well capitalized financiers). This provides the foundation for another cycle, assuming that all the required factors (displacement, monetary expansion, appetite for speculation) are present.

2.2 Conceptual Framework

Let us now focus on the empirical question of the existence and detection of financial bubbles. But what are really bubbles? The term "bubble" is widely used but rarely clearly defined. Following Case and Shiller [13], the term "bubble" refers to a situation in which excessive public expectations of future price increases cause prices to be temporarily elevated. For instance, during a housing price bubble, homebuyers think that a home that they would normally consider too expensive for them is now an acceptable purchase because they will be compensated by significant further price increases. They will not need to save as much as they otherwise might, because they expect the increased value of their home to do the saving for them. First-time homebuyers may also worry during a housing bubble that if they do not buy now, they will not be able to afford a home later. Furthermore, the expectation of large price increases may have a strong impact on demand if people think that home prices are very unlikely to fall, and certainly not likely to fall for long, so that there is little perceived risk associated with an investment in a home.

What is the origin of bubbles? In a nutshell, speculative bubbles are caused by "precipitating factors" that change public opinion about markets or that have an immediate impact on demand, and by "amplification mechanisms" that take the form of price-to-price feedback, as stressed by Shiller [57]. Consider again the example of a housing bubble. A number of fundamental factors can influence price movements in housing markets. On the demand side, demographics, income growth, employment growth, changes in financing mechanisms or interest rates, as well as changes in location characteristics such as accessibility, schools, or crime, to name a few, have been shown to have effects. On the supply side, attention has been paid to construction costs, the age of the housing stock, and the industrial organization of the housing market. The elasticity of supply has been shown to be a critical factor in the cyclical behavior of home prices. The cyclical process that we observed in the 1980s in those cities experiencing boom-and-bust cycles was caused by the general economic expansion, best proxied by employment gains, which drove demand up. In the short run, those increases in demand encountered an inelastic supply of housing and developable land, inventories of for-sale properties shrank, and vacancy declined. As a consequence, prices accelerated. This provided an amplification mechanism as it led buyers to anticipate further gains, and the bubble was born. Once prices overshoot or supply catches up, inventories begin to rise, time on the market increases, vacancy rises, and price increases slow down, eventually encountering downward stickiness. The predominant story about home prices is always the prices themselves [57, 59]; the feedback from initial price increases to further price increases is a mechanism that amplifies the effects of the precipitating factors. If prices are going up rapidly, there is much word-of-mouth communication, a hallmark of a bubble. The word-of-mouth can spread optimistic stories and thus help cause an overreaction to other stories, such as ones about employment. The amplification can also work on the downside as well.

Another vivid example is the proposition offered close to the peak of the Internet bubble that culminated in 2000, that better business models, the network effect, first-to-scale advantages, and real options effect could account rationally for the high prices of dot-com and other New Economy companies [50]. These interesting views expounded in early 1999 were in synchrony with the bull market of 1999 and preceding years. They participated in the general optimistic view and added to the strength of the herd. Later, after the collapse of the bubble, these explanations seemed less attractive. This did not escape US Federal Reserve chairman Alan Greenspan, who said [27]:

> Is it possible that there is something fundamentally new about this current period that would warrant such complacency? Yes, it is possible. Markets may have become more efficient, competition is more global, and information technology has doubtless enhanced the stability of business operations. But, regrettably, history is strewn with visions of such new eras that, in the end, have proven to be a mirage. In short, history counsels caution.

In this vein, as mentioned above, the buzzword "new economy" so much used in the late 1990s was also hot in the 1960s during the "tronic boom" before a market crash, and during the bubble of the late 1920s before the Oct. 1929 crash. In this latter case, the "new" economy was referring to firms in the utility sector. It is remarkable how traders do not learn the lessons of their predecessors!

Positive feedback occurs when an action leads to consequences which themselves reinforce the action and so on, leading to virtuous or vicious circles. We propose the hypotheses that (1) bubbles may be the result of positive feedbacks and (2) the dynamical signature of bubbles derives from the interplay between fundamental value investment and more technical analysis. The former can be embodied in nonlinear extensions of the standard financial Black–Scholes model of log-price variations [2, 14, 30, 61].

The mechanisms for positive feedbacks in financial markets include (1) technical and rational mechanisms (option hedging, insurance portfolio strategies, trend following investment strategies, asymmetric information on hedging strategies) and (2) behavioral mechanisms (breakdown of "psychological Galilean invariance" [60], imitation). We stress here particularly the second mechanism which, we believe, dominates. First, it is actually "rational" to imitate when lacking sufficient time, energy and information to make a decision based only on private information and processing, that is, most of the time. Second, imitation has been documented in psychology and in neuro-sciences as one of the most evolved cognitive processes, requiring a developed cortex and sophisticated processing abilities. It seems that imitation has evolved as an evolutionary advantageous trait, and may even have promoted the development of our anomalously large brain (compared with other mammals) [19]. Furthermore, we learn our basics and how to adapt mostly by imitation all through our life. Imitation is now understood as providing an efficient mechanism of social learning. Experiments in developmental psychology suggest that infants use imitation to get to know people, possibly applying a like-me test (people who I can imitate and who imitate me). Imitation is found in highly social living species which show, from a human observer point of view, intelligent behavior and signs for the evolution of traditions and culture (humans and chimpanzees, whales and dolphins, parrots). In non-natural agents such as robots, imitation is a principal tool for easing the programming of complex tasks or endowing groups of robots with the ability to share skills without the intervention of a programmer. Imitation plays an important role in the more general context of interaction and collaboration between software agents and human users.

Humans are perhaps the most social mammals and they shape their environment to their personal and social needs. This statement is based on a growing body of research at the frontier between new disciplines called neuro-economics, evolutionary psychology, cognitive science, and behavioral finance [10, 15, 25]. This body of evidence emphasizes the very human nature of humans with its biases and limitations, opposed to the previously prevailing view of rational economic agents optimizing their decisions based on unlimited access to information and to computation resources.

Imitation, in obvious or subtle forms, is a pervasive activity of humans. In the modern business, economic and financial worlds, the tendency for humans to imitate leads in its strongest form to herding and to crowd effects. Imitation is a prevalent form in marketing with the development of fashion and brands. We hypothesize that financial bubbles are footprints of perhaps the most robust trait of humans and the most visible imprint in our social affairs: imitation and herding (see [59], and references therein).

2.3 Finite-Time Singular Behavior of Bubbles

This understanding of imitation and herding has led us to propose that one of the hallmarks of a financial bubble is the faster-than-exponential growth of the price of the asset under consideration. It is convenient to model this accelerated growth by a power law with a so-called finite-time singularity [63]. This feature is nicely illustrated by the price trajectory of the Hong-Kong Hang Seng index from 1970 to 2000, as shown in Fig. 10. The Hong Kong financial market is repeatedly rated as providing one of the most pro-economic, pro-entrepreneurship and free market-friendly environment in the world, and thus provides a textbook example of the behavior of weakly regulated liquid and striving financial markets. In Fig. 10, the logarithm of the price $p(t)$ is plotted as a function of the time (in linear scale), so that an upward trending straight line qualifies as exponential growth with a constant

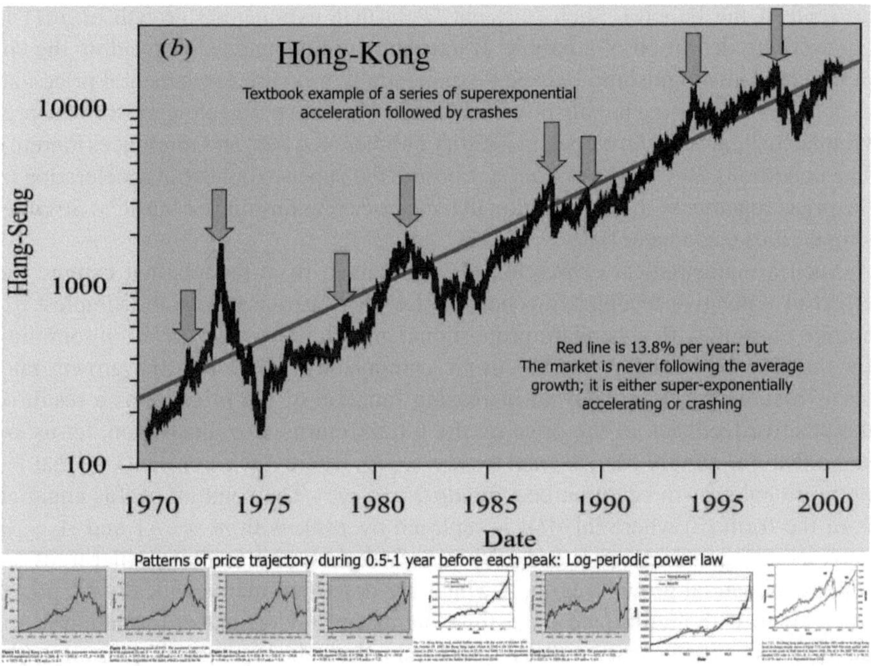

Fig. 10 Trajectory of the Hong-Kong Hang Seng index from 1970 to 2000. The vertical log-scale together with the linear time scale allows one to qualify an exponential growth with constant growth rate as a straight line. This is indeed the long-term behavior of this market, as shown by the best linear fit represented by the *solid straight line*, corresponding to an average constant growth rate of 13.8% per year. The eight *arrows* point to eight local maxima that were followed by a drop of the index of more than 15% in less than 3 weeks (a possible definition of a crash). The eight *small panels at the bottom* show the upward curvature of the log-price trajectory preceding each of these local maxima, which diagnose an unsustainable bubble regime, which culminates at the peak before crashing. Reproduced from [62]

growth rate equal to the slope of the line: the straight solid line corresponds indeed to an approximately constant compounded growth rate of the Hang Seng index equal to 13.8% per year. However, the most striking feature of Fig. 10 is not this average behavior, but the obvious fact that the real market is never following and abiding to a constant growth rate. One can observe a succession of price run-ups characterized by growth rates ... growing themselves: this is reflected visually in Fig. 10 by transient regimes characterized by strong upward curvature of the price trajectory. Such an upward curvature in a linear-log plot is a first visual diagnostic of a faster than exponential growth (which of course needs to be confirmed by rigorous statistical testing). Such a price trajectory can be approximated by a characteristic transient finite-time singular power law of the form

$$\ln[p(t)] = A + B(t_c - t)^m, \quad \text{where} \quad B < 0, \ 0 < m < 1, \tag{2}$$

and t_c is the theoretical critical time corresponding to the end of the transient run-up (end of the bubble). Such transient faster-than-exponential growth of $p(t)$ is our working definition of a bubble. It has the major advantage of avoiding the conundrum of distinguishing between exponentially growing fundamental price and exponentially growing bubble price, which is a problem permeating most of the previous statistical tests developed to identify bubbles (see [46] and references therein). The conditions $B < 0$ and $0 < m < 1$ ensure the super-exponential acceleration of the price, together with the condition that the price remains finite even at t_c. Stronger singularities can appear for $m < 0$ [26].

Such a mathematical expression (2) is obtained from models that capture the effect of a positive feedback mechanism. Let us illustrate it with the simplest example. Starting with a standard proportional growth process $dp/dt = rp$ (omitting for the sake of pedagogy the stochastic component), where r is the growth rate, let us assume that r is itself an increasing function of the price p, as a result of the positive feedback of the price on the future returns. For illustration, let us assume that r is simply proportional to p ($r = cp$, where c is a constant), so that the proportional growth equation become $dp/dt = cp^2$. The solution of this equation is of the form (2) where $\ln[p(t)]$ is replaced by $p(t)$, with $m = -1$ and $A = 0$, corresponding to a divergence of $p(t)$ at t_c. Many systems exhibit similar transient super-exponential growth regimes, which are described mathematically by power law growth with an ultimate finite-time singular behavior: planet formation in solar systems by runaway accretion of planetesimals, Euler equation of inviscid fluids, general relativity coupled to a mass field leading to formation of black holes in finite time, Zakharov equation of beam-driven Langmuir turbulence in plasma, rupture and material failures, nucleation of earthquakes modeled with the slip-and-velocity weakening Ruina–Dieterich friction law, models of micro-organisms interacting through chemotaxis aggregating to form fruiting bodies, Mullins–Sekerka surface instability, jets from a singular surface, fluid drop snap-off, the Euler rotating disk, and so on. Such mathematical equations can actually provide an accurate description of the transient dynamics, not too close to the mathematical singularity where new mechanisms come into play. The singularity at t_c mainly signals a change of regime. In the present context, t_c is the end of the bubble and the beginning of a new market phase, possible a crash or a different regime.

Such an approach may be thought at first sight to be inadequate or too naive to capture the intrinsic stochastic nature of financial prices, whose null hypothesis is the geometric random walk model [47]. However, it is possible to generalize this simple deterministic model to incorporate nonlinear positive feedback on the stochastic Black–Scholes model, leading to the concept of stochastic finite-time singularities [2, 21, 22, 61]. Still much work needs to be done on this theoretical aspect.

Coming back to Fig. 10 , one can also notice that each burst of super-exponential price growth is followed by a crash, here defined for the eight arrowed cases as a correction of more than 15% in less than 3 weeks. These examples suggest that the non-sustainable super-exponential price growths announced a "tipping point" followed by a price disruption, i.e., a crash. The Hong-Kong Hang Seng index provides arguably one of the best textbook example of a free market in which bubbles and crashes occur repeatedly: the average exponential growth of the index is punctuated by a succession of bubbles and crashes, which seem to be the norm rather than the exception.

More sophisticated models than (2) have been proposed to take into account the interplay between technical trading and herding (positive feedback) versus fundamental valuation investments (negative mean-reverting feedback). Accounting for the presence of inertia between information gathering and analysis on the one hand and investment implementation on the other hand [30] or between trend followers and value investing [20], the resulting price dynamics develop second-order oscillatory terms and boom-bust cycles. Value investing does not necessarily cause prices to track value. Trend following may cause short-term trend in prices, but also cause longer-term oscillations.

The simplest model generalizing (2) and including these ingredients is the so-called log-periodic power law (LPPL) model ([59] and references therein). Formally, some of the corresponding formulas can be obtained by considering that the exponent m is a complex number with an imaginary part, where the imaginary part expresses the existence of a preferred scaling ratio λ describing how the continuous scale invariance of the power law (2) is partially broken into a discrete scale invariance [58]. The LPPL structure may also reflect the discrete hierarchical organization of networks of traders, from the individual to trading floors, to branches, to banks, to currency blocks. More generally, it may reveal the ubiquitous hierarchical organization of social networks recently reported [74] to be associated with the social brain hypothesis [19]. The simple implementation of the LPPL model that we use in Sect. 3 reads

$$\ln[p(t)] = A + B(t_c - t)^m \left[1 + C \cos\left(\omega \log(t_c - t) + \phi\right)\right], \quad \text{with } 0 < m < 1, \text{ and } B < 0. \qquad (3)$$

The constant A is by construction equal to $\ln[p(t_c)]$. The two key parameters are the exponent m, which characterizes the strength of the super-exponential acceleration of the price on the approach to the critical time t_c, and ω, which encodes the discrete hierarchy of accelerated "impulse-retracting" market wave patterns associated with the super-exponential acceleration. Specifically, the preferred scaling ratio encoding the accelerated oscillations is given by $\lambda \equiv e^{\frac{2\pi}{\omega}}$ [58].

Examples of calibrations of financial bubbles with one implementation of the LPPL model are the eight super-exponential regimes discussed above in Fig. 10: the eight small insets at the bottom of Fig. 10 show the LPPL calibration on the Hang Seng index on the bubble phase that preceded each peak. Preliminary tests suggest that the LPPL model provides a good starting point to detect bubbles and forecast their most probable end [59]. Rational expectation models of bubbles a la Blanchard and Watson implementing the LPPL model [32, 37, 38] have shown that the end of the bubble is not necessarily accompanied by a crash, but it is indeed the time where a crash is the most probable. But crashes can occur before (with smaller probability) or not at all. That is, a bubble can land smoothly, approximately one-third of the time, according to preliminary investigations [37]. Therefore, only probabilistic forecasts can be developed. Probability forecasts are indeed valuable and commonly used in daily life, such as in weather forecasts.

3 A 15-Year History of the 2007–???? Financial and Economic Crisis

Using the general framework for bubbles and crashes outlined in Sect. 2, we now present the evidence on the five successive bubbles that developed over the last 15 years. We suggest that these five bubbles reveal the belief in the "perpetual money machine" that characterized this epoch, as discussed in Sect. 1.3.

Each bubble excess was thought and felt as "solved" by the following excess... leading to a succession and combination of mutually reinforcing unsustainable financial bubbles, preparing the ground for the instabilities that have been unravelling since 2007. The evidence presented in this section is useful to fully appreciate that the present crisis and economic recession are to be understood as the "hangover" and consolidation phase following this series of unsustainable excesses.

One should conclude that the extraordinary severity of this crisis is not going to be solved by the same implicit or explicit "perpetual money machine" thinking, that still characterize most of the proposed solutions. "The problems that we have created cannot be solved at the level of thinking that created them." said Albert Einstein.

We start by presenting the analysis using the LPPL model (3) presented in Sect. 2.3 of a global index obtained as follows. Starting from time series of emerging market equity indices, freight indices, soft commodities, base and precious metals, energy, and currencies, a principal component analysis (PCA) yields a set of principal components that are thought to correspond to common factors acting on these time series. The first principal component, which explains the largest fraction of the covariance of these time series, is shown in Fig. 11, together with its fit with

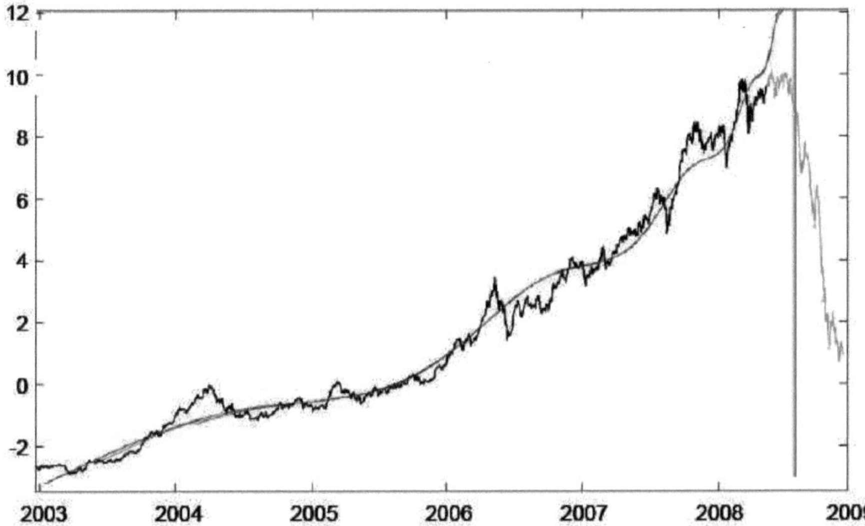

Fig. 11 First component obtained from a principal component analysis performed on a data set containing emerging markets equity indices, freight indices, soft commodities, base and precious metals, energy, and currencies. Source: Peter Cauwels, Fortis Bank – Global Markets

the LPPL model (3). It is striking to observe the overall super-exponential behavior, with a clear change of regime occurring mid-2008. The following subsections allows us to decompose this overall process into bubble components.

3.1 First Phase: The ITC "New Economy" Bubble (1995–2000)

The nature of the ITC bubble is striking when comparing the price trajectories of two indices constructed on the 500 companies forming the S&P 500 index. The Internet stock index is an equally weighted portfolio of 100 firms related to the Internet. The non-Internet stock price index is an equally weighted portfolio made of the remaining 400 "brick-and-mortar" companies. Figure 12 shows that the non-Internet stock price index remained basically flat from 1998 to 2002, while exhibiting fluctuation of roughly $\pm 20\%$ over this period. In contrast, the Internet stock index was multiplied by a factor 14 from 1998 to its peak in the first quarter of 2000, and then shrunk with a great crash followed by a jumpy decay to below its initial value at the end of 2002. The contrast between the behavior of these two indices over the same 4-years interval cannot be more shocking.

The super-exponential nature of the Nasdaq composite index allows us to diagnose this period before 2000 as an unambiguous bubble as first reported by Johansen and Sornette [34], according to the definition presented in Sect. 2.3. Figure 13 shows that the logarithm of the Nasdaq composite index indeed increased with an

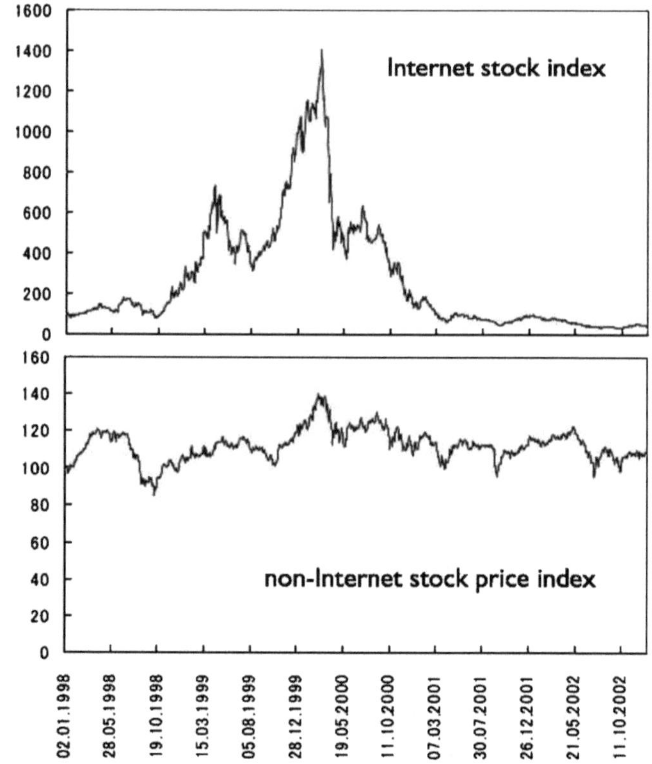

Fig. 12 The Internet stock index and non-Internet stock index which are equally weighted as explained in the text. Comparison of the index levels of the Internet index and the non-Internet Stock index for the period 2 Jan. 1998 to 31 Dec. 2002. The two indexes are scaled to be 100 on 2 Jan. 1998. Courtesy of Taisei Kaizoji

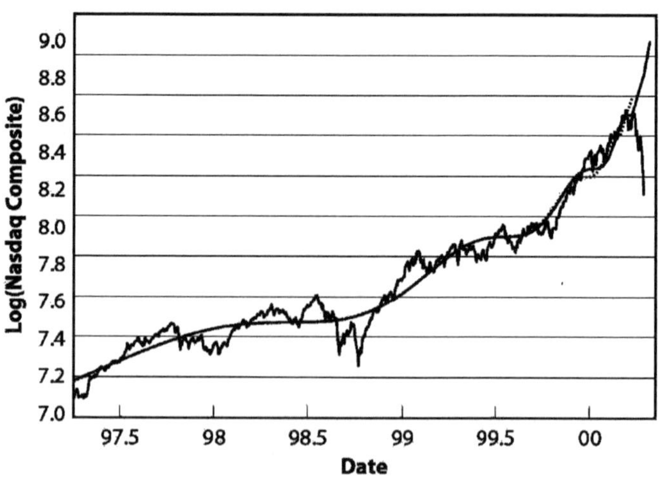

Fig. 13 Calibration of the LPPL model (3) to the Nasdaq Composite Index from early 1997 to the end of 1999. Reproduced from [34]

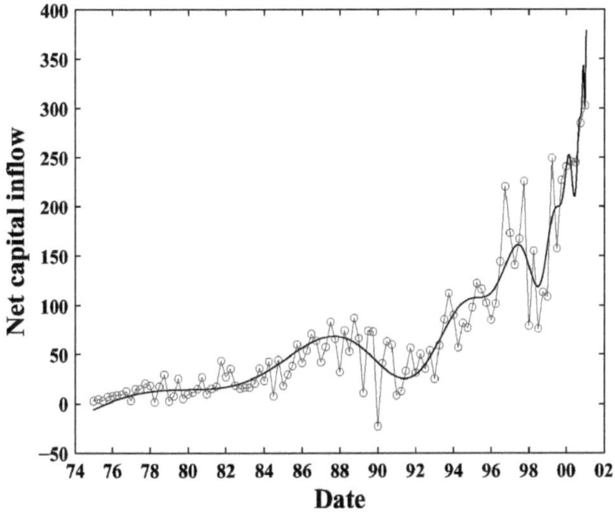

Fig. 14 Foreign capital inflow in the USA during the ICT bubble, illustrating the growth of the euphoria phase in scenario (1) of Sect. 2.1. The *smoothed curve* shows the fit of the net capital inflow by an extension of the LPPL model (3) using higher-order log-periodic components presented in [65]

overall upward curvature, signaling a super-exponential growth. The calibration of the LPPL model (3) to the Nasdaq index is excellent (see [34] for details and statistical tests).

As explained in the scenario (1) in Sect. 2.1, a typical bubble goes through a period of euphoria. This euphoria is characterized by an irresistible attraction, in particular, to foreign investors, who cannot wait to be part of the celebration. This pattern is vividly observed in the case of the ICT bubble in Fig. 14, which shows the flux of foreign capital inflow to the USA. This inflow almost reached 400 billion dollars per year at the peak. A significant part of this foreign capital was invested in the US market to profit from the return opportunities it provided until 2000. The smoothed curve shows that the net capital inflow can also be well-fitted by the LPPL model (3), yielding values for the exponent m and log-frequency ω, which are consistent with those obtained for other bubbles [31].

3.2 Second Phase: Slaving of the Fed Monetary Policy to the Stock Market Descent (2000–2003)

To fight the recession and the negative economic effects of a collapsing stock market, the Fed engaged in a pro-active monetary policy (decrease of the Fed rate from 6.5% in 2000 to 1% in 2003 and 2004). Figure 15 shows this decrease of the Fed rate and compares it with the behavior of two US market indices, the S&P 500 and the Nasdaq composite indices.

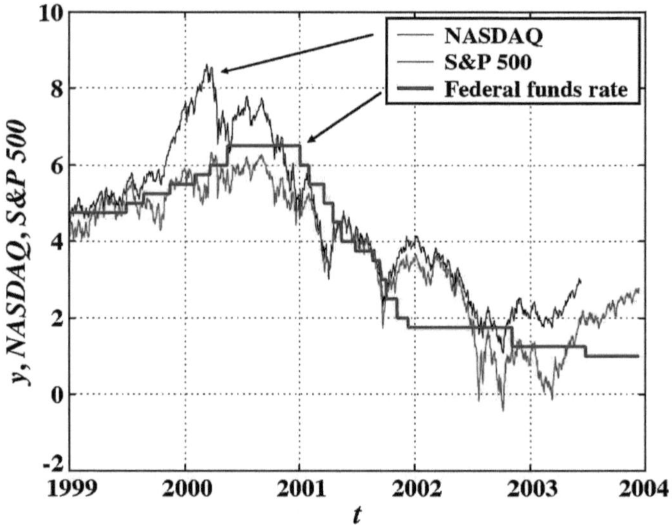

Fig. 15 Comparison of the Federal funds rate, the S&P 500 Index $x(t)$, and the NASDAQ composite $z(t)$, from 1999 to mid-2003. To allow an illustrative visual comparison, the indices have been translated and scaled as follows: $x \to 5x - 34$ and $z \to 10z - 67$. Reproduced from Zhou and Sornette [71]

It is quite apparent that the Fed rate decreased in parallel to the US stock market. But did it lead it or lag behind it? According to common wisdom, the Federal Reserve control of the leading rate indicator is supposed to influence the stock markets. A decrease of the Fed rate makes borrowing cheaper, giving more leverage to firms to invest for the future. As a consequence, this should lead to anticipations of larger future growth, and hence to larger present market values. Hence, logically, the Fed rate drops should precede the market losses.

We check this prediction by showing in Fig. 16 the cross-correlation between the returns of the S&P 500 index and the increments of the Federal funds rate as a function of time lag. The remarkable result is that the Fed rate decreased with a robustly determined lag of about 1–2 months behind the on-going loss of the S&P 500. This reverse causality suggests that the Fed monetary policy has been influenced significantly by (or "slaved" to) the vagaries of the stock market.

In 2003, Fed Chairman A. Greenspan argued that the Fed needed to set low interest rates to prevent the US economy from deteriorating so much that it would follow a deflationary spiral, often referred to as a "liquidity trap" [43], a situation in which conventional monetary policy loses all traction. Greenspan's critics continue to debate about the influence of the exceptionally low Fed rates in 2002 and 2003 that are thought to have caused an extraordinary real estate bubble... that led eventually to the 2007–???? crisis and recession.

Recently, economist Nick Rowe presented a piece of evidence [54] that is reproduced in Fig. 17, which illuminates this debate. The growth rate of the non-farm business productivity is compared with the real Federal fund rate. It is apparent that

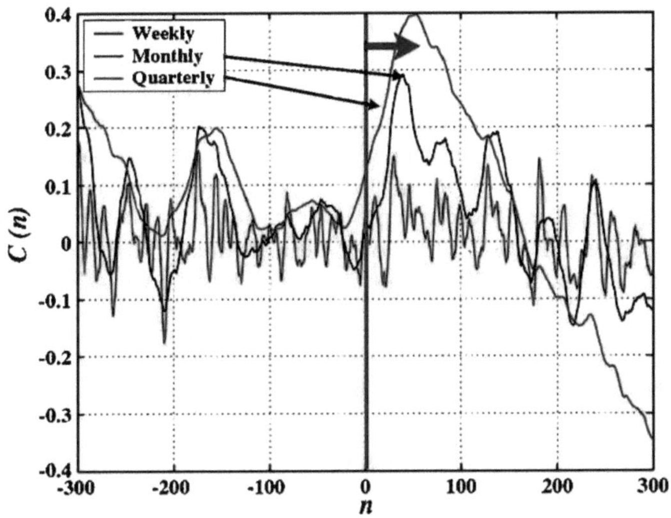

Fig. 16 "Causal Slaving" of the US Treasury Bond Yield by the Stock Market Antibubble of August 2000. The cross-correlation coefficient $C(n)$ between the increments of the logarithm of the S&P 500 Index and the increments of the Federal funds rate is shown as a function of time lag n in days. The three curves corresponds to three different time steps used to calculate the increments: weekly (noisiest curve), monthly and quarterly (curve showing the largest peak). A positive lag n corresponds to having the Federal funds rate changes posterior to the stock market returns. The *arrow* points to this lag. Reproduced from Zhou and Sornette [71]

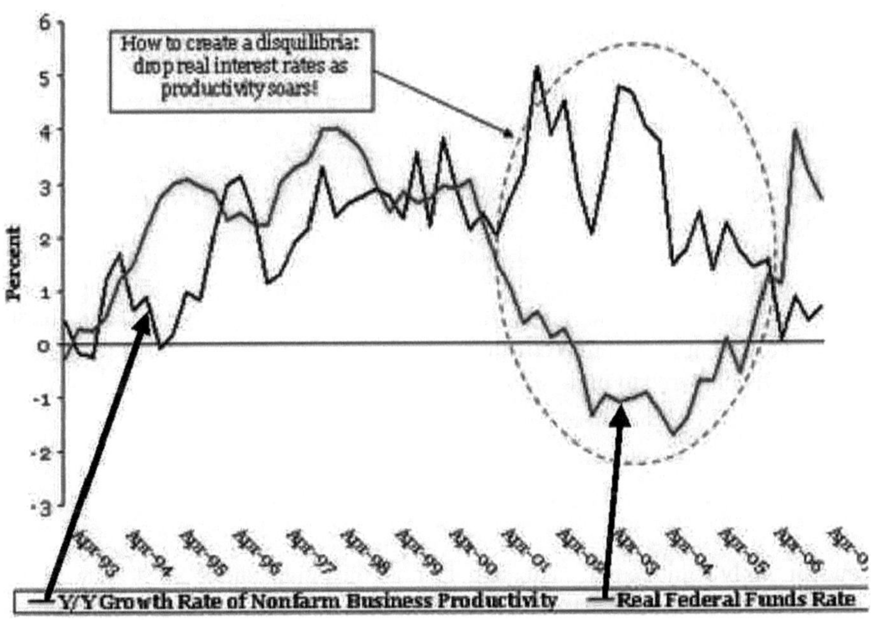

Fig. 17 Growth rate of the non-farm business productivity compared with the real Federal fund rate. Reproduced from [54]

the Fed rate was pushed down at the time of a surge of productivity gains, not really a deteriorating economy. This combines with the previous evidence of Fig. 16 to support the view that the Federal Reserve has been too obnubilated by the stock market signals. This is additional evidence that, even in the higher spheres of finance, the stock market is taking over in shaping economic and strategic decisions. We mentioned in the introduction of Sect. 1 that the impact of financial markets has been growing to basically dominate strategic decision at the firm level (see also [9] for a dramatic example of this trend for the case Royal Ahold firm in the Netherland). It seems that the monetary authorities are also infected by the same stock market virus.

3.3 Third Phase: Real-Estate Bubbles (2003–2006)

The pro-active monetary policy of the Federal Reserve described in the previous subsection, together with expansive Congressional real-estate initiatives, fueled what can now be rated as one of the most extraordinary real-estate bubbles in history, with excesses on par with those that occurred during the famous real-estate bubble in Japan in the late 1980s.

As Alan Greenspan himself documented in a scholarly paper researched during his tenure as the Federal Research Chairman at that time [28], the years from 2003 to 2006 witnessed an extraordinary acceleration of the amount of wealth extracted by Americans from their houses as shown in Fig. 18, which parallels the accelerated house price appreciation shown in Fig. 19. The negative effects on consumption and income due to the collapse of the first ICT bubble were happily replaced by an enthusiasm and a sense of riches permeating the very structure of US society.

In June 2005 (proof from the arXiv submission http://arxiv.org/abs/physics/0506027), Zhou and Sornette [72] issued a diagnostic that about two-fifths of the states of the USA were developing real estate bubbles. Zhou and Sornette predicted a peak for most of the US real estate bubbles in mid-2006 [72]. The validity of this prediction can be checked in Fig. 20.

It should be noted that the real estate bubble has not been confined to the USA but was active in many (but not all) countries. Exceptions include Germany, Japan, Switzerland and The Netherlands. But the critical time of the peak of the bubble has been different in different countries. For instance, it was mid-2004 for the UK bubble [70] compared to mid-2006 for the US bubble.

3.4 Fourth Phase: MBS, CDOs Bubble (2004–2007)

Concomitantly with the real estate bubble, both the public and Wall Street were the (sometimes unconscious) actors of a third bubble of subprime mortgage-backed securities (MBS) and complex packages of associated financial derivatives, as already described in Sect. 1.2.2 and now shown in Figs. 21 and 22.

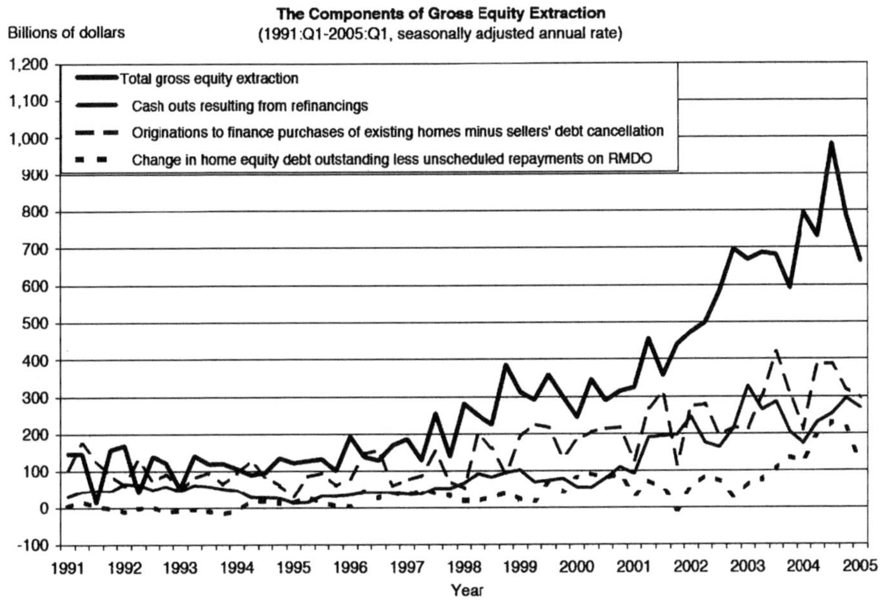

Fig. 18 Quantification of gross equity extraction by homeowners from their houses, showing the accelerated growth that spilled over to the economy by fueling consumption. This figure shows that, over the past decade and a half, equity extraction has been closely correlated with realized capital gains on the sale of homes. Source: Greenspan and Kennedy [28]

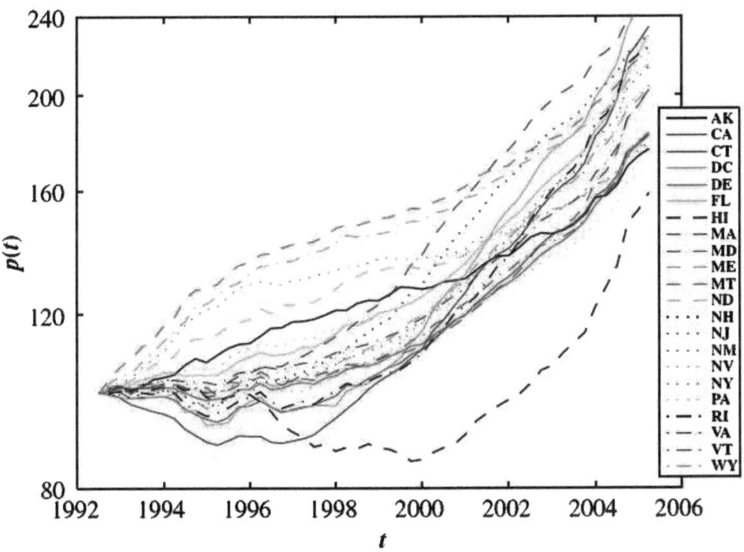

Fig. 19 (Color online) Quarterly average HPI (house price index) in the 21 states and in the district of Columbia (DC) that were found to exhibit a clear faster-than-exponential growth. For better comparison, the 22 house price indices have been normalized to 100 at the second quarter of 1992. The corresponding states symbols are given in the legend. Reproduced from Zhou and Sornette [72]

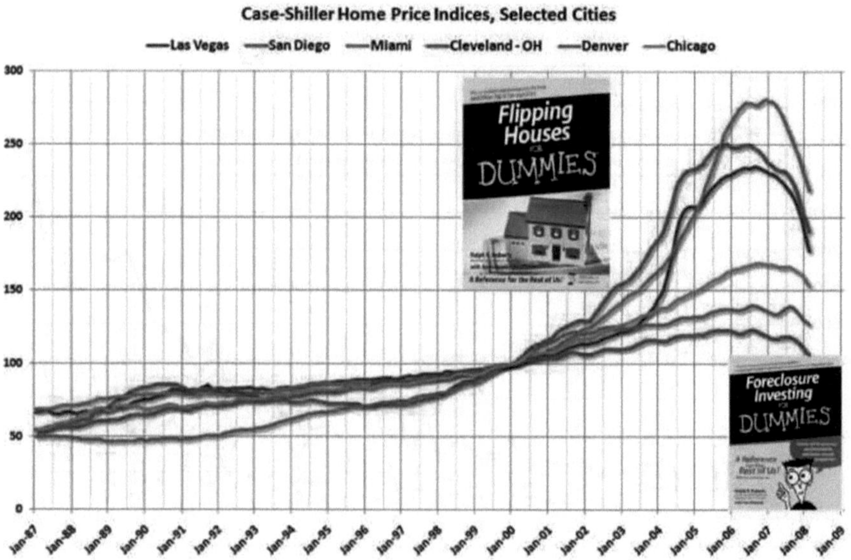

Fig. 20 (Color online) Year-over-year price changes for the Case–Shiller composite 10 and 20 indices (through February 2008), and the Case–Shiller and OFHEO National price indices (through Q4 2007). Adapted from http://calculatedrisk.blogspot.com. The pictures of the two books are put here to emphasize the dominating sentiment in each phase (see [53] for a study of how book sales reflect market bubbles)

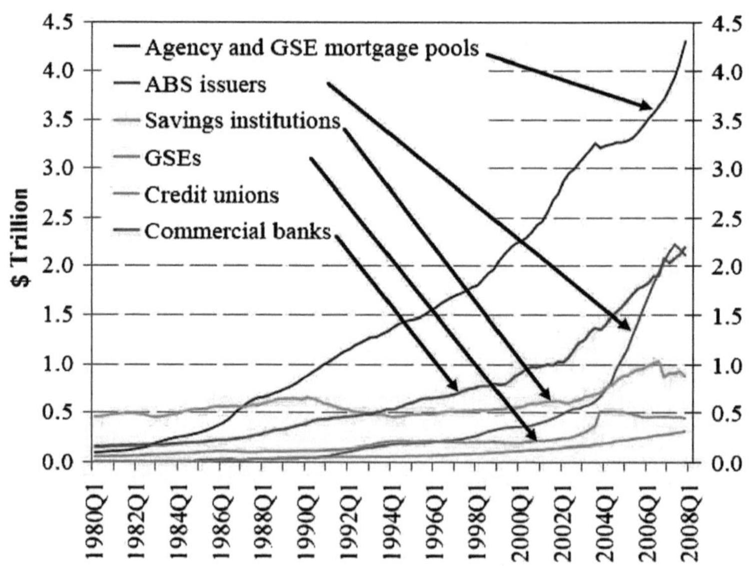

Fig. 21 Total holdings of US home mortgages by type of financial institution. Source: Hyun Song Shin, Princeton University

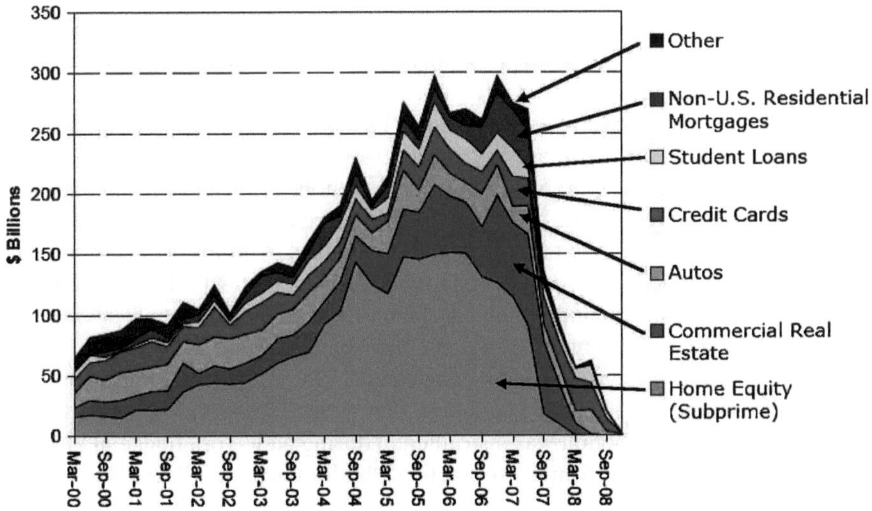

Fig. 22 New issuance of asset backed securities in previous three months. Source: JP Morgan

The growth of the MBS derivatives is exemplified in Figs. 21 and 22, which respectively show (1) the total holding of mortgage related securities of different financial institutions and (2) the accelerated rate of new issuance of asset backed securities (ABS) until the peak in March 2007, when the first signs of accelerating loan payment defaults started to be felt on the MBS.

These two Figs. 21 and 22 clearly illustrate the MBS bubble and its bursting. In addition, as pointed out by many astute observers, many of the MBS were "fragile" as they were linked to two key unstable processes: the value of houses and the loan rates. The "castles in the air" of bubbling house prices promoted a veritable eruption of investments in MBS, these investments themselves pushing the demand for and therefore the prices of houses – until the non-sustainability of these mutually as well as self-reinforcing processes became apparent.

But to be clear; these financial instruments were great innovations which, in normal times, would indeed have provided a win-win situation: more people have access to loans, which become cheaper because banks can sell their risks to the supposed bottomless reservoirs of investors worldwide with varying appetites for different risk-adjusted returns.

The problem is that the MBS and collateral debt obligations (CDO) constituted new types of derivatives. Their complexity together with the lack of historical experience may have provided the seed for unrealistic expectations of low risks and large returns. Actually, this is part of a larger debate on the role of financial derivatives.

Many financial economists hold that derivatives serve a key role of making markets more complete, in the sense that more states of the world can be hedged by a corresponding asset. As a consequence, financial markets become more efficient and stable. Perhaps the most influential proponent of this view has been Alan Greenspan himself. For more than a decade, Greenspan has fiercely objected whenever derivatives have come under scrutiny in Congress or on Wall Street. "What we have found over the years in the marketplace is that derivatives have been an extraordinarily useful vehicle to transfer risk from those who shouldn't be taking it to those who are willing to and are capable of doing so," Mr. Greenspan told the Senate Banking Committee in 2003. "We think it would be a mistake" to more deeply regulate the contracts, he added. "Not only have individual financial institutions become less vulnerable to shocks from underlying risk factors, but also the financial system as a whole has become more resilient." – Alan Greenspan in 2004.

Others disagree. The well-known financier G. Soros avoids using derivatives "because we do not really understand how they work." Felix G. Rohatyn, the investment banker whose action was instrumental during New York financial turmoils in the 1970s, described derivatives as potential "hydrogen bombs." And, in the 2002 Berkshire Hathaway annual report, Warren E. Buffett observed that derivatives were "financial weapons of mass destruction, carrying dangers that, while now latent, are potentially lethal."

These statements have recently been given theoretical support in new out-of-equilibrium models of financial markets in which it is found that, paradoxically, on the one hand the proliferation of financial instruments tends to make the market more complete and efficient by providing more means for risk diversification, while at the same time this proliferation of financial instruments erodes systemic stability as it drives the market to a critical state characterized by large susceptibility, strong fluctuations and, enhanced correlations among risks [8, 48, 49].

3.5 *Fifth Phase: Stock Market Bubble (2004–2007)*

The exuberant real-estate market and MBS bubbles spilled over to the stock market. Figure 23 shows the S&P 500 index (in logarithmic scale) as a function of time. A clear upward overall upward curvature can be observed, which is characteristic of a super-exponential growth. The LPPL calibration confirms the existence of bubble characteristics.

3.6 *Sixth Phase: Commodities and Oil Bubbles (2006–2008)*

As explained in Sect. 3.3 and shown in Fig. 18, the growth of the real-estate bubble, of the MBS bubble, and of the stock market bubble led to a huge extraction of wealth or, in other words, the creation of a lot of wealth and of money. Both money creation and wealth increase led to higher demand in all things that can be consumed. In fact,

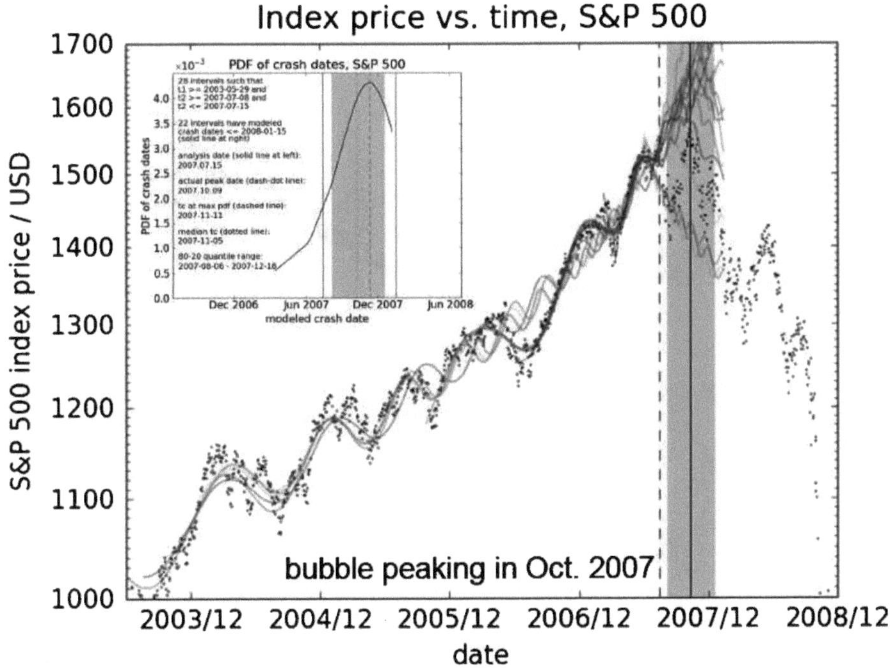

Fig. 23 S&P 500 index (in logarithmic scale) shown as dots as a function of time. The *dashed vertical line* shows the last observed time t_{last} used to perform the calibration of the LPPL model (3) and the *different smoothed curves* correspond to different estimations obtained with distinct time windows extending no later that t_{last}. The *grey zone* corresponds to the 80% confidence interval for the predicted critical time t_c of the end of the bubble. The *inset* shows the probability density function of the predicted t_c's

the demand has been accelerating on basic commodities, which developed clear bubble characteristics, as shown in Figs. 24 and 25.

Oil prices exhibited a record rise, whose starting phase can be traced to 2003 according to our analysis [64], followed by a spectacular crash in 2008. The peak of $145.29 per barrel was set on July 3, 2008 and a recent low of $40.81 was scraped on December 5, 2008, a level not seen since 2004. On May 27, 2008, we addressed the question of whether oil prices were exhibiting a bubble-like dynamics, which may be symptomatic of speculative behavior, using our techniques based on statistical physics and complexity theory [64]. Thorough analysis of our May 27, 2008 experiment predicted a peak within a 80% confidence interval between May 17, 2008 and July 14, 2008. The actual observed "crash", where prices began a long downward trend, began on the last day of this period.

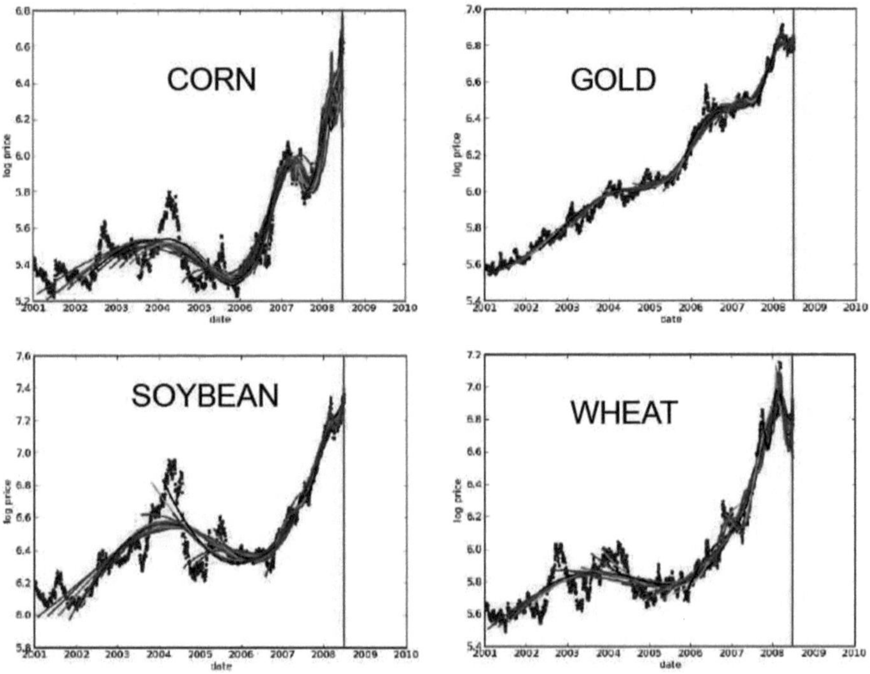

Fig. 24 Calibration of the LPPL model (3) to the time series of four commodities (corn, gold, soybean and wheat) expressed in US dollars

4 Thoughts on Resolution of the Crisis and Its Aftermath

4.1 Summary

We have presented evidence that the fundamental cause of the unfolding financial and economic crisis lies in the accumulation of at least five bubbles whose interplay and mutual reinforcement has led to an illusion of the "perpetual money machine" allowing financial institutions to extract wealth from an unsustainable artificial process.

The path from MBS problem to a global World recession and to potentially even higher risks can be outlined as follows: drop in confidence → drop in demand → major recession → interaction between a deep recession and a weakened financial system → increased risk of trade wars → collapse of global commodity prices and, thus, revenues for low-income countries → global instability.

In March 2007, the first visible signs of a worrisome increase in default rates appeared, followed in the summer of 2007 by the startling realization by the financial community that the problem would be much more serious. In parallel with an acceleration of the default of homeowners and a deteriorating economic outlook, the second part of 2007 and first part of 2008 saw a rapid increase, punctuated by

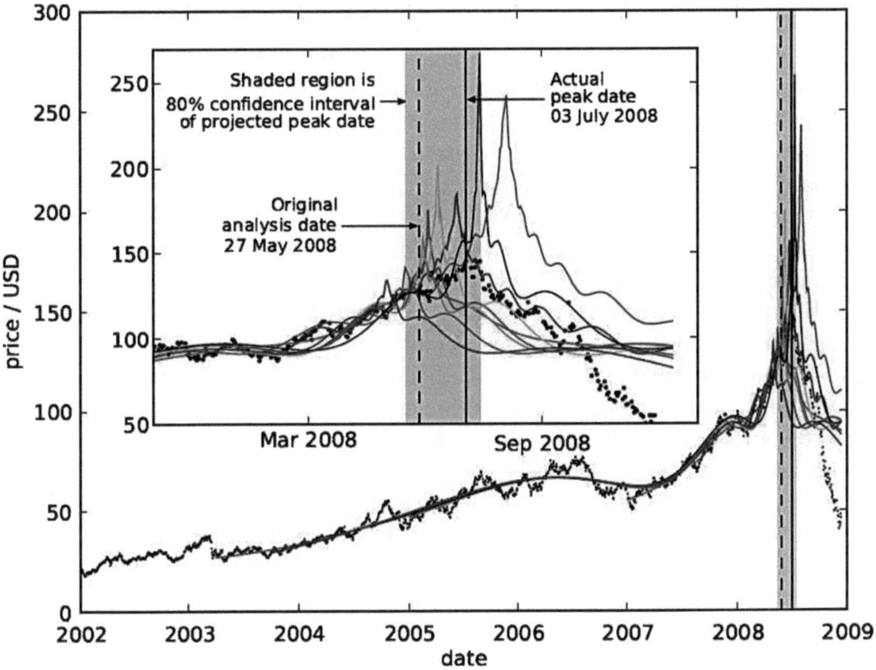

Fig. 25 Typical result of the calibration of the simple LPPL model to the oil price in US dollars in shrinking windows with starting dates t_{start} moving up towards the common last date t_{last} = May 27, 2008. Reproduced from Sornette et al. [64]

dramatic bankruptcies, in the estimated cumulative losses facing banks and other investment institutions; from initial guesses in the range of tens of billions, to hundreds of billions, then to more than a trillion of US dollars. The US Federal Reserve and US Treasury stepped up their actions in proportion to the ever-increasing severity of the uncovered failures, unaware of the enormity of the underlying imbalances, or unwilling to take the measures that would address the full extent of the problem, which only now (one hopes) has revealed its enormous systemic proportions. Let us be blunt: Government has been blissfully unaware of the predictable effect of the piling up these bubbles, as each one appeared to displace the problems induced by the previous one. Monetary easing, the injection of liquidity, successive bailouts – all address symptoms and ignore the sickness.

The sickness is the cumulative excess liability present in all the sectors of the US economy: debts of US households as a percentage of disposable income at around 130%, those of US banks as a percentage of GDP currently around 110%, US Government debt at 65% of GDP, corporate debts at 90% GDP, state and local government debts at 20% GDP, unfunded liabilities of Medicare, Medicaid, and Social Security in the range of three to four times GDP. Such levels of liabilities in the presence of the bubbles have produced a highly reactive unstable situation in which sound economic valuation becomes unreliable, further destabilizing the system. This sickness has only been worsened by measures that disguise or deny it, like the misapplied innovations of the financial sector, with their flawed incentive structures, and the de facto support of an all-too willing population, eager to believe that wealth extraction could be a permanent phenomenon. On the sustainable long term, the growth of wealth has to be equal to the actual productivity growth, which is about 2–3% in real value on the long term in developed countries.

Acting on the "theory" that crises can be stopped if confidence is restored by stopping the hemorrhage of MBS losses, and on the basis of the diagnostic that the financial crisis was a "liquidity" problem, central banks and governments have actively intervened to combat the observed swing in risk taking shown by banks. One set of measures amount basically to attempt stopping the devaluation of the MBS assets held by financial institutions. Another set of measures tries to remove the so-called "toxic" assets and park them in vehicles mainly owned by the government and the Federal Reserve, until their values rebound, one hopes. Most measures attempt to encourage more consumption and spending.

The evidence discussed above suggests strongly that this approach constitutes a fundamental error because it misses the crucial point about the cause of the "losses." The losses are not just the downturn phase of a business cycle or the liquidity response to an unlucky exogenous shock. They express a simple truth that is too painful to accept for most, that previous gains were not real, but just artificially inflated values that have bubbled in the financial sphere, without anchor and justification in the real economy. In the last decade, banks, insurance companies, Wall Street as well as Main Street – we all have lured ourselves into believing that we were richer. But this wealth was just the result of a series of self-fulfilling bubbles: in the USA and in Europe, we had the Internet bubble (1996–2000), the real-estate bubble (2002–2006), the MBS bubble (2002–2007), an equity bubble (2003–3007), and a commodity bubble (2004–2008), each bubble alleviating the pain of the previous bubble or supporting and justifying the next bubble. In a word, the overgrowth of the "financial economy" compared with that of the real economy is reminiscent of the fabled frog seized by a jealous desire to equal the ox in size [1].

The painful consequence of this brutal truth is that trying to support the level of valuation based on these bubbles is tantamount to continue supporting the "perpetual money machine", which was the cause of the problem. Worse, it misuses scarce taxpayer resources, increasing long-term debts and liabilities, which are already at dangerous levels in many countries.

Encouraging over-spending to solve a crisis due to over-spending is an ill-advised approach.

There are no silver bullets, but the following concepts should constitute part of a basis for a pragmatic approach to the crisis.

4.2 Trust! Why It Has Been Lost and How to Regain It

The on-going credit crisis and panic shows that financial price and economic value are based fundamentally on trust; not on fancy mathematical formulas, not on subtle self-consistent efficient economic equilibrium; but on trust in the future, trust in economic growth, trust in the ability of debtors to face their liabilities, trust in financial institutions to play their role as multipliers of economic growth, trust that your money in a bank account can be redeemed at any time you choose. Usually, we take these facts for granted, as an athlete takes for granted that her heart will continue to pump blood and oxygen to her muscles. But what if she suffers a sudden heart attack? What if normal people happen to doubt banks? Then, the implicit processes of a working economy – all we take for granted – starts to dysfunction and spirals into a global collapse.

Because of the failed governance, the crisis has accelerated, now in a burst of such intensity that it has forced coordinated actions among all major governments. While their present involvement may restore short-term confidence, much more is needed to address the depth of the problem. At one level, the loss of trust between financial institutions stems from the asymmetric information on their suddenly escalading counter-party risks, making even the most solid bank be perceived as a potential candidate for default, leading to credit paralysis. This can be addressed by financial measures and market forces. At a deeper level, people in the street have lost confidence, by observing a succession of problems, timidly addressed by decision makers proposing ad hoc solutions to extinguish the fire of the day (Bear Stearns, Fannie, Freddie, AIG, Washington Mutual, Wachovia), with no significant result and only more deterioration.

Nothing can resist loss of trust, since trust is the very foundation of society and economy. That people haven't yet made a run on the banks is not, given today's insurance policies against the catastrophes of the past, sufficient indication to the contrary. In fact, there has been already invisible run on the banks, as electronic and wire transfers have been accelerating in favor of government-backed Treasury bills. A significant additional impediment to the restoration of public trust is that the Fed, Treasury and concerted government actions are perceived as supporting the notion that "gains are private while losses are socialized."

Present actions attempt to stabilize the financial sector by making governments, therefore taxpayers, the lenders and buyers of last resort. But collectively people are more intelligent than governments and decision makers think. They know that governments, in particular in the West, have not saved (counter-cyclically a la Keynes) during the good years,[2] and they thus wisely doubt their prudence during the bad ones. They suspect that their governments will eventually extract the needed capital from them. They suspect intuitively that the massive measures taken to support the financial world will do little to help general economies of the USA, Europe and the rest of the world.

---

[2] In this respect, note the information from Reuters, Santiago, March 27, 2009, reporting that Chile's President Michelle Bachelet unwittingly embarrassed British Prime Minister Gordon Brown when she said Chile had put aside money during good economic times to help it through the downturn. "I would say that because of our decision during ... the good times in copper prices, we decided to save some of the money for the bad times and I would say that policy today is producing good results."

How to restore trust? This is a long, tedious and fragile process. We suggest that governing bodies must for once play to the intelligence of the crowd. What needs to be done is to explain truthfully (is this possible?) the true cause of the problems: the more than one decade of excesses and the successive and inter-related bubbles, the fact that the liabilities are enormous and that the budget has in the end to be balanced, that accelerating borrowing on the future cannot be a sustainable strategy. As humans, we are more inspired to trust when failures are acknowledged than when blame is shifted to someone else. This is the core reason why going to the fundamental source of the problems may be part of the solution in restoring confidence in the existence of a solution at minimal cost.

Second, the issue of fairness is essential for restoring confidence and support. There is an absolute need to rebuild that confidence, and this applies also to the regulators. This requires new strong regulations to deal with conflicts of interest, moral hazard, and to enforce the basic idea of well-functioning markets in which investors who took risks to earn profits must also bear the losses. For instance, to fight the rampant moral hazard that fueled the bubbles, share-holders should be given "clawback" permission, that is, the legal right to recover senior executive bonus and incentive pay, that proved to be ill-founded. In addition, many of the advisors and actors of the present drama have vested interest and strong conflict of interests. This particularly the case in the USA, for the Fed, the Treasury and the major banks acting on behalf or with the approval of the Treasury. An independent (elected?) body would be one way to address this problem, ensuring separation of interest and power.

4.3 Short-term: Melting the Cash Flow Freeze

The most immediate issue is to address the cash flow freeze imposed by banks, with their newfound overly restrictive lending rules, on companies and households. This cash flow problem bears the seed of a spiraling recession of catastrophic amplitude, which has no fundamental reason to develop, except as an unwanted consequence of pro-cyclical feedbacks aggravating a necessary correction that should only be confined to the financial sphere. Here, the central banks and governments should show creativity in ensuring that small and medium size companies have access to monthly liquidity, to allow them to continue producing and hiring. This is the issue that has been by far the most under-estimated and which requires a fast and vigorous solution.

In addition to providing lending facilities to banks conditional on serving their natural multiplier role in the economy, special governmental structures could be created with a finite lifetime, with the mandate to provide liquidity to the real economy, bypassing the reluctant banks. Note that this procedure should not necessarily be used to bailout some large badly managed companies in some industry sectors, when in obvious need of restructuring. Crises are often opportunities for restructuring, which provide increased benefits in the future as some cost in the present.

B. Lietaer has proposed a different approach based on the use of a complementary currency to help make the network of financial interactions between companies more robust [45], in analogy with ecological networks which derive their resilience from the multiple network levels they are built on. As a first candidate for this complementary currency could be a professionally run business-to-business system of payments on the model of the WIR system,[3] which has been successfully operational for 75 years in Switzerland, involving a quarter of all the businesses in that country. This system has been credited by J. Stodder as a significant counter-cyclical stabilizing factor that may contribute to the proverbial stability of the Swiss economy [66].

4.4 Long-term: Growth Based on Returning to Fundamentals and Novel Opportunities

Long-term economic stimulation programs are needed on a large scale, probably a few percent of GDP, with pragmatic adaptive tuning as the crisis unfolds. They should focus on the fundamentals of wealth growth: infrastructure, education and entrepreneurship, with the goal of promoting productivity growth and the creation of new real economic sources of wealth. Many studies demonstrate for instance a direct impact of machinery equipment on economic growth.

Similarly, by many metrics, the quality of education in the USA and to a lesser degree in Europe has been degrading in the recent decades. This crisis is an opportunity to go back to the fundamentals of the roots of long-term sustainable wealth creation. These stimulation programs offer an opportunity to adapt and develop new infrastructure which are more energy and pollution efficient, thus promoting the development of new industry sectors such as wind energy, electricity storage, nuclear waste processing and recycling and so on.

Given growing evidence that mankind is facing global challenges for its sustainability on the finite Earth, the financial-rooted crisis offers a chance for using its associated political capital to make bold choices to steer an environmentally friendly economic development. Governments are best in their role of risk takers investing in long-term R&D projects that provide the support for innovations that industry can build upon to provide increased prosperity in the future.

[3] The Swiss Economic Circle (Wirtschaftsring-Genossenschaft or WIR) is an independent complementary currency system in Switzerland that serves small and medium-sized businesses. It was founded in 1934 by businessmen Werner Zimmermann and Paul Enz as a result of currency shortages after the stock market crash of 1929 and the great recession. "Its purpose is to encourage participating members to put their buying power at each other's disposal and keep it circulating within their ranks, thereby providing members with additional sales volume." Cited from Wikipedia.

4.5 The Financial Sphere, Bubbles and Inflation

One has to accept the need for an abrupt deflation of the financial sphere. And for the future, mechanisms should be designed to control the over-growth of the financial economy to ensure better stability. When functioning properly, the financial world provides many services such as efficient access to funding for firms, governments and private people. Furthermore, it works as an effective storage of value, which should reflect the "real economy." But the extraordinary growth of the component of wealth associated with the financial world has been artificial and based on multipliers amplifying the virtual fragile components of wealth. Objective measures and indicators can be developed to quantify the ratio of wealth resulting from finance compared with the total economy. For instance, when it is assessed that, on average, about 40% of the income of major US firms result from financial investments, this is clearly a sign that the US economy is "building castles in the air." In the academic literature, this is related to the concept of "financialization",[4] according to which profit making occurs increasingly through financial channels, and shareholder value tends to dominate corporate governance.

The way we think of inflation also needs to be re-evaluated. For instance, a house price appreciation does not just mean that you are more wealthy as a homeowner; it also implies that you need more dollars or euros to buy one unit of habitation compared to units of food, vacation or university tuition. From this vantage, it is part of inflation. While already considered in present measure of the so-called consumer price index (CPI), its weight and impact need to be re-evaluated. We propose that real-estate and equity indices should be incorporated as constituents of inflation metrics, of course with adequate consideration for the hedonic gains.[5] A financial ratio index could be created to follow a broader definition of inflation useful for central bank monitoring, which includes the growth of total fixed assets, working capital, excess supply of money and so on.

With such tools, monetary policy with inflation targets will provide natural partial control over some of the asset bubbles at the origin of the present financial crisis. Guidelines could be drawn to flag warning signals to central banks and governments when the ratio of the financial wealth compared with the real economy value grows above a bracket that could be defined from a consensus among economists and actions could be taken to moderate the growth of this ratio. These indicators should be the key targets of modern central banks.

Central banks and governments should step in to support financial institutions, but only under fair conditions that ensure that stockholders and lower priority debt holders support the consequences of the losses, avoiding the privatization of gains and socialization of losses. Different technical mechanisms have been proposed by financial economists, which serve this goal, safeguarding the interest of the taxpayers on the long term.

---

[4] See http://en.wikipedia.org/wiki/Financialization.

[5] Governments use so-called hedonic regression in computing their CPI to take quality changes into account.

As a final point on the issue of the size of the financial sphere, the first author is a happy professor teaching financial economics to a growing corpus of students in a World-renowned technical university. We are however worried by the growing flood of civil, mechanical, electrical and other engineers choosing to become "defectors" and work in finance: Is this another bubble in the making? Finance will not solve the many problems mentioned above. Creativity and entrepreneurship occurring in the real economy and the real world need to be better rewarded.

4.6 Recipes for a More Robust and Sustainable World

The present crisis is illustrating the accelerating fragility of society. We believe that this is just a foreshock of much more serious jolts to come on times scales of just one or two decades. In this respect, we refer to Johansen and Sornette [35] and Sornette [59, Chap. 10], in which the analysis of Earth's human population, its economic output and global stock market capitalization suggest a transition to a completely new regime circa 2050, over a time scale of several decades around that date.

However, now is an opportunity to build a more resilient World. Recipes are known. They involve the need for variety, redundancy, compartments, sparseness of networks and understanding the consequences of the unavoidable delays and synchronization of actions. This "robustness" approach is well exemplified by a conservative investment approach based on (1) understanding the vehicles and firms in which one invests (which contrasts with the opaqueness of the MBS investments) and (2) keeping capital even under extraordinarily adverse conditions. This strongly contrasts with standard financial practices based on estimated likelihoods for meeting obligations and short-term gains. This requires fundamentally new design in infrastructures and in regulations. The task is complex, but realizing and formulating it is a major step that should be followed by a vigorous program at the international level, based on multidisciplinary task forces that are well-funded and empowered with authority. Leading countries should start at their domestic level to nucleate the process internationally.

Beyond the immediate concerns, we need to keep in mind the big picture, that this time is a unique opportunity for change and improvement. The present crisis should be exploited to start developing a genuine culture of risks, which should be obligatory training for managers in governments, in regularity bodies, and in financial institutions. One little discussed reason for the present crisis was indeed the lack of adequate education of top managers on risks in all its dimensions and implications. This is also the time that a culture of risk should start permeating the public at large. In the twenty-first century, "linear" and "equilibrium" thinking should be replaced by a growing appreciation of the inter-connectivity and feedbacks of the complex systems we deal with, which creates shocks – with opportunities.

References

1. Gibbs L (trans) (2002) Aesop's fables. Oxford University Press, Oxford
2. Andersen JV, Sornette D (2004) Fearless versus fearful speculative financial bubbles. Physica A 337(3–4):565–585
3. Wikipedia (n.d.) Basel II accord. http://en.wikipedia.org/wiki/Basel_II
4. Berns G, Capra CM, Moore S, Noussair C (2009) Neural mechanisms of social influence in consumer decisions. Working paper
5. Blanchard OJ (2008) The crisis: basic mechanisms, and appropriate policies. http://ssrn.com/abstract=1324280
6. Borgatti SP, Mehra A, Brass DJ, Labianca G (2009) Network analysis in the social sciences. Science 323:892–895
7. Boss M, Elsinger H, Summer M, Thurner S (2003) An empirical analysis of the network structure of the Austrian interbank market. Fin Stab Rep 7:77–87
8. Brock WA, Hommes CH, Wagener FOO (2008) More hedging instruments may destabilize markets. CeNDEF Working paper 08-04, University of Amsterdam
9. Broekstra G, Sornette D, Zhou W-X (2005) Bubble, critical zone and the crash of Royal Ahold. Physica A 346:529–560
10. Camerer CF (2003) Behavioral game theory: experiments in strategic interaction. Princeton University Press, Princeton
11. Campbell JY, Cocco JF (2005) How do house prices affect consumption? Evidence from micro data. NBER Working Paper No. 11534, August 2005
12. Cannata F, Quagliariello M (2009) The role of Basel II in the subprime financial crisis: guilty or not guilty? CAREFIN Research Paper No. 3/09. http://papers.ssrn.com/sol3/papers.cfm?abstract_id=1330417
13. Case KE, Shiller RJ (2003) Is there a bubble in the housing market. Brookings Pap Econ Act 2:299–362
14. Corcos A, Eckmann J-P, Malaspinas A, Malevergne Y, Sornette D (2002) Imitation and contrarian behavior: hyperbolic bubbles, crashes and chaos. Quant Finan 2:264–281
15. Damasio A (1994) Descartes' error. Putman Adult, New York
16. Demyanyk YS (2009) Quick exits of subprime mortgages. Federal Reserve Bank of St. Louis Rev 91(2):79–93
17. Demyanyk Y, van Hemert O (2009) Understanding the subprime mortgage crisis. Rev Fin Stud 1305 (forthcoming). http://ssrn.com/abstract=1020396
18. Doms M, Furlong F, Krainer J (2007) Subprime mortgage delinquency rates. Working Paper Series 2007-33, Federal Reserve Bank of San Francisco. http://www.frbsf.org/publications/economics/papers/2007/wp07-33bk.pdf
19. Dunbar RIM (1998) The social brain hypothesis. Evol Anthropol 6:178–190
20. Farmer JD (2002) Market force, ecology and evolution. Ind Corp Change 11(5):895–953
21. Fogedby HC (2003) Damped finite-time-singularity driven by noise. Phys Rev E 68:051105
22. Fogedby HC, Poukaradzez V (2002) Power laws and stretched exponentials in a noisy finite-time-singularity model. Phys Rev E 66:021103
23. Freixas X, Parigi BM, Rochet J-C (2000) Systemic risk, interbank relations, and liquidity provision by the central bank. J Money Credit Banking 32(3):611–638
24. Garber PM (2000) Famous first bubbles: the fundamentals of early manias. MIT Press, Cambridge, MA
25. Gintis H, Bowles S, Boyd R, Fehr E (eds) (2005) Moral sentiments and material interests. MIT Press, Cambridge, MA
26. Gluzman S, Sornette D (2002) Classification of possible finite-time singularities by functional renormalization. Phys Rev E 66:016134
27. Greenspan A (1997) Federal Reserve's semiannual monetary policy report, before the Committee on Banking, Housing, and Urban Affairs, U.S. Senate, February 26, 1997
28. Greenspan A, Kennedy J (2008) Sources and uses of equity extracted from homes. Oxford Rev Econ Policy 24(1):120–144

29. Helbing D (ed) (2008) Managing complexity: insights, concepts, applications. Understanding complex systems. Springer, Heidelberg
30. Ide K, Sornette D (2002) Oscillatory finite-time singularities in finance, population and rupture. Physica A 307(1–2):63–106
31. Johansen A (2003) Characterization of large price variations in financial markets. Physica A 324(1–2):157–166
32. Johansen A, Ledoit O, Sornette D (2000) Crashes as critical points. Int J Theor Appl Fin 3(2):219–255
33. Johansen A, Sornette D (1998) Stock market crashes are outliers. Eur Phys J B 1:141–143
34. Johansen A, Sornette D (2000) The Nasdaq crash of April 2000: yet another example of log-periodicity in a speculative bubble ending in a crash. Eur Phys J B 17:319–328
35. Johansen A, Sornette D (2001) Finite-time singularity in the dynamics of the world population and economic indices. Physica A 294(3–4):465–502
36. Johansen A, Sornette D (2001) Large stock market price drawdowns are outliers. J Risk 4(2):69–110
37. Johansen A, Sornette D (2006) Shocks, crashes and bubbles in financial markets. Brussels Econ Rev (Cahiers economiques de Bruxelles) 49(3/4). Special issue on nonlinear analysis. http://papers.ssrn.com/abstract=344980
38. Johansen A, Sornette D, Ledoit O (1999) Predicting financial crashes using discrete scale invariance. J Risk 1(4):5–32
39. Kaizoji T, Sornette D (2010) Market bubbles and crashes. In: Encyclopedia of quantitative finance. Wiley, New York (in press). http://www.wiley.com//legacy/wileychi/eqf/ (long version of the paper at http://arXiv.org/abs/0812.2449)
40. Keys BJ, Mukherjee T, Seru A, Vig V (2008) Did securitization lead to lax screening? Evidence from subprime loans. Athens Meetings Paper, European Finance Association, December 2008. http://papers.ssrn.com/sol3/papers.cfm?abstract_id=1093137
41. Kindleberger CP (2005) Manias, panics, and crashes: a history of financial crises, 5th edn. Wiley, New York
42. Krugman P (2008) The return of depression economics. W.W. Norton, New York
43. Krugman PR, Dominquez KM, Rogoff K (1998) It's baaack: Japan's slump and the return of the liquidity trap. Brookings Pap Econ Act 1998(2):137–205
44. Laherrère J, Sornette D (1998) Stretched exponential distributions in nature and economy: "fat tails" with characteristic scales. Eur Phys J B 2:525–539
45. Lietaer B, Ulanowicz R, Goerner S (2008) White paper on the options for managing systemic bank crises (November 2008). http://www.lietaer.com
46. Lux T, Sornette D (2002) On rational bubbles and fat tails. J Money Credit Banking 34(3): 589–610
47. Malkiel BG (2003) A random walk down Wall Street, 8th edn. W.W. Norton, New York
48. Marsili M (2008) Eroding market stability by proliferation of financial instruments. Working paper (November 21, 2008). Available at SSRN: http://ssrn.com/abstract=1305174
49. Marsili M, Raffaelli G, Ponsot B (2008) Dynamic instability in generic model of multi-assets markets. Working paper (November 21, 2008). Available at SSRN: http://ssrn.com/abstract=1305205
50. Mauboussin MJ, Hiler R (1999) Rational exuberance? Equity research report of Credit Suisse First Boston, January 26, 1999
51. Philippon T, Reshef A (2009) Wages and human capital in the U.S. financial industry: 1909–2006. Working Paper 14644, National Bureau of Economic Research. http://www.nber.org/papers/w14644
52. Poser NS (2009) Why the SEC failed: regulators against regulation. Brooklyn J Corp Fin Comm Law 3(Spring). Brooklyn Law School, Legal Studies Paper No. 132. http://papers.ssrn.com/sol3/papers.cfm?abstract_id=1348612
53. Roehner BM, Sornette D (2000) "Thermometers" of speculative frenzy. Eur Phys J B 16: 729–739
54. Rowe N (2009) No, Greenspan was not right. http://macromarketmusings.blogspot.com/2009/02/no-greenspan-was-not-right.html

55. Serrano MA, Boguna M, Vespignani A (2007) Patterns of dominant flows in the world trade web. J Econ Interact Coord 2(2):111–124
56. Sheffrin H (2005) A behavioral approach to asset pricing. Academic, New York
57. Shiller RJ (2000) Irrational exuberance. Princeton University Press, New York
58. Sornette D (1998) Discrete scale invariance and complex dimensions. Phys Rep 297(5): 239–270
59. Sornette D (2003) Why stock markets crash (critical events in complex financial systems). Princeton University Press, New York
60. Sornette D (2008) Nurturing breakthroughs: lessons from complexity theory. J Econ Interact Coord 3:165–181
61. Sornette D, Andersen JV (2002) A nonlinear super-exponential rational model of speculative financial bubbles. Int J Mod Phys C 13(2):171–188
62. Sornette D, Johansen A (2001) Significance of log-periodic precursors to financial crashes. Quant Fin 1(4):452–471
63. Sornette D, Takayasu H, Zhou W-X (2003) Finite-time singularity signature of hyperinflation. Physica A 325:492–506
64. Sornette D, Woodard R, Zhou W-X (2009) The 2006–2008 oil bubble and beyond. Physica A 388:1571–1576
65. Sornette D, Zhou W-X (2004) Evidence of fueling of the 2000 new economy bubble by foreign capital inflow: implications for the future of the US economy and its stock market. Physica A 332:412–440
66. Stodder J (2000) Reciprocal exchange networks: implications for macroeconomic stability. Paper presented at the International Electronic and Electrical Engineering (IEEE) Engineering Management Society (EMS), August 2000, Albuquerque, New Mexico
67. Taylor JB (2009) How government created the financial crisis. Wall Street Journal, February 9, 2009
68. Taylor JB (2009) Getting off track: how government actions and interventions caused, prolonged, and worsened the financial crisis. Hoover Institution Press, Stanford
69. Vasiliki S, Veldkamp L (2009) Ratings shopping and asset complexity: a theory of ratings inflation. NBER working paper 14761. http://www.nber.org/papers/w14761
70. Zhou W-X, Sornette D (2003) 2000–2003 Real estate bubble in the UK but not in the USA. Physica A 329:249–263
71. Zhou W-X, Sornette D (2004) Causal slaving of the U.S. treasury bond yield antibubble by the stock market antibubble of August 2000. Physica A 337:586–608
72. Zhou W-X, Sornette D (2006) Is there a real-estate bubble in the US? Physica A 361:297–308
73. Zhou W-X, Sornette D (2008) Analysis of the real estate market in Las Vegas: bubble, seasonal patterns, and prediction of the CSW indexes. Physica A 387:243–260
74. Zhou W-X, Sornette D, Hill RA, Dunbar RIM (2005) Discrete hierarchical organization of social group sizes. Proc R Soc London 272:439–444

Global and Local Approaches Describing Critical Phenomena on the Developing and Developed Financial Markets

Dariusz Grech

Abstract We define and confront global and local methods to analyze the financial crash-like events on the financial markets from the critical phenomena point of view. These methods are based respectively on the analysis of log-periodicity and on the local fractal properties of financial time series in the vicinity of phase transitions (crashes). The log-periodicity analysis is made in a daily time horizon, for the whole history (1991–2008) of Warsaw Stock Exchange Index (WIG) connected with the largest developing financial market in Europe. We find that crash-like events on the Polish financial market are described better by the log-divergent price model decorated with log-periodic behavior than by the power-law-divergent price model usually discussed in log-periodic scenarios for developed markets. Predictions coming from log-periodicity scenario are verified for all main crashes that took place in WIG history. It is argued that crash predictions within log-periodicity model strongly depend on the amount of data taken to make a fit and therefore are likely to contain huge inaccuracies. Next, this global analysis is confronted with the local fractal description. To do so, we provide calculation of the so-called local (time dependent) Hurst exponent H_{loc} for the WIG time series and for main US stock market indices like DJIA and S&P 500. We point out dependence between the behavior of the local fractal properties of financial time series and the crashes appearance on the financial markets. We conclude that local fractal method seems to work better than the global approach – both for developing and developed markets. The very recent situation on the market, particularly related to the Fed intervention in September 2007 and the situation immediately afterwards is also analyzed within fractal approach. It is shown in this context how the financial market evolves through different phases of fractional Brownian motion. Finally, the current situation on American market is analyzed in fractal language. This is to show how far we still are from the end of recession and from the beginning of a new boom on US financial market or on other world leading stocks.

D. Grech (✉)
Institute of Theoretical Physics, University of Wrocław, 50-204 Wrocław, Poland
e-mail: dgrech@ift.uni.wroc.pl

M. Takayasu et al. (eds.), *Econophysics Approaches to Large-Scale Business Data and Financial Crisis*, DOI 10.1007/978-4-431-53853-0_7, © Springer 2010

1 Introduction

One believes that the analogy of the financial market with complex dynamical systems could be a very fruitful idea in description of various financial events including crashes (see, e.g. [1–17]). The similarity between financial crashes and phase transitions in complex systems has a long history. This idea, mentioned by Mandelbrot already in [1], was further developed more than a decade ago [2, 3]. Since then many authors have been analyzing crashes in financial world with the use of phase transition language. The main aim of this article is to present, review and discuss in some details two main philosophies developed so far to apply complexity in finance.

The first approach we call the global one. Its aim is to observe well defined, repeatable structure in financial time series before the phase transition point occurs. This point will be denoted on time axis as t_c (the crash point). Complexity may help to find some candidate functions to fit this repeatable structure. If $p(t)$ is the price of a given stock (or the market index value) at time t and $\tau = t_c - t$ is the time left to the crash ($t_c > t$), one expects that $p(t)$ should explode at t_c according to known complexity law

$$\frac{dp(\tau)}{d\tau} \sim \tau^{-\alpha} \tag{1}$$

with some positive, real exponent α.

The above equation is not the only one we may consider,[1] but it is the simplest one describing the price evolution at large time scales. It has two qualitatively different solutions. If $\alpha \neq 1$ one arrives with the power-law divergent solution

$$p(t) = A - B(t_c - t)^m, \quad B > 0 \tag{2}$$

with four free parameters $(A, B, t_c, m \equiv 1 - \alpha)$. This approach was originally introduced in finance by Sornette et al. [2,9].

Another point of view is presented by Vandewalle et al. in [10]. The latter paper suggests to take the particular value $\alpha = 1$ what leads to log-divergent solution of (1)

$$p(t) = A - B\ln(t_c - t), \quad B > 0 \tag{3}$$

instead of power-law divergent solution shown in (2).

The latter approach was shown to be very successful in determination of crash point in October 1987 and October 1997 on US market [11]. The same authors proposed also in [10] to apply the log-divergent fit not to the pure $p(t)$ signal itself but to the exponentially detrended $\widetilde{p}(t)$ signal, where:

$$\widetilde{p}(t) = p(t) - D\exp(rt). \tag{4}$$

They considered the background signal $C_{bg} = D\exp(rt)$ subtracted from $p(t)$ as the "natural evolution" of the stock market out of any euphoric phenomena. Such

[1] Sornette proposed recently faster than exponential, bubble-like growth of price, replacing the RHS of (1) by p^d, $d > 0$.

subtraction is a classical step used in phase transitions if one searches for critical exponents of the unbiased complex system. In the case of economic index, C_{bg} has the meaning of long-lasting natural increasing trend with a rate $r \ll 1$ per annum.

Once one approaches the critical time ($t \to t_c$), the short scale behavior of complex systems close to the phase transition point should be revealed. It makes the formulae in (2) or in (3) a bit more complicated. The relative change of price $\Delta p(\tau)/p(\tau)$, according to critical system behavior, should be independent on the chosen time scale. Thus

$$\frac{\Delta p(\tau)}{p(\tau)} = \Re e f\left(\frac{\Delta\tau}{\tau}\right), \tag{5}$$

where the function f is an arbitrary (also complex in general) function of $\Delta\tau/\tau$ (in the case of complex function one takes its real part as it is shown on RHS of (5)).

Expanding the RHS of (5) into Taylor power series in $\Delta\tau/\tau$ and leaving only leading linear term, we arrive with

$$\frac{\Delta p(\tau)}{p(\tau)} = (m + i\omega)\frac{\Delta\tau}{\tau}, \tag{6}$$

where m, ω are real constants.

The solution of (6) leads to so called generalized power-law:

$$p(\tau) \sim \left(\frac{\tau}{\tau_0}\right)^m \cos\left[\omega \ln\left(\frac{\tau}{\tau_0}\right) + \varphi\right] \tag{7}$$

with some constants τ_0 and φ.

The full price evolution comes then as a long-time scale behavior decorated with short-time scale solutions, i.e. we get the form of power-law terms decorated with so called log-periodic oscillations:

$$p(t) = A - B(t_c - t)^m[1 + C\cos(\omega\ln(t_c - t) + \varphi] \tag{8}$$

or the form of log-divergent terms decorated with log-periodic oscillations:

$$p(t) = A - B\ln(t_c - t)[1 + C\cos(\omega\ln(t_c - t) + \varphi]. \tag{9}$$

It is straightforward to show that the subsequent moments t_n ($n = 1, 2, 3, \ldots$) when the signal is in the same phase, satisfies the following relation

$$\frac{t_{n-1} - t_n}{t_n - t_{n+1}} = \lambda, \tag{10}$$

where λ is some constant. Thus, the period of the signal is not constant any longer and it decreases geometrically with time.

An alternative explanation for log-periodicity in financial time series may come out also from the so called discrete scale invariance (DSI) [18] observed for some complex systems. Let $\Phi(t)$ is an observable near a critical point t_c. Under the change of time scale $t \to \lambda t$, one expects the power-law behavior

$$\Phi(\lambda t) = \lambda^m \Phi(t). \tag{11}$$

However, in the case of DSI, it happens only for infinite but countable set of $\lambda's$. Such condition can be justified for the general solution of (11) written as

$$\Phi(t) = t^m P\left(\frac{\ln(t)}{\ln(\lambda^*)}\right), \tag{12}$$

where P is an arbitrary periodic function and λ^* is the parameter characteristic for the system. Indeed, one gets (11) satisfied under the time rescaling transformation $t \to \lambda t$ only if

$$P\left(\frac{\ln(\lambda t)}{\ln(\lambda^*)}\right) \equiv P\left(\frac{\ln(\lambda)}{\ln(\lambda^*)} + \frac{\ln(t)}{\ln(\lambda^*)}\right) = P\left(\frac{\ln(t)}{\ln(\lambda^*)}\right) \tag{13}$$

what requires $\lambda = k \times \lambda^*$ with $k = 1, 2, \ldots$ and P of period 1. The solution given by (7) is then obtained directly as a first order Fourier expansion of RHS of (12) with $\omega = 2\pi / \ln(\lambda)$.

There is a number of papers suggesting that λ^* parameter should be unique for all complex financial systems. There are empirical arguments that for well established and grown financial markets there exists the preferential scaling $\lambda^* \sim 2$ [15, 17, 19]. It should apply both to the leading pattern in the signal (called a bubble) as well as to the related substructures (called sub-bubbles) the main pattern is decorated with due to the fractal nature of the signal. Moreover, the scenario of (12) enables to generalize periodicity replacing *cosine* by other simple periodic functions, e.g. cosine modulus, saw-like functions, etc. They sometimes give better description of oscillations departure from the average trend [15].

Other mechanistic founded global approaches to periodicity in finances have also been developed. It is especially worth to mention, e.g., those based on the analogy with viscoelastic materials properties [20]. The periodic evolution of a stock index before and immediately after the crash is described within this approach by Mittag-Leffler generalized exponential function superposed with various types of oscillations.

The global approach seems to be interesting and encouraging. The main difficulty in its application lies in the fractal structure of financial time series. Due to this fractal nature we never know if oscillations or even the leading shape of the price index described theoretically by (8) and (9) are connected with the main bubble (i.e. the structure of time series being formed from the beginning of increasing trend till the crash point t_c) or with some mini-bubbles appearing as second order or higher order corrections to these equations. Usually, it is difficult to separate data connected with the main bubble and its mini-bubble corrections before an extreme event (crash) happens. Such distinction becomes explicitly clear only after the event already had happened.

The other philosophy of complex phenomena applied to finances is therefore to study the local scaling properties of financial time series in order to distinguish whether the involved stochastic process can be long-memory correlated or not. Several techniques have been proposed in literature to attack this problem. Their common aim is to calculate the Hurst exponent H [21,22] of the system. It is known that for long-memory correlated process called fractional Brownian motion (fBM), the value of H exponent ranges from $0 < H < 1/2$ for negative persistence (negatively autocorrelated signal) to $1/2 < H < 1$ for positive persistence (positively autocorrelated signal). The particular case $H = 1/2$ corresponds to the signal with fully uncorrelated increments called integer Brownian motion or shortly Brownian motion.

The Hurst exponent is related to the fractal dimension D_f of one dimensional time series by the well-known relation $D_f = 2 - H$, so the search for H is equivalent to the search for fractal properties of time series. There are various techniques to calculate H exponent for the given time series. Let us mention the Rescaled Range Analysis (R/S) [21], Detrended Fluctuation Analysis (DFA) [23], Detrended Moving Average Analysis (DMA) [24, 25] or the power spectrum analysis [26]. The accurate and fast algorithm enabling to extract H from the given time series is served by DFA, applied in the following chapters of this article. DFA is very well described in the literature (see, e.g. [27–31]) with detailed discussion of various effects on DFA results like the effects of trends [31], non-stationarities [32], or nonlinear filters [33]. So far, the technique was widely applied to various topics – from DNA structures [23, 34], through stellar X-ray binary systems [35], up to finances [27,36 40]. Briefly, the method contains the following steps: (1) time series of length N is divided into non-overlapping boxes of length τ each, (2) in each box the linear trend (found within this box) is subtracted from the signal, (3) in each box the root-mean-square fluctuation $F^2(\tau)$ of the detrended signal is calculated and then $F^2(\tau)$ is averaged over all boxes of size τ, (4) the procedure is repeated for all box sizes τ $(1 < \tau < N)$.

The edge part of time series is usually not covered by any box. We propose to overcome this difficulty as follows. If the remaining part of time series has the length $\tau/2 \leq \Delta L < \tau$, we cover it by an additional box of size τ partly overlapping the preceding data. If $\Delta L < \tau/2$, we do not take the part of data contained in ΔL into account. Such recipe is particularly useful in the "local" version of DFA described later on in this chapter.

The power-law relation

$$F(\tau) \sim \tau^H \tag{14}$$

is expected from DFA, where H stands for the Hurst exponent. The maximal scale $\tau^\star$ for which the scaling holds, indicates the length (time-scale) of correlation in the system or the scaling range.

It turns out that the "local" version of DFA is very promising. It has been applied for the first time in financial crash analysis in [16]. The "local" DFA is nothing else but DFA applied to small subseries of a given set of data. We expect positive autocorrelations in time series if financial system relaxes (i.e., just after the critical moment t_c). Thus $H(t)$ should reach then the value $H > 1/2$ corresponding to the persistent signal. It means however, that for some time before the crash $(t < t_c)$

the system is antipersistent in order to reproduce the observed mean Hurst exponent value $\langle H \rangle \sim 1/2$ for large time limit. Therefore, $H(t)$ called also the "local" Hurst exponent $H_{\rm loc}$, fluctuates between $H < 1/2$ and $H > 1/2$ revealing some structure. This structure should be well recognized and compared with real events on the financial market, with hope to make possible predictions just from its form. The alternative explanation for such fluctuation of H and for its drop before a crash is based on market psychology. As far as the financial market approaches closer and closer the crash regime, investors become more nervous and sceptic about index or price rise in forthcoming sessions. Therefore, the stock index starts to fluctuate more and more around its mean drift. Such fluctuations can be translated quantitatively as substantial decrement of "local" H value. Financial time series becomes more "jagged".

The time evolution of $H_{\rm loc}$ is shown in a case of WIG signal in its 1991–2007 history at the bottom part of Fig. 1. The local H exponent calculated for part of the Dow Jones (DJIA) index is shown in Fig. 2. The latter figure shows also how the $H_{\rm loc}$ evolution changes for various length of time subseries (time-window lengths) the Hurst exponent is calculated for. In particular, the effect of trends in $H_{\rm loc}$ starts to be seen if it is calculated over small enough time-window (around 210 trading days). Once the time window length is bigger, only statistical fluctuations in $H_{\rm loc}$

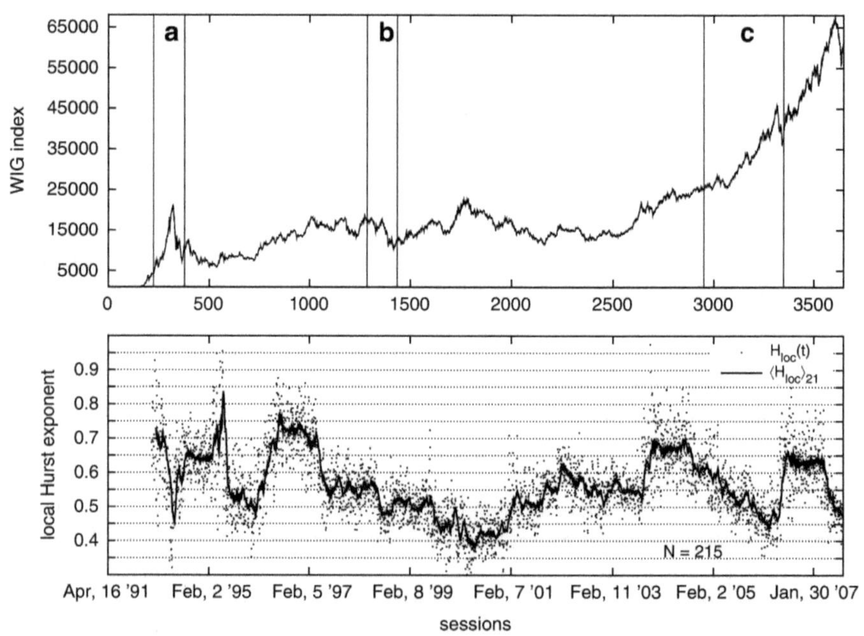

Fig. 1 The closure day WIG time series history April 1991–May 2007 (*top*) and its corresponding local Hurst exponent evolution (*bottom*). The time-dependent Hurst exponent has been calculated in the observation box of $N = 215$ sessions. The *solid line* represents the moving average of $H_{\rm loc}$ (marked as *dots*) calculated for one trading month back (21 sessions). Three main crash periods discussed in the main text are marked within the *vertical lines* (**a**), (**b**), (**c**), and correspond to the crashes or rupture point described in Table 1. (Taken from [41])

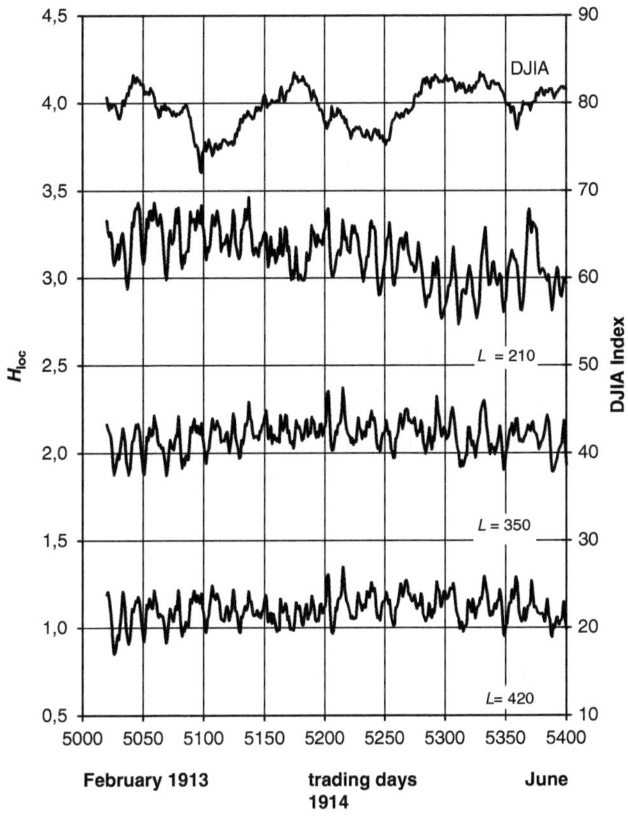

Fig. 2 An example of time dependent $H_{\rm loc}$ evolution for three choices of time-window length $N = 210, 350, 420$ calculated for DJIA signal in the period: Feb. 1913 – June 1914. $H_{\rm loc}$ values are artificially multiplied by 2 and then displaced along the vertical axis with a shift 0.5, 1.5, 3.0 respectively to make the differences in $H_{\rm loc}$ evolution in different time windows more noticeable. DJIA index values are also shown on the top. (Taken from [16])

signal are visible. We may expect that advantage in using $H_{\rm loc}$ over other methods is that it actually measures the temporal fluctuations on the market. Therefore, the local method seems to be more precise or at least more adequate to apply in open complex systems with rapidly changing boundary conditions. It is just a case of financial market.

The $H_{\rm loc}$ values are usually widely spread out so one needs to use some kind of data filter to extract information from these values and to eliminate statistical fluctuations. The easiest choice for such a filter is the moving average. We applied in this article the moving average $\langle H \rangle_5$ of last five sessions (one trading week) and the moving average $\langle H \rangle_{21}$ of last 21 sessions (one trading month) to limit the possible statistical noise. It turns out that although fluctuations in $H_{\rm loc}$ do depend on the time-window length N for which DFA is applied, the main pattern of $H_{\rm loc}$ containing its time evolution trends, does not depend on N [42]. The optimum choice of N seems to be a matter of intuition, statistics and (or) economic regards. If N is too large H_{loc} loses its "local" character, if N is too small we deal with not sufficient statistics causing huge fluctuations in H_{loc} values what makes H_{loc} data hardly readable. For such reasons we suggested in [16] to take $N < 240$, i.e. the time-window length less than one trading year. The good choice for N was suggested in [16] to be around 10 trading months (215 daily sessions). This warranties the statistical uncertainty in H_{loc} below 10% and at least 15–20 different box sizes τ within the scaling range (see [42]). For smaller N we are stuck with bigger statistical uncertainty in H_{loc} values, mainly because of insufficient (≤ 10) number of points in the scaling range of (14) used to make a regression fit.

The Polish market represented by the full WIG index accommodating around 300 companies is a good example of large but still emerging European stock market. The criticality based methods described above have not been compared so far for such emerging markets. On the other hand, we would like to know if local, fractal approach based on the calculation of H_{loc} values may give us a signal of forthcoming extreme events in finance. The latter question should be asked for developed as well as for developing markets. The following chapters of this article deal with the above problems and review main findings published in literature so far.

2 Log-Periodicity of Developing Markets: The Case of WIG Index

The most interesting extreme financial events on the Polish stock market are shown between the vertical lines in Fig. 1 (top part) and are also described in more details in Table 1. Some of them are real crashes, while others, like the event of July 1998, terminate the long-lasting increasing trend on the market. The events of the latter kind are sometimes called rupture points. They are also interesting for practical purposes for obvious reasons.

We shall pay attention at one big crash in March 1994 and several minor crashes that had taken place in the WIG history (1991–2007): in July 1998 and May 2006. The reverse of long-lasting increasing trend on the stock market in July 2007 will also be analyzed.

Table 1 The main crashes on the Polish financial market. The total percentage drop in WIG and its duration time is shown as well as the percentage drop in the first three sessions after the critical point t_c. All dates indicate the beginning of a trend reverse event

Date	Initial 3 sessions drop	Total relative drop (duration time)
17.03.94	11%	65% (41 sessions)
22.07.98	4%	39% (30 sessions)
12.05.06	5%	21% (24 sessions)

Let us examine first the log-periodicity indicated by (8) and (9). We have preformed fits according to [12], using initially only the power-law (2) or log-divergent (3) functions and thus establishing A, B, t_c, m parameters. These parameters were used in the next step to fit other ones included in log-periodic oscillations ((8), (9)). The nonlinear fit performed as the minimization of the variance of the signal with respect to the assumed log-periodic function was also done, as suggested in [2]. The latter one allows to express A, B, C as a function of other parameters, so the number of free fitting parameters drops twice. The fit was done for the signal containing at least $\sim$95% of the data points before any of the crashes (rupture points). Two mentioned fitting methods have shown agreement in results of the fit. The difference between two methods in fitted critical time t_c can be found as $\Delta t_c \leq 5$ trading days.

One may consider exponentially detrended signal according to (4) recipe as well as the pure not detrended WIG signal. Below are the results of our findings.

The first discovery was that the exponentially detrended signal (see (4)) does not work at all for developing stocks. One gets for the 1994 crash the log-divergent and power-law fits with goodness only $R^2 \sim 0.5$. The fit is much better ($R^2 \sim 0.9 - 0.95$) if the pure (not detrended) signal is considered. The same happens for the May 2006 rupture point.

The 1994 crash turns out to be very well predicted by the critical phenomena approach – equally well by power-law shaped signal (8) with the critical exponent $m \sim 0.08$ and by log-divergent solution (9) (see Fig. 3 and Table 2). Contrary, the second crash in July 1998 has not been detected at all by log-periodicity phase transition methods. This probably should not be very surprising. In fact, the 1998

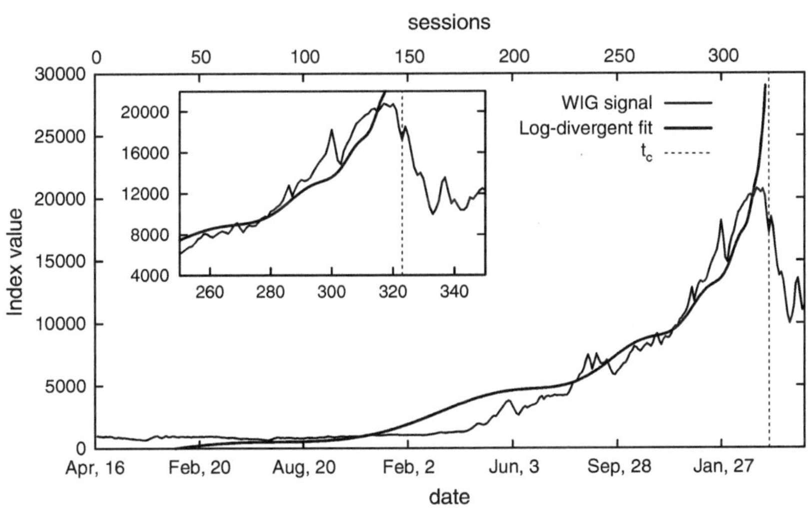

Fig. 3 Log-divergent best fit of log-periodic oscillations (see (9)) to the first financial crash on Polish stock market in March 1994. Fitting parameters are $A = (3.205 \pm 0.124) \times 10^4$, $B = 5785 \pm 239$, $C = 115.3 \pm 39$, $\omega = 8.6 \pm 0.1$, $\phi = 5.0 \pm 0.1$, $t_c = 322.6 \pm 1.8$ with $R^2 = 0.903$. The expected crash moment t_c is given as the session number (trading day). The real crash started at $t_c^{\text{real}} = 322$ (see also Table 2). (Taken from [41])

Table 2 Log-periodic oscillation fit to data preceding major crash-like events on the Polish stock market with assumed power-law divergent and logarithm divergent prices. Dates for all events are scaled in sessions numbers (trading days), t_c is the predicted and t_c^{real} is the real date of event. Scaling factor λ for all crash-like events is also pointed out

		Power-law div. fit		Log. div. fit		
Beginning of event	t_c^{real}	t_c	$t_c^{\mathrm{real}} - t_c$	t_c	$t_c^{\mathrm{real}} - t_c$	λ
March 17, 1994	322	321 ± 2	1 ± 2	323 ± 2	-1 ± 2	2.08
July 22, 1998	1,362	Not detected	Not detected	Not detected	Not detected	Not detected
May 12, 2006	3,320	$3{,}278 \pm 1$	42 ± 1	$3{,}272 \pm 3$	48 ± 3	1.53
without 1st event		$3{,}379 \pm 5$	-59 ± 5	$3{,}331 \pm 2$	-11 ± 2	1.23

drop in WIG was rather a long-lasting decreasing trend what can be interpreted as the relaxation of the system without going through the phase transition state itself.

The third (mini) crash of May 2006 is predicted to happen (see Table 2) two trading months before the actual event took place if the log-periodic pattern was fitted to *all available data including the first crash.* Once we have made a fit only to data after the first crash relaxation period, i.e. since February 2005, the predicted phase transition moment t_c was behind the actual event (see data in Table 2). More precise determination of t_c is offered in this case by the log-divergent fit (2 trading weeks difference between predicted and actual event) than by power-law fit (see the last row in Table 2). The power law fit leads to almost 60 trading days error. The goodness of fit was found better for $m = 0$, what additionally favors the log-divergent fit over the power-law one. This conclusion agrees with the statement of [11] for developed stock markets.

It is also interesting to calculate the scaling factor λ in order to see whether it is unique as for developed markets. The calculated λ varies between 1.2 and 2.1 what does not support the idea of its universality on the Polish stock market. This also indicates that the same common scenario of log-periodic oscillation for already developed and yet developing stock markets is a very rough approximation. An analysis of far-east emerging markets leads to similar conclusions (see, e.g [43]).

Finally, we may look at the last main rupture point in WIG index that took place on July 6, 2007. The log-periodic fit to this bubble is shown in Fig. 4. The best result is served by log-divergent curve (see Fig. 4 and Table 3) predicting very precisely the date of extreme event. However, the fitting results were found to be very sensitive to the number of data (the final input data length) one takes into account (see Fig. 5). The similar event was predicted to happen earlier when smaller number of data was available. The predicted transition point t_c tends to increase with the number of input data and has no clear stability region, so this prediction seems to be rather accidental. The further analysis is here more than welcome – also for more developed financial systems than the Polish developing stock exchange market. Therefore, one may conclude that minor crashes on the Polish market, contrary to big events (only one such event took place so far, i.e. March 1994 crash), are not very precisely predicted by the critical phenomena description based on the log-periodicity assumption. It is interesting to check how much other methods like,

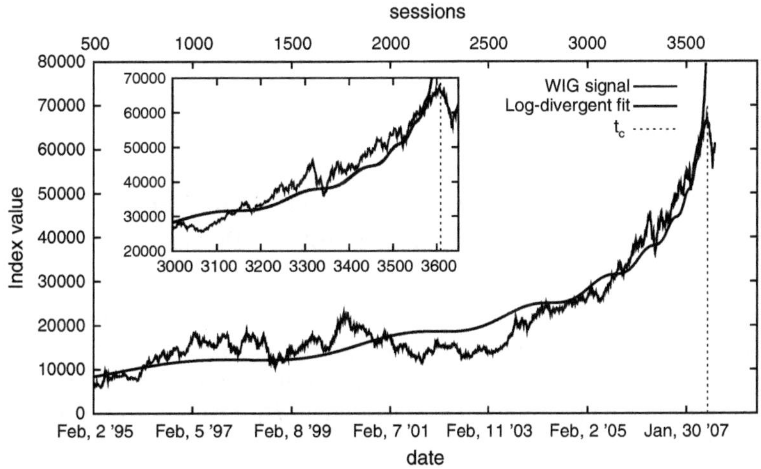

Fig. 4 Log-divergent best fit of log-periodic oscillation to the rupture point of July 2007 for WIG. Fit is done for data taken in the period Feb. 2, 1995 – June 28, 2007 and the parameters of fit are: $A = (1.057 \pm 0.014) \times 10^5$, $B = (1.21 \pm 0.02) \times 10^4$, $C = 150.6 \pm 25$, $\omega = 11.67 \pm 0.23$, $\phi = 55.61 \pm 1.77$, $t_c = 3609 \pm 3$ with $R^2 = 0.947$. The real rupture point t_c^{real} took place on July 6, 2007, i.e. session #3609. See also Table 3. (Taken from [41])

Table 3 Log-periodic fit to the recent major rupture point in WIG data. Shown are results of analysis including first crash data (i.e. from the beginning of WIG quotations) and without first crash data (i.e. when analysis starts in Feb. 1995 after relaxation of the system connected with the first crash). Power-law-divergent (8) and log-divergent (9) cases are studied

	Without 1st crash		With 1st crash	
Data analysed till	Power-law div. fit	Log. div. fit	Power-law div. fit	Log. div. fit
3,575 (May 19, 2007)	3,741 ± 28	3,579 ± 2	3,580 ± 3	3,580 ± 4
3,600 (June 28, 2007)	3,700 ± 10	3,609 ± 3	3,659 ± 11	3,620 ± 2

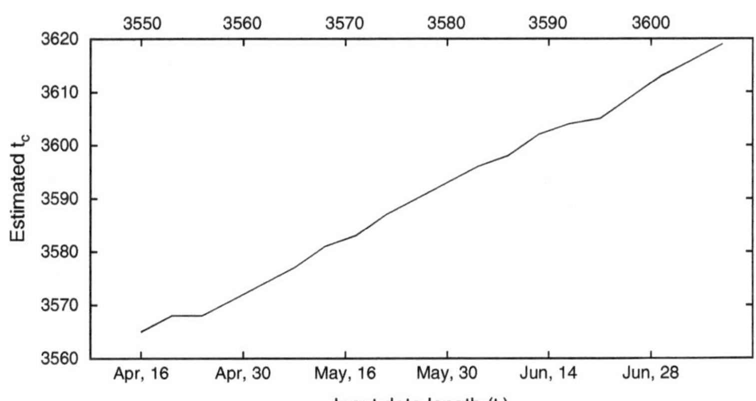

Fig. 5 Evolution of the predicted t_c for WIG as a function of the last point (input data length) for data analyzed after the first crash relaxation of the system, i.e. from Feb. 1995 till t_f. (Taken from [41])

e.g. the one based on Mittag-Leffler generalized exponent function decorated with periodicity, proposed in [20], are sensitive to the number of data a fit is made to. However, it is beyond the scope of this article.

3 Local Fractal Properties of Financial Time Series in the Vicinity of Crashes and Rupture Points

Let us proceed now to analyze extreme events on the stock market, looking at the local fractal properties of financial time series. All the events considered in the preceding chapter for Polish WIG index as well as main crashes on US market in its whole history will be investigated now looking at H_{loc} evolution properties around the critical point t_c.

Let us first focus our attention on the most important events in US market history. These are crashes in September 1929, October 1987, July 1998 and the termination of long-term prosperity in September 2001.

Let us start with the 1929 crash shown in details with its local fractal characteristic in Fig. 6. The DJIA signal was in clear increasing mode for about 4 months – from session ♯9760 up to ♯9840. Simultaneously, a very clear decreasing trend in H_{loc} values is visible. It had started about 1 month before the DJIA index reached its maximal value on September 3, 1929. The H_{loc} reached a deep and clearly seen local minimum $H_{\text{loc}} \sim 0.45$ two weeks before the crash.

To see if it was a chance we may check other crashes in US. The fractal scenarios of crashes in 1987 and 1998 are shown in Figs. 7 and 8, respectively. The 1987 crash reveals also clear decreasing trend in H_{loc} for the 1-year period preceding the crash moment (see Fig. 7), despite DJIA signal was still rising at that time. These two contradicting trends seem to indicate that investors were gradually more sceptic about the market future, despite the market index was rising. The H_{loc} gains again deep minimum $H_{\text{loc}} \simeq 0.43$ just before the October crash. The percentage drop between maximal and minimal H_{loc} value in decreasing trend was about the same as for 1929 crash.

The same phenomena occurs for 1998 crash illustrated with its fractal evolution in Fig. 8. The market was much more "nervous" at that time what is reflected by more jagged index plot and smaller H_{loc} values forming decreasing trend. H_{loc} reaches its deep minimum $H_{\text{loc}} \sim 0.35 - 0.4$ in the half of 1998. The crash actually occurred on July 17, 1998.

Finally, the total display of the period 1995–2003 for DJIA signal history with corresponding evolution of H_{loc} values is shown in Fig. 9. Some important events are marked here like: the beginning of 1998 crash (July 17, 1998), the end of the longest economic boom in US history after 107 months of stock market expansion (January 14, 2000), or the terrorist attack in New York (September 11, 2001). The decreasing trend in H_{loc} exponents is being observed from January 2000 till September 10, 2001 – just one day before terrorist attack. The local Hurst exponent gets values as low as $H_{\text{loc}} \sim 0.4$ just before this attack with average values around 0.45. If one

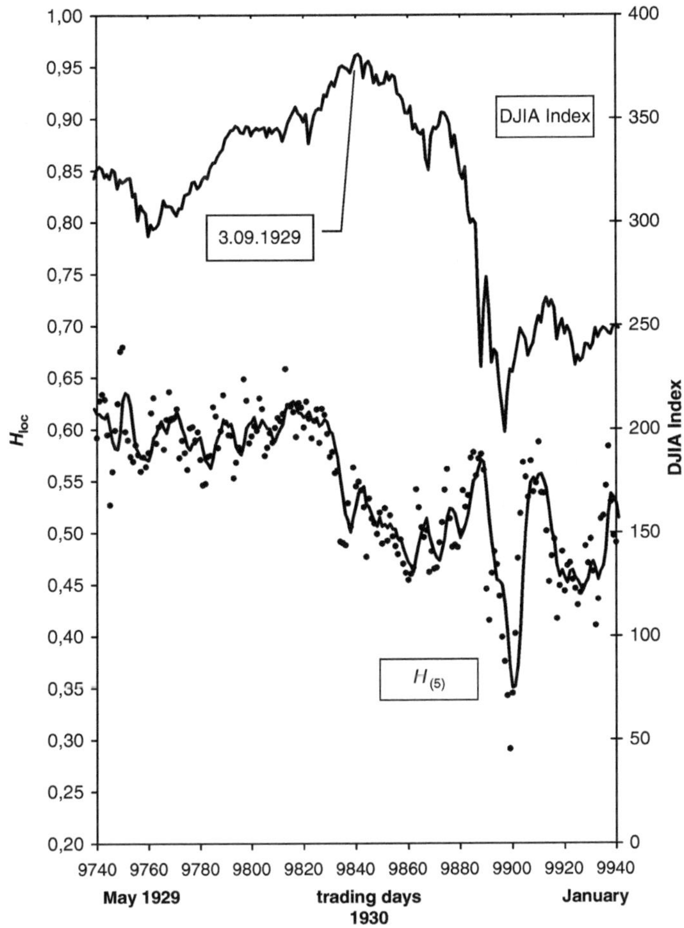

Fig. 6 The 1929 crash on US market seen from the fractal point of view. All local H_{loc} values (marked as *dots*) are calculated session by session for $N = 215$ session time-window. The moving average H_5 of H_{loc} calculated over last 5 days is marked as *solid line*. The significant drop in H_{loc} values is seen before the moment of crash – between sessions ♯9820 and ♯9860. (Taken from [16])

believes that previous findings were not accidental then we should state that the huge drop in DJIA signal on September 11, 2001 would have occurred, even if the terrorist attack did not take place!

It is interesting to see if the characteristic pattern of H_{loc} evolution before a crash will be repeated for developing markets. We did this task for WIG signal [41,42] analyzing crashes investigated in the previous chapter with log-periodic scenario. The zoomed evolution of time dependent $H_{\text{loc}}(t)$ for these crashes is shown in Figs. 10–13, respectively. The decreasing trend of H_{loc} before the critical point t_c is also evident from these plots. It lasts for many sessions before the extreme event occurs. One may summarize the common characteristic pattern of H_{loc} plots before the crash moment in the form of the following necessary conditions to be

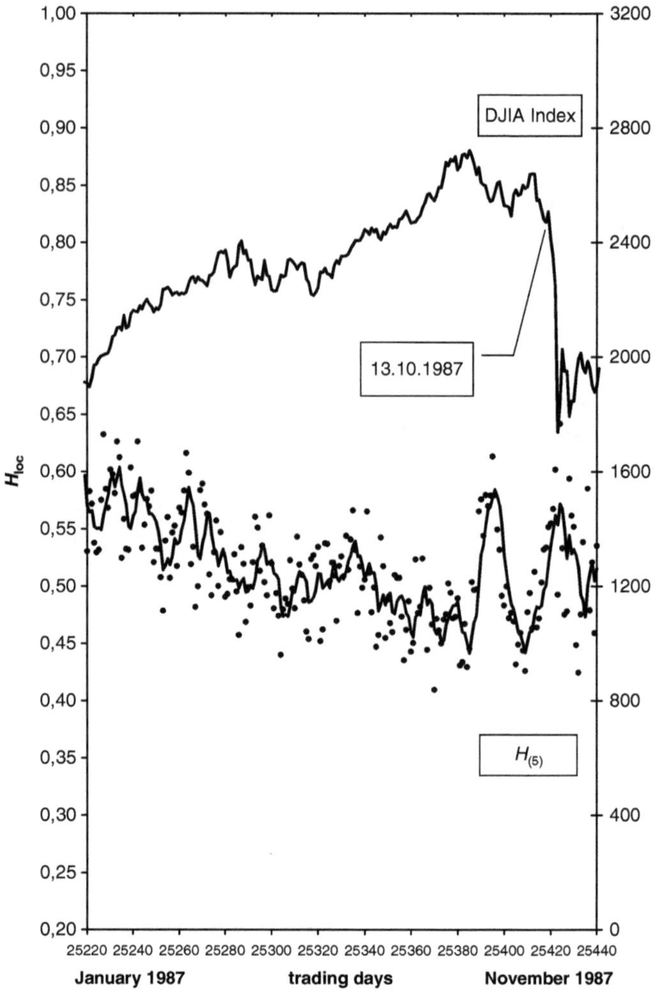

Fig. 7 The history of 1987 crash. The moving 1-week average H_5 explains the decreasing trend formation in $H_{\rm loc}$. (Taken from [16])

simultaneously satisfied (*signal to sell*) if the rupture point in the increasing index signal is expected soon, i.e. usually within one to two trading weeks [42]:

1. $H_{\rm loc}(t)$ is in decreasing trend (hence $\langle H_{\rm loc}\rangle_5 < \langle H_{\rm loc}\rangle_{21}$ except for small fluctuations).
2. $\langle H_{\rm loc}\rangle_{21} \lesssim 0.5$.
3. $\langle H_{\rm loc}\rangle_5 \lesssim 0.45$.
4. Minima of $H_{\rm loc}(t)$ (for not necessary consecutive sessions) satisfy $H_{\rm loc}^{min}(t) \lesssim 0.4$.

Contrary, we expect the strong *signal to buy* on the market if all the above conditions are not satisfied.

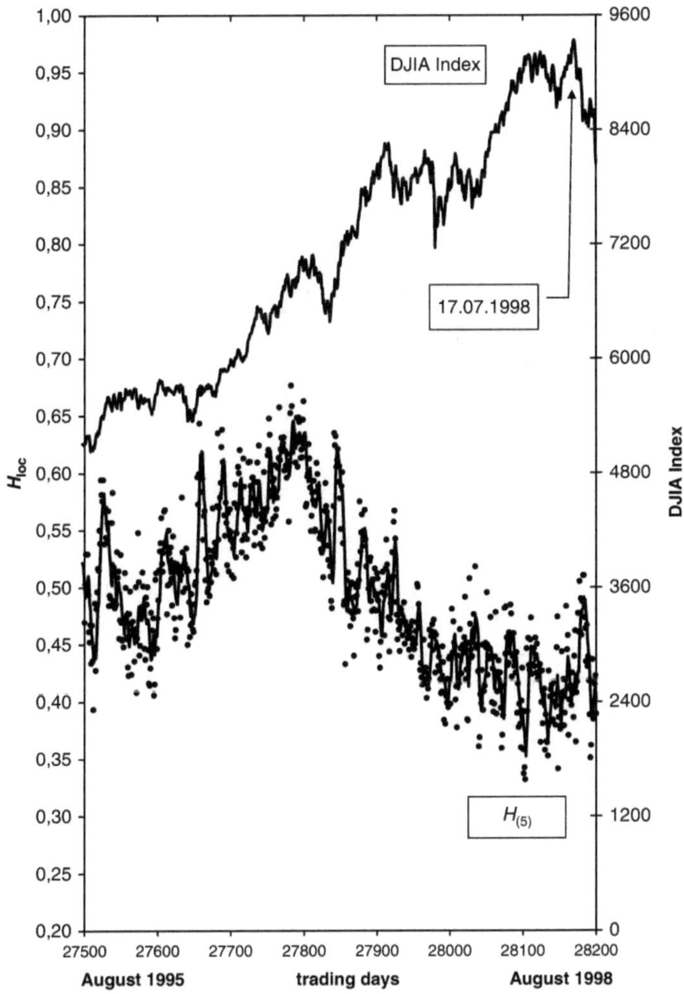

Fig. 8 The history of 1998 crash. The decreasing trend in H_{loc} is much deeper than before. The strong decreasing trend with deep minimum $H_{loc} \sim 0.35 - 0.4$ in the half of 1998 seems to predict the July 1998 crash. (Taken from [16])

One may obviously ask if the characteristic pattern in H_{loc} is truly related to autocorrelations in marked signal. This can be done by random shuffling the logarithmic returns $r_i = \ln(P_i/P_{i-1})$ where P_i is the index value in the end of i-th session. We did the shuffling of the return signal for all crashes discussed in this article for US and Polish markets. The time dependent Hurst exponent did not reveal its previous structure after shuffling. An example is seen in Fig. 14 for March 1994 crash in WIG. The similar results were obtained also for German DAX index in [44]. It is worth to stress that July 1998 crash, previously not seen by log-periodicity, could have been detected by well developed H_{loc} crash structure described above. This H_{loc} crash pattern was visible as early as one week before the event took place (see Fig. 11).

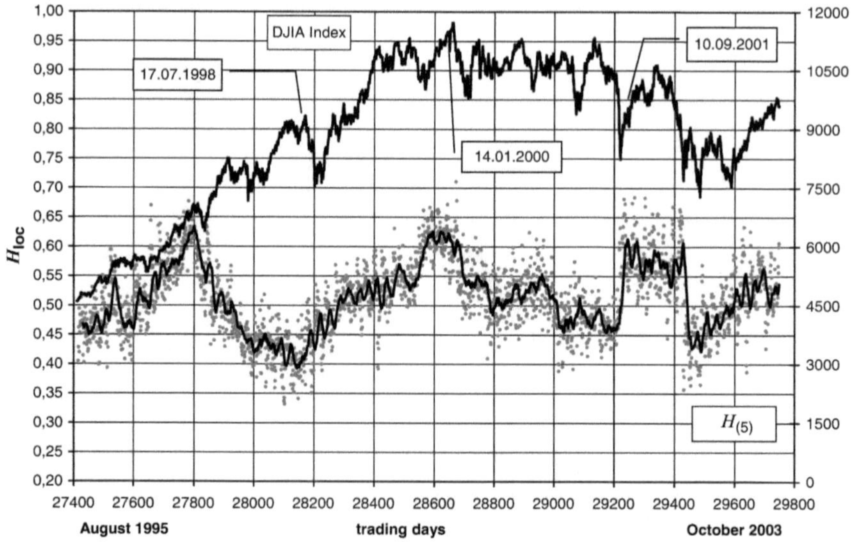

Fig. 9 The total display of local fractal properties of DJIA index in the period 1995–2003. Some important historical events are marked. (Taken from [16])

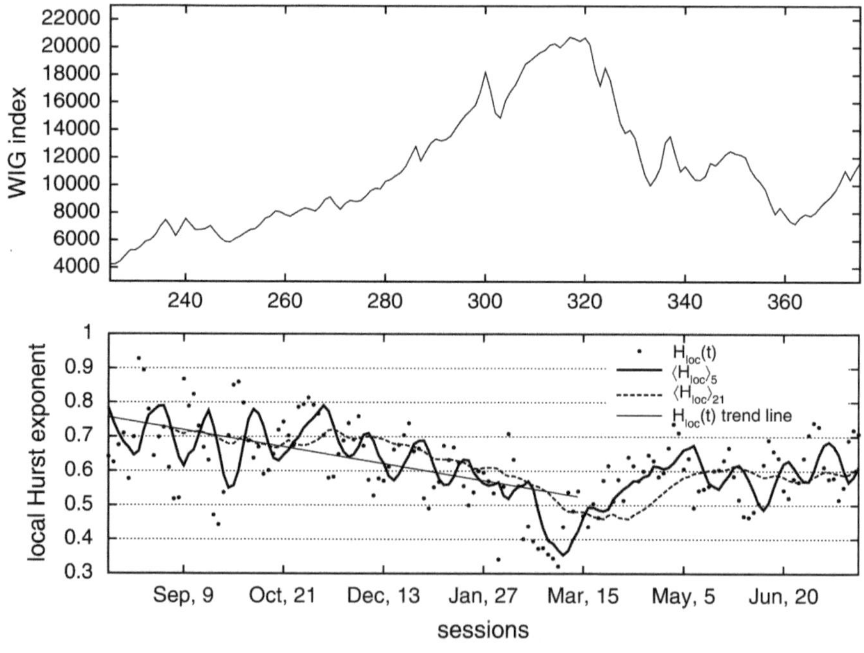

Fig. 10 The H_{loc} pattern before and after the first crash on the Polish stock market in March 1994. The dots represent H_{loc} values, *solid lines* indicate the 1-week $\langle H_{\text{loc}}\rangle_5$ and 1-month $\langle H_{\text{loc}}\rangle_{21}$ moving averages of $H_{\text{loc}}(t)$. The line-fit of the decreasing trend for $H_{\text{loc}}(t)$ before the crash is also shown. The relaxation of the system (increasing $\langle H\rangle_5$) immediately after the crash point (March 17, 1994) is seen. (Taken from [41])

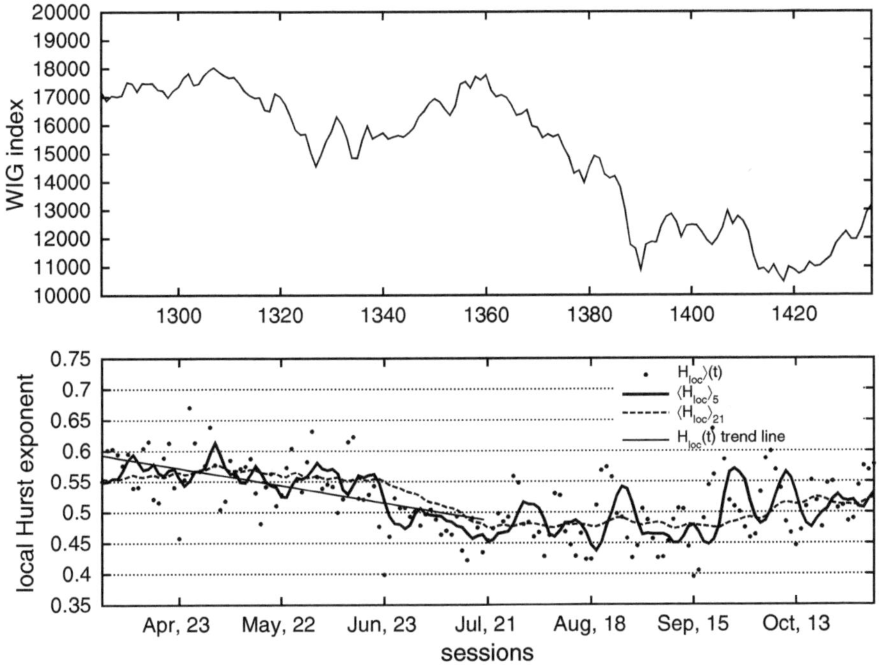

Fig. 11 The same as in Fig. 10 but for the July 1998 rupture point of WIG signal

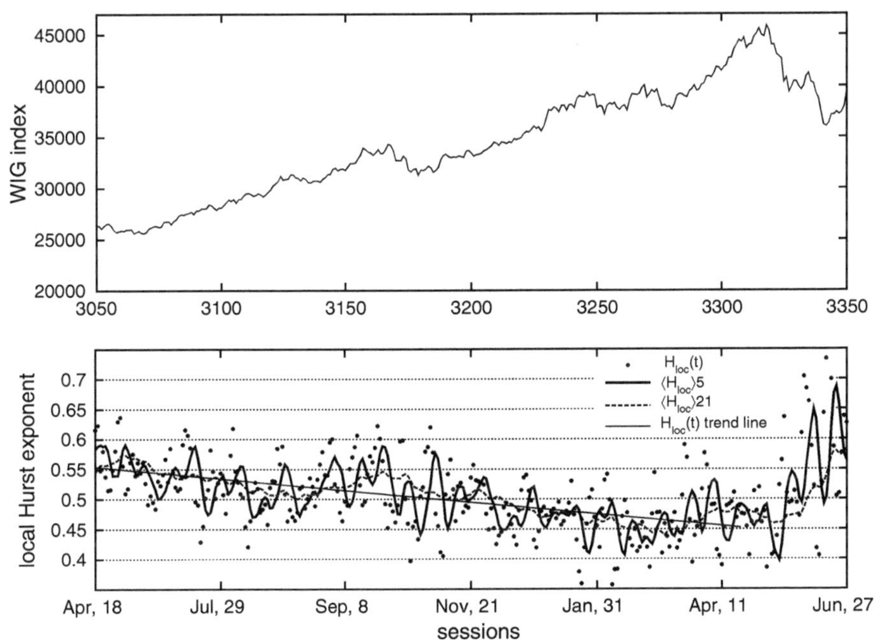

Fig. 12 The same as in Fig. 10 but for the May 2006 crash of WIG index. (Taken from [41])

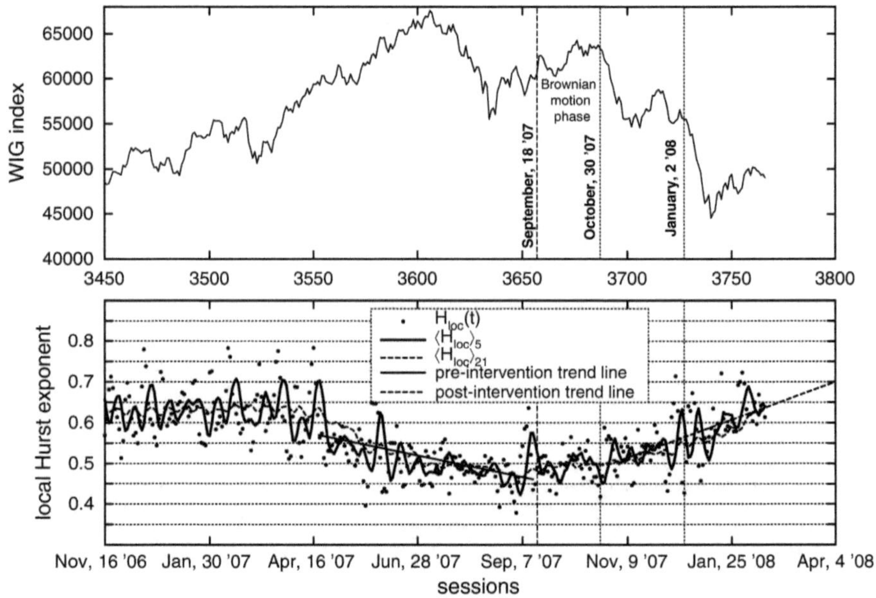

Fig. 13 Recent WIG history and the corresponding H_{loc} pattern. Data terminate on Feb. 29, 2008. *Dashed thick line* indicates the moment of Fed intervention; two *thin dashed lines* indicate sessions with local strong minima in H_{loc} values, i.e. Oct. 30, 2007 and Jan. 2, 2008. These sessions terminate two major positive corrections in WIG signal during its long-lasting decreasing trend started yet on July 6, 2007 (session #3609). H_{loc} pattern shows also the relaxation of the system after Oct. 30, 2007 (see post-intervention H_{loc} trend-line) and the Brownian-like motion phase on the stock market between the intervention of Sept. 18, 2007 and the end of October 2007. (Taken from [41])

Finally, let us notice that the main rupture point of increasing trend of WIG signal in July 2007 had also H_{loc} pattern indicating the extreme event is coming. However, the decreasing trend in H_{loc} did not terminate on July 6, 2007 (see Fig. 13). In other words the relaxation of the system did not started yet and one could have expected the major crash is still ahead. We predicted in [42] that it should happen no later than mid-September 2007 if the H_{loc} was still in decreasing mode with the same dropping rate.

The reason for such statement was that $\langle H_{\text{loc}}(t)\rangle$ average was calculated to reach $\sim$0.4 value at that time (see Fig. 8 in [42]). What happened next is an interesting problem to be discussed within local fractal approach, because for the first time one could notice how exterior influence on the market is seen from the fractal point of view. It is the subject of next chapter.

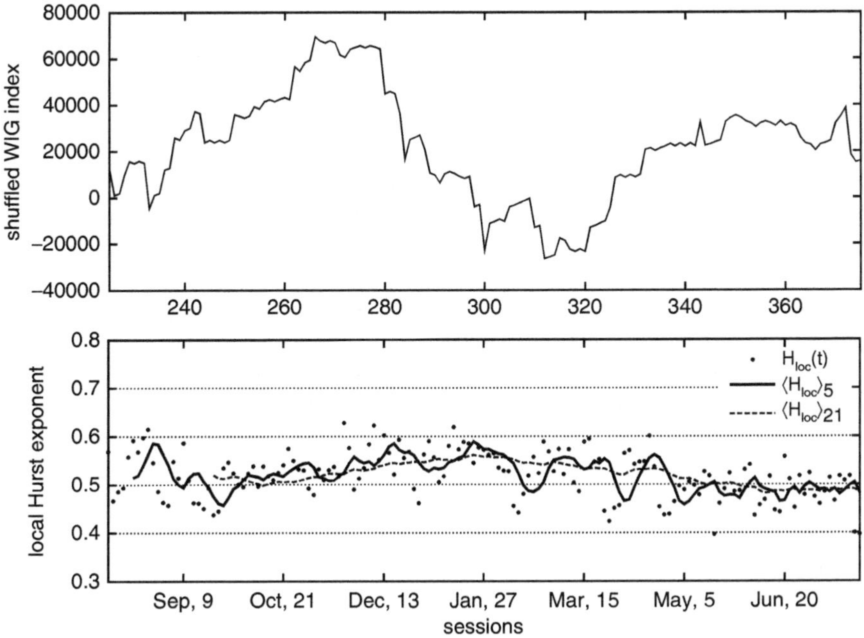

Fig. 14 The local fractal structure of integrated shuffled returns in WIG time series corresponding to the same period as in Fig. 10. H_{loc} evolution is very much different from the one shown in Fig. 10 and does not indicate any extreme event coming. (Taken from [42])

4 Exterior Interventions and the Current Situation Seen in Fractal Analysis of Stock Indices

The crash pattern conditions (1)–(4) for WIG signal were indeed fully formed in the first trading week of September 2007 as seen in Fig. 13. This should have caused the crash formation no later than within two trading weeks. However, the first signatures of difficulties with US economy and dangerous situation at that time on US mortgage credits market forced the intervention of the US Federal Reserve central bank (Fed). Fed decreased significantly the credit rate levels and pushed therefore the US stock exchange up. Other leading world financial markets obviously followed this trend. This intervention is a good example of exterior influence on the financial complex dynamical system. We will see in this chapter how this influence is visible in the case of WIG, DJIA and S&P 500 behavior and in their fractal structures.

If the intervention did not take place, the crash in WIG signal should have started in our opinion [42] no later than in the third trading week of September 2007, i.e. about two weeks from the moment the described above crash pattern was formed. We observe (Fig. 13) that the H_{loc} exponent changed its trend immediately after Fed intervention from decreasing to horizontal one. Looking at the particular values of the local Hurst exponent one recognizes that Fed intervention led market into

Brownian-like integer motion with $0.45 \leq H \leq 0.55$. This kind of Brownian evolution was continuing for about 6 trading weeks (Sept. 18 – Oct. 30, 2007). However, the external action did not prevent the crash-like event. It postponed only the major rupture point in time. Indeed, the main rupture point took place on Oct. 30, 2007 instead of ~Sept. 20, 2007 suggested by H_{loc} pattern. The financial market started then to relax what can be confirmed just by looking at the increasing post-intervention trend of H_{loc}. It is likely the first observation of changes in local fractal properties of financial stock time series caused by the external influence on the complex financial system.

Notice that even if stock market is in decreasing trend (bearish period), an appearance of strong consecutive minima in H_{loc} pattern indicate moments when positive corrections in market index values terminate. In the case of WIG signal it happened twice after Fed intervention: on Oct. 30, 2007 and on Jan. 2, 2008. These phenomena are marked as vertical thin lines in Fig. 13. Both corrections were followed by the substantial drop of WIG index (14% and 20% respectively) in three trading weeks after each rupture point and were linked to $H_{\text{loc}} \sim 0.4$.

One observes also that minima of H_{loc} in bearish period seem to be slightly higher than in bullish period (compare with previous figures). Thus the condition (4) virtually works even in the bearish phase, and may help to predict the interior structure (rupture points) during relaxation of the system.

It is interesting to see the hypothetical simulated evolution of the WIG index if the Fed intervention in September 2007 did not take place, by simple cut-off the data corresponding to Brownian motion phase in WIG signal (i.e. between Sept. 18, 2007 and Oct. 30 2007). Indeed, the H_{loc} pattern can be calculated (see Fig. 15) for such simulated signal, and compared with the one previously found for the real WIG evolution (Fig. 13). We see that the system relaxes now immediately. The H_{loc} pattern does not reveal the Brownian motion intermediate phase and H_{loc} enters the increasing trend with $H_{\text{loc}} > 0.5$ just after Sept. 18, 2007 what can be translated as the beginning of the relaxation phase on the market. Such immediate relaxation after crash was also very significant for other typical crashes in WIG history (see Figs. 10, 12).

The similar phenomenon can be found for developed stocks. Let us look at DJIA and S&P 500 behavior and corresponding fractal structure of these indices during and just after the Fed intervention. This is shown in Figs. 16, 17, respectively. The moment of Fed intervention stops for several months the decreasing trend in H_{loc} (up to May 2008). However, both markets are still in antipersistent mode after intervention ($\langle H_{\text{loc}} \rangle_{5,21}$ are below 0.5), contrary to Brownian motion revealed by WIG. Simultaneously, DJIA index seemed to be closer to Brownian motion generated dynamics (Hurst values were closer to 0.5) than S&P 500 index. The latter one revealed some antipersistency at the same time (Hurst exponents closer to 0.45).

Finally, we may consider how far markets are still from the end of recession. One possible view at this question is to consider the actual fractal structure of leading financial time series in the world – e.g. DJIA or S&P 500 indices. The long-lasting increasing trend on stocks corresponds to long-range memory (persistence) of financial time series. Thus $H_{\text{loc}} > 0.5$ is expected for the established

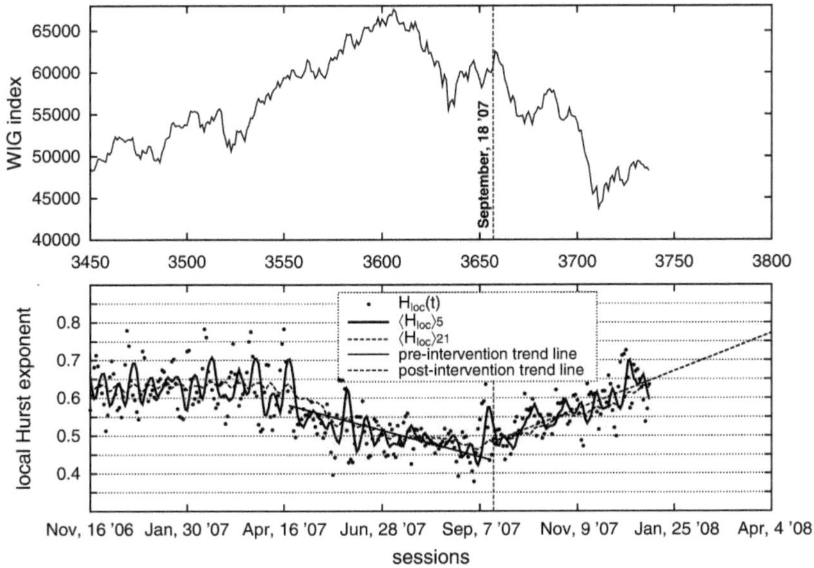

Fig. 15 Hypothetical evolution of WIG index if the Fed intervention in September 2007 did not take place. In that case the rupture point happens around Sept. 18, 2007 and the system relaxes immediately, i.e. H_{loc} pattern does not reveal the Brownian motion phase shown in Fig. 9. Note that H_{loc} pattern is *calculated* from the new hypothetical data and *it is not taken* from the joined parts shown in Fig. 13. (Taken from [41])

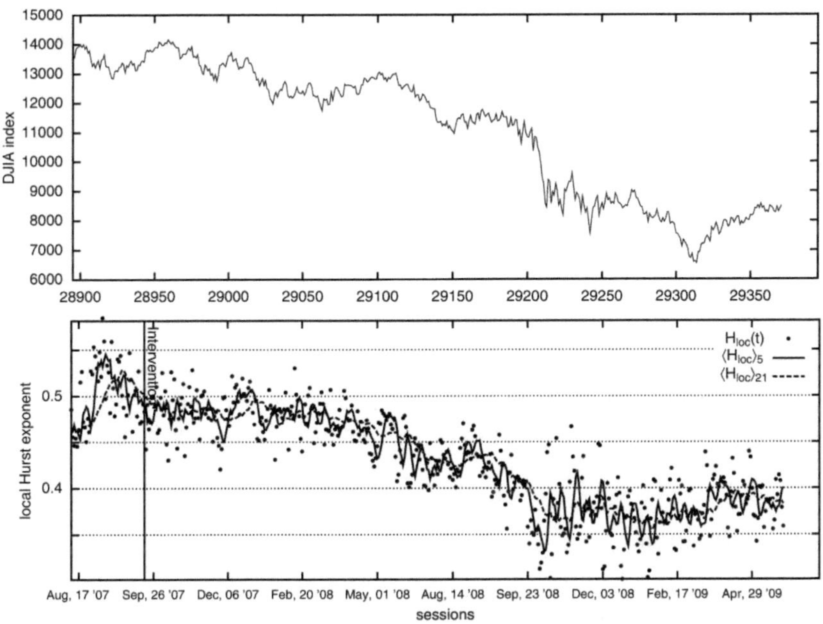

Fig. 16 An influence of Fed intervention on DJIA index evolution and its local fractal properties. A distance to long-lasting increasing trend is indicated with H_{loc} values terminating around $H_{loc} \sim 0.35 - 0.42$

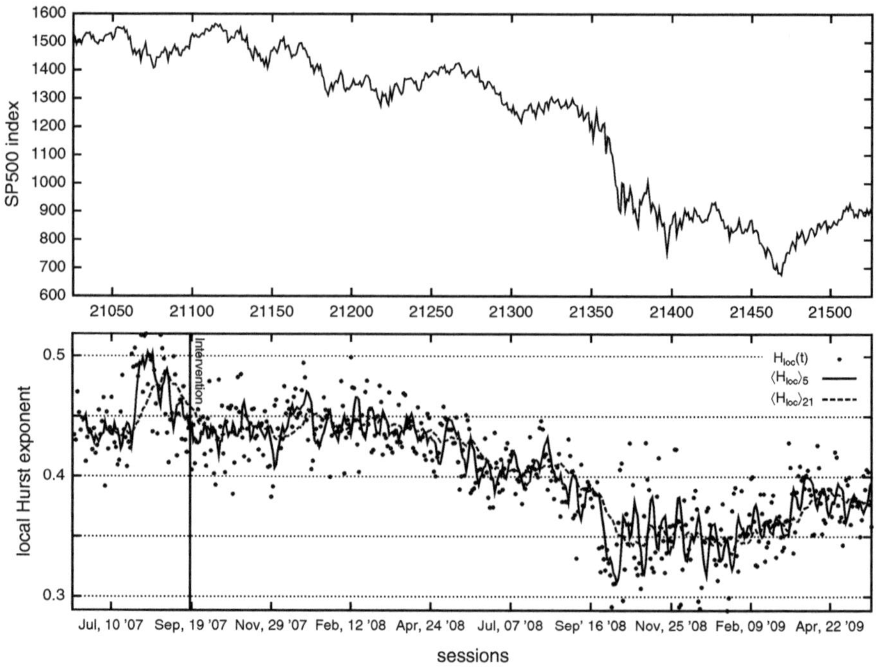

Fig. 17 The same as in Fig. 16 but for S&P 500 index

bullish phase on stocks. Looking at actual H_{loc} values for DJIA and S&P 500 indices in Figs. 16, 17, based on data taken up to June 2009, one observes, however, that local Hurst exponents are on the average still below 0.4 indicating antipersistence. A time-distance ΔT when H_{loc} should pass to $\sim$0.5 region is then quite significant. It can be roughly estimated as

$$\Delta T \sim \frac{1/2 - \langle H_{\mathrm{loc}} \rangle_{21}}{\kappa} \tag{15}$$

where κ is the average (15) increasing rate of $\langle H(t)_{\mathrm{loc}} \rangle_{21}$ in last few months and $\langle H(t)_{\mathrm{loc}} \rangle_{21}$ is the actual value of monthly moving average for H_{loc}. Substituting actual data at the moment this article is being written (August 2009), one estimates the beginning of boom at least 10 months ahead, i.e. late spring 2010, of course under condition that the average increasing rate of κ remains unchanged.

5 Conclusions

Concluding, we state that with the use of local Hurst exponent one is able to identify various phases of financial market evolution in the language of complex system phenomena. We can generalize these pattern into repeatable scheme revealing the major forthcoming events, particularly interesting for investors like, e.g. crashes, rupture points, beginning of bullish periods, etc. It seems there exists a rather straightforward connection between the H_{loc} pattern and phase transitions (crashes or rupture points) on the market. This connection is caused by the intrinsic organization of the complex system. The presented scheme works equally well for the developed and for the developing markets. Recently, an interest in application of local H values to establish actual trends on financial markets is more and more encouraging [45–47].

Contrary, the log-periodicity does not work so well for emerging markets. Especially, it cannot be used as the well confirmed phase transition indicator for the young stock markets in the same way as for the old ones. It suffers also from the problem of dependence on the number of data one takes backward in time into account. Further research is more than welcome in this subject, particularly related to, e.g. multi-fractal and multi-scaling properties of financial time series in the vicinity of crash points.

Acknowledgement Author wishes to thank Ewa Sochocka for her outstanding support and encouragement he experienced preparing this article.

References

1. Mandelbrot BB (1963) J Business 36:349
2. Sornette D, Johansen A, Bouchaud J-P (1996) J Phys I (France) 6:167
3. Feigenbaum JA, Freund PGO (1996) Int J Mod Phys B 10:3737
4. Takayasu H, Miura H, Hirabayashi T, Hamada K (1992) Physica A 184:127
5. Bouchaud J-P, Sornette D (1994) J Phys I (France) 4:863
6. Mantegna RN, Stanley HE (1995) Nature 376:46
7. Liu Y, Cizeau P, Meyer M, Peng C-K, Stanley HE (1997) Physica A 245:437
8. Bak P, Paczuski M, Shubik M (1997) Physica A 246:430
9. Sornette D, Johansen A (1997) Physica A 245:411
10. Vandewalle N, Boveroux Ph, Minguet A, Ausloos M (1998) Physica A 255:201
11. Vandewalle N, Ausloos M, Boveroux Ph, Minguet A (1998) Eur Phys J B 4:139
12. Vandewalle N, Ausloos M, Boveroux Ph, Minguet A (1999) Eur Phys J B 9:355
13. Johansen A, Sornette D (2000) Eur Phys J B 17:319
14. Ausloos M, Ivanova K, Vandewalle N (2002) Crashes: symptoms, diagnoses and remedies. In: Takayasu H (ed) Empirical sciences of finacial fluctuations. The advent of econophysics. Tokyo, Japan, Nov. 15–17, 2000 Proceedings. Springer, Berlin, pp 62–76 (arXiv: cond-mat/0104127)
15. Drożdż S, Grummer F, Ruf F, Speth J (2003) Physica A 324:174
16. Grech D, Mazur Z (2004) Physica A 336:133
17. Bartolozzi M, Drożdż S, Leinweber DB, Speth J, Thomas AW (2005) Int J Mod Phys C 16:1347
18. Sornette D (1998) Phys Rep 297:239
19. Drożdż S, Ruf F, Speth J, Wójcik M (1999) Eur Phys J B 10:589
20. Kozłowska M, Kasprzak A, Kutner R (2008) Int J Mod Phys C 19:453
21. Hurst HE (1951) Trans Am Soc Civ Eng 116:770
22. Mandelbrot BB, Wallis JR (1969) Water Resour Res 5(2):321
23. Peng C-K, Buldyrev SV, Havlin S, Simons M, Stanley HE, Golberger AL (1994) Phys Rev E 49:1685
24. Allesio E, Carbone A, Castelli G, Frappietro V (2002) Eur Phys J B 27:197

25. Carbone A, Castelli G (2003) Proc SPIE 5114:407
26. Geweke J, Porter-Hudak S (1983) Jour Time Ser Anal 4:221
27. Vandewalle N, Ausloos M (1997) Physica A 246:454
28. Ausloos M, Vandewalle N, Boveroux Ph, Minguet A, Ivanova K (1999) Physica A 274:229
29. Viswanathan GM, Peng C-K, Stanley HE, Goldberger AL (1997) Phys Rev E 55:845
30. Kantelhardt JW, Koscielny-Bunde E, Rego HHA, Havlin S, Bunde A (2001) Physica A 295:441
31. Hu K, Ivanov PCh, Chen Z, Carpena P, Stanley HE (2001) Phys Rev E 64:011114
32. Chen Z, Ivanov PCh, Hu K, Stanley HE (2002) Phys Rev E 65:041107
33. Chen Z, Hu K, Carpena P, Bernaola-Galvan P, Stanley HE, Ivanov PCh (2005) Phys Rev E 71:011104
34. Buldyrev SV, Dokholyan NV, Golberger AL, Havlin S, Peng C-K, Stanley HE, Viswanathan GM (1998) Physica A 249:430
35. Moret MA, Zebenda GF, Nogueira E, Pereira MG (2003) Phys Rev E 68:041104
36. Vandewalle N, Ausloos M (1998) Phys Rev E 58:6832
37. Ausloos M, Ivanova K (2001) Int J Mod Phys C 12:169
38. Ausloos M, Ivanova K (2000) Physica A 286:353
39. Ausloos M, Ivanova K (2001) Eur Phys J B 20:537
40. Oświęcimka P, Kwapień J, Drożdż S, Rak R (2005) Acta Phys Pol B 36:2447
41. Czarnecki L, Grech D, Pamuła G (2008) Physica A 387:6801
42. Grech D, Pamuła G (2008) Physica A 387:4299
43. Drożdż S, Kwapień J, Oświęcimka P, Speth J (2008) Acta Phys Pol A 114:539. arXiv: 0802.4043v1 [physics.soc-ph]
44. Carbone A, Castelli G, Stanley HE (2004) Physica A 344:267
45. Cajueiro DO, Tabak BM (2004) Physica A 336:521
46. Eom C, Choi S, Oh G, Jung W-S (2008) Physica A 387:4630
47. Eom C, Oh G, Jung W-S (2008) Physica A 387:5511

Root Causes of the Housing Bubble

Taisei Kaizoji

Abstract In this chapter we investigate root causes of the recent US housing bubble which has been caused a serious downturn in US economic growth since autumn of 2008. We propose a simple model of housing markets in order to indicate the possible determinants of recent housing prices. Utilizing the model, we verify a number of hypotheses which have been proposed in the recent literature on the housing bubbles. We suggest that the main causes of the housing bubble from 2000 to 2006 are (1) non-elastic housing supply in the metropolitan areas, and (2) declines in the mortgage loan rate and the housing premium by the massive mortgage credit expansion. We also suggest that these factors were strongly influenced by policies that governments and the Federal Reserve Board performed.

1 Introduction

The principal cause of the current financial and economic crises is the housing bubble. The housing bubble in USA was the biggest financial bubble in the recent history in USA. The unprecedented US housing bubble began to inflate in the first quarter of 1998 and continued till the second quarter of 2006. Then the housing prices began falling, and as of July 2009 the housing price still has been falling. The housing market meltdown was triggered by a dramatic rise in mortgage delinquencies and foreclosures for subprime residential mortgage loans.

Figure 1 shows the nominal housing price index and the real housing price index which is adjusted for inflation from 1890 to 2009. From 1998 to a peak in 2006, the nominal housing price index increased by 130% and the inflation adjusted housing prices index increased by 80%. The growth rate of the nominal housing price index had averaged 2.9% between 1890 and 1997, and averaged 9.3% between 1998 and 2007. The real housing price index increased by an average of 6.8% a year from

T. Kaizoji (✉)
International Christian University, 3-10-2 Osawa, Mitaka, Tokyo 181-8585, Japan
e-mail: kaizoji@icu.ac.jp

M. Takayasu et al. (eds.), *Econophysics Approaches to Large-Scale Business Data and Financial Crisis*, DOI 10.1007/978-4-431-53853-0_8, © Springer 2010

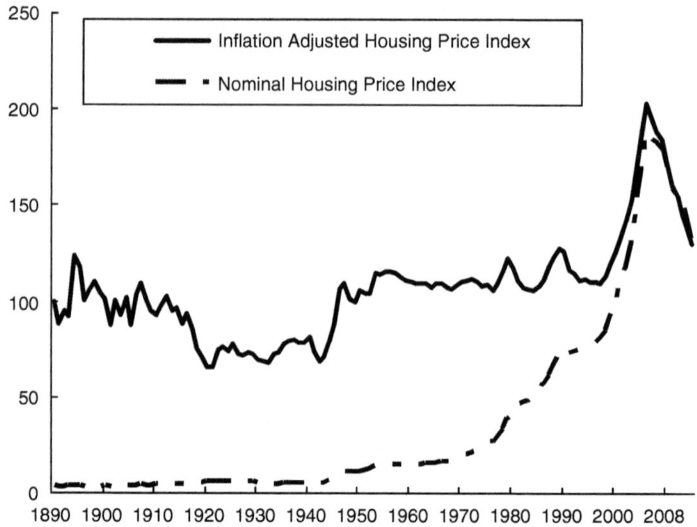

Fig. 1 The US housing price index (data was provided by Shiller [15] as updated by the author)

1997 to 2006 while the index increased by an average of 0.086% a year from 1890 to 1997. The nominal housing price index decline by 4.2% from the first quarter of 2007 to the first quarter of 2009 while the real housing price declines by 4.9% per a quarter for the same period.

The housing bubble had an immeasurable negative impact on the US economy because they caused misallocation of resources, and furthermore, the housing market crash destroyed a large amount of wealth. What is most important things that we learned in order to prevent the disaster? This chapter investigate the reasons why the housing bubble was caused through verifying hypotheses which have been proposed in the recent literature on the housing bubbles. To give a theoretical background of the determinants of housing prices, we propose a simple model of housing markets which is based on the classical economics.

2 A Simple Model of Housing Markets

In this session we propose a simple of housing markets and derive a theoretical housing price. We formalize the price-to-rent ratio in the framework of the household's utility maximization. We assume that the household makes a optimal decision regarding consumption of goods and housing services for the present. The utility of the household who owns a house is given by the following form:

$$U^H\left(c^H, s^H\right) = \alpha \log c^H + \beta \log s^H, \tag{1}$$

where α and β are the parameters which express the household of preference. The household's budget constraint is

$$c^H + \tilde{p}s^H = y^H. \tag{2}$$

c^H denotes a homeowner's consumption, s^H his demand for housing, and y^H his real disposable income. $\tilde{p}$ denotes the user cost of home ownership, known in the housing literature as the imputed rent (see [8, 12, 13]). It is defined as follows:

$$\tilde{p} = p(i + \tau + f - \pi), \tag{3}$$

where:

(i) i is the after-tax nominal interest rate.
(ii) τ is the property tax rate on owner-occupied houses.
(iii) f is the recurring holding costs.
(iv) π is the expected capital gain (or loss).
(v) p is the house price.

i is the cost of foregone interest that the homeowner could have earned on an alternative investment. f consists of depreciation, maintenance and the risk premium on residential property.

Hence, the household's utility maximization problem is formulated as

$$\max U\left(c^H, s^H\right),$$

$$\text{s.t.}\ \ c^H + \tilde{p}s^H = y^H.$$

The problem is solved by forming the Lagrangian

$$L\left(c^H, s^H, \lambda\right) = U\left(c^H, s^H\right) + \lambda\left[y^H - c^H - \tilde{p}s^H\right]. \tag{4}$$

The first-order conditions to find the critical points of the Lagrangian function are

$$\begin{aligned}\frac{\partial L}{\partial c^H} &= \frac{\alpha}{c^H} - \lambda = 0,\\ \frac{\partial L}{\partial s^H} &= \frac{\beta}{s^H} - \lambda = 0,\\ \frac{\partial L}{\partial \lambda} &= y^H - c^H - \tilde{p}s^H = 0.\end{aligned}$$

Consumption c and the demand for housing services s are obtained as

$$c^H = \alpha y^H, \quad s^H = \beta\frac{y^H}{\tilde{p}}. \tag{5}$$

Secondly, let us consider the utility maximization problem of the household who rents a house. We assume that the household, who rents a house, has the same utility function as that of the household who owns a house.

$$U^R\left(c^H, s^H\right) = \alpha \log c^R + \beta \log s^R. \tag{6}$$

The renter's budget constraint is

$$c^R + ws^R = y^R. \tag{7}$$

Solving the first-order conditions, consumption c^R and the demand for housing services s^R by the household who rents a house are obtained as

$$c^R = \alpha y^R, \quad s^R = \beta \frac{y^R}{w}. \tag{8}$$

c^R denotes a renter's consumption, s^R his demand for housing, and y^R his real disposable income. Aggregate demand for purchasing houses is

$$\sum_{i=1}^{n_1} s_i^H = \frac{\beta y^H}{\tilde{p}} n_1, \tag{9}$$

where n_1 denotes the number of households who own a house. The aggregate demand for leasing houses is

$$\sum_{j=1}^{n_2} s_j^R = \frac{\beta y^R}{w} n_2, \tag{10}$$

where n_2 denotes the number of households who rent a house.

We assume that the supply of housing for owing and renting is constant: N_1 and N_2. In equilibrium, the demand has to be equal to supply,

$$\frac{\beta y^H}{\tilde{p}} n_1 = N_1, \quad \text{and} \quad \frac{\beta y^R}{w} n_2 = N_2. \tag{11}$$

Arranging (11), the price-to-rent ratio is written as

$$\frac{p}{w} = \frac{n_1}{n_2} \frac{N_2}{N_1} \frac{y^H}{y^R} \frac{1}{(i + \tau + f - \pi)}. \tag{12}$$

We introduce the new variable,

$$s = \frac{(n_1 - n_2)}{2N} = \frac{1}{2N} \sum_{j=1}^{2N} s_j, \quad (-1 \le s \le 1), \tag{13}$$

where $n_1 + n_2 \equiv 2N$. The variable s, indicates the *homeownership rate*. Then we can rewrite the price-to-rent ratio as follows:

$$\frac{p}{w} = \frac{(1+s)}{(1-s)} \frac{N_2}{N_1} \frac{y^H}{y^R} \frac{1}{(i+\tau+f-\pi)}. \tag{14}$$

As many literatures show, when the following conditions are hold: $n_1 = n_2$, $N_1 = N_2$ and $y^H = y^R$, the expected annual cost of owning a house equals that of renting at an equilibrium in the housing market, that is, $p = w$ (see [9, 12]).[1]

2.1 The Price-to-Rent Ratio

The price-to-rent ratio is a measure which is commonly used to get an indication of over- or under-valuation in terms of the cost of owning versus renting a house. If this ratio rises above its long-term average, then it could be an indication that prices are overvalued.

Figure 2 shows the ratio of price to rent from the first quarter of 1960 to the first quarter of 2009. The figure shows that the price-to-rent ratio ranged between 16.5 and 20 times between 1960 and 1995, but the price-to-rent ratio rapidly increased after 1995, and reached an historic high of 32 times by year-end 2006. Since the first quarter of 2007, the price to rent ratio decreased from its historical high, and the latest data suggest that the ratio has continued to decrease [3].

A related finding is that price-to-rent ratio differs across cities and the differences are persistent over time: cities with high price-to-rent ratios tend to remain high, and cities with low price-to-rent ratio remain low (see [9]). The price-to-rent ratios in 2006 are substantially above their long-term averages. In the metropolitan areas with the largest house price increases, these ratios exceed their long-term averages significantly. The unprecedented and steep increase in the price-to-rent ratio lead us to conclude that the housing markets in the metropolitan areas experienced a sizeable bubble over the 1995–2006 period. Davis et al. [3] forecasts that assuming nominal rents were to increase by 4% per year, about the average since 2001, a decline in nominal house prices of about 3% per year would bring the price-to-rent ratio up to its historical average, 5%, by mid-2012.

[1] They implicitly assume low-cost arbitrage between owning and renting and the supply of housing were perfectly elastic. In reality, mortgage origination fees, broker commissions and moving costs make it expensive to switch back and forth between owning and renting. As we show below, the housing supply is often regulated by the local governments.

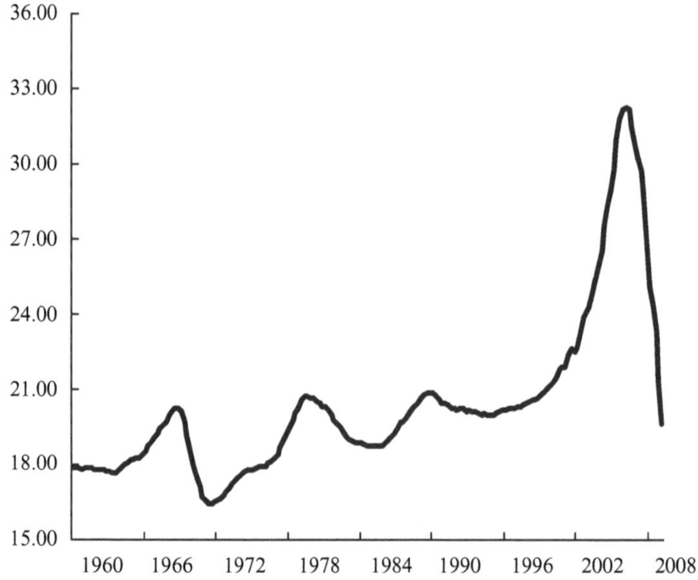

Fig. 2 The price-to-rent ratio for the aggregate stock of owner-occupied housing (data was provided by Davis et al. [3])

3 Determinants of Housing Prices

Although rapid derivation of the price-to-rent ratio from the long-term average is necessarily evidence of overvaluation of housing price, the ratio of prices to rent by itself is not a sufficient metric to indicate possible causes of the housing bubble. To address this issue, it is necessary to relate the price-to-rent ratio to their underlying determinants in the right-hand side of (14). We focus attentions on the following: (1) income, y^H, (2) housing supply, N_1, (3) the user cost of home ownership, $\tilde{p}$, and (4) homeownership rate, s.

3.1 Price-to-Income Ratio

Another measure that is commonly used to assess whether housing prices are unsustainably high is the price-to-income ratio, which provides a measure of local housing costs relative to the local ability to pay (see, for examples, [4] and [9]). Himmelberg et al. [9] investigates the price-to-income ratio computed as the OFHEO price index divided by an index of median per-capita income based on data from the US Bureau of Economic Analysis. They show that at the national level the growth in real house prices was outpaced by the high growth in incomes (see also [10]). However, at the metro area level, price-to-income ratios have generally increased the most in cities where house price growth has been highest such as Boston, New York, San Francisco and San Diego. Price-to-income in 2004 is above its peak value in many of these cities, and as much as 40–50% above its 25-year average. But in Houston, Dallas and other southern cities, price-to-income is well below its 25-year average. Gyourko et al. [7] refers to cities with high long-run rates of house price growth such as San Francisco, Boston, New York, and Los Angeles as *superstar cities*, and show that differences in house price and income growth rates across metropolitan areas have led to an ever-widening gap in housing values and incomes between the typical and highest-priced locations. They argue that the new home buyers in superstar cities are high-income households who have moved from other parts of the country, so that those cities can experience above-average house price growth over a very long horizon under tight supply constraints.

3.2 Elasticity of Housing Supply

In principle, if the housing market is perfect competitive, and the supply of housing is perfectly elastic, house prices would be determined solely by construction costs [11]. In practice, the housing prices have derived from the housing cost in many metropolitan area over the long-run period. Glaeser et al. [6] investigates the ratio of the average house price to the estimated physical construction cost for 102 metropolitan areas in each Census year from 1950–2000. They found that house prices grew relative to construction costs in most of the metropolitan statistical areas. The rise in the gap between price and construction cost throughout parts of the Northeast and the West coast over the past three decades. Therefore, changes in construction costs explain neither the overall rise in real house prices nor cross-sectional differences in appreciation rates across markets. Several studies on the US regional housing markets have found that the low supply elasticity of housing units is an important factor behind the recent larger price increases in some urban markets, and have argued that housing supply regulation is an important determinant of house prices [5–7]. Glaeser et al. [6] argue that differences in appreciation rates of housing across metropolitan areas reflect tight housing supply regulation combined with an increasing number of households who want to live in the area. They also point out that US homebuilders have faced increasing difficulty in obtaining regulatory approval for the construction of new homes in some states, (for notable examples, California, Massachusetts, New Hampshire, and New Jersey) and in Washington, DC. An additional factor has been the increased ability of established residents to block new projects. The supply constraint on new construction have push up prices relative to construction cost, and this is considered to be able to explain partially house price overheating in housing markets in the metropolitan area.

3.3 The User Cost of Home Ownership

Campbell et al. [2] uses Campbell and Shiller's [1] dynamic Gordon growth model to decompose the log rent-price ratio, ν_t, into three components: the expected present value of real interest rates (the real 10-year Treasury), $\tilde{r}_t$, and the expected present value of the housing premium, σ_t, and the expected present value of real rent growth, $\tilde{w}_t$:

$$\nu_t = \kappa + \tilde{r}_t + \sigma_t - \tilde{w}_t. \tag{15}$$

which is an alternative to the rent-to-price ratio proposed by Portaba [12]. Campbell et al. [2] shows that both of real interest rates and housing premia have fallen from 1997 to 2005 in almost all the metropolitan areas, and almost all of the decline in the rent-to-price ratio (inversely, increase in the price-to-rent ratio) is attributable to a steep decline in the risk premium paid to housing over and above a 10-year Treasury bond, and a decline in the expected present values of real interest rate. Especially they indicate that the housing premium plays a relatively more important role, and can explain about 65% of the recent run-up of house prices relative to rents since 1997.

3.4 Policies

Why did both of the real interest rate and the risk premium paid to housing declined sharply over the 1997–2005 period?

First, the Federal Reserve Board sharply lowered its sort-term interest rates in response to the recession caused by bursting internet bubbles in 2000. The Fed Funds rate dropped from 6.5% at the end of 2000 to 1% in mid-2003. The Federal Reserve's monetary policy in this period was overly accommodative compared to the optimal policy suggested by the so-called Taylor rule [16, 17]. Banks, other financial institutions eagerly expanded credit through loans and investments in debt and derivative securities. As a result, monetary policy also helped hold mortgage rates at a low level. Taylor [17] argue that this significant deviation may have been a cause of the boom and bust in US housing market.

Second, federal policymakers have promoted home-ownership as the fulfillment of the American dream. Especially the Clinton administration encouraged home-buying among financially marginal and minority households. To purse this aim, the administration performed the following policies: (1) enacting preferential income tax policies to reduce the cost of home ownership relative to renting, (2) pressing depository institutions and mortgage banks to lower their credit standards and reduce down payment requirements, and (3) promoting exotic alternatives to traditional fixed-rate fully amortizing residential mortgage loans, such as interest-only residential mortgage loans and negatively amortizing residential mortgage loans. (For a detailed documentation, see Saxton [14].) These policies were intended to help financially marginal and minority households that could not qualify for traditional mortgage loans under normal credit standards to buy homes. The Bush administration left these Clinton administration policies in place.

3.5 Home Ownership Rate

The federal government's policies encouraged many households to buy housing during the bubble. Thereby, the home ownership rate increased to a peak of 69.0% in 2004 while the rate had averaged 64.3% of all households from 1982 to 1997. These policies also designed to help financially marginal and minority households generated many households who buy homes through unaffordable subprime residential mortgage loans. However, many of these new home owners had poor credit histories, and were unprepared or unable to discharge their mortgage obligations.

4 Concluding Remarks

In this chapter we study the reasons why the housing price was inflated excessively in USA. To this aim, we propose a simple model of housing market. The main causes of the housing bubble are summarized as

1. Non-elastic housing supply in the metropolitan areas
2. Declines in the interest rate and the housing premium

First, inflation of the housing prices in highest-price areas, especially New York City and California does not reflect physical costs of construction, but zoning strictness and other restrictions on building by government regulations. These findings suggests that policy makers of local governments possibly prevented the housing bubble through appropriate zoning reform in the metropolitan area.

Second, Federal policymakers adopted a number of policies to promote home ownership and help financially marginal and minority households to buy homes with unaffordable subprime residential mortgage loans. After all, these policies made households vulnerable to foreclosure after the housing bubble burst, and caused a serious downturn in US economic growth.

Finally, the Federal Reserve's monetary policy which was overly accommodative from the second quarter of 2002 through the third quarter of 2006 lowered the cost of funds for financial institutions, and encouraged them to expand credit aggressively by extending loans. The policies created a massive credit expansion and stimulated the demand for housing among households, so that US housing prices soared. In conclusion, Federal policymakers could also have adopted a number of policies to preventing housing bubble.

References

1. Campbell J, Shiller R (1998) Stock prices, earnings and expected dividends. J Fin 43:661–676
2. Campbell S, Davis M, Gallin J, Martin R (2009) What moves housing markets: a variance decomposition of the rent-price ratio. J Urban Econ 66(2):90–102
3. Davis MA, Lehnert A, Martin RF (2008) The rent-price ratio: for the aggregate stock of owner-occupied housing. Rev Income Wealth 54(2):279–284
4. Girouard N, Kennedy M, van den Noord P, Christophe Andre C (2006) Recent house price developments: the role of fundamentals. OECD Economics Department Working Papers, No. 475, OECD Publishing
5. Glaeser E, Gyourko J (2003) The impact of building restrictions on housing affordability. Federal Reserve Bank of New York Economic Policy Review, June 2003
6. Glaeser E, Gyourko J, Saks R (2005) Why have housing prices gone up? Harvard Institute of Economic Research Discussion Paper, No. 2061
7. Gyourko J, Mayer C, Sinai T (2004) Superstar cities. Wharton Working Paper, July 2004
8. Hendershott P, Slemrod J (1983) Taxes and the user cost of capital for owner-occupied housing. AREUEA J 10(4):375–393
9. Himmelberg C, Mayer C, Sinai T (2005) Assessing high house prices: bubbles, fundamentals and misperceptions. NBER Working Paper 11643
10. McCarthy J, Peach R (2004) Are home prices the next bubble? Federal Reserve Bank of New York Economic Policy Review, December 2004
11. Muth RF (1960) The demand for nonfarm housing. In: Harberger AC (ed) The demand for durable goods. University of Chicago Press, Chicago
12. Poterba J (1984) Tax subsidies to owner-occupied housing: an asset market approach. Q J Econ 99:729–752
13. Poterba J (1992) Taxation and housing: old questions, new answers. Am Econ Rev 82(2): 237–242
14. Saxton J (2008) The U.S. housing bubble and the global financial crisis: housing and housing-related finance. Joint Economic Committee, United States Congress, May 2008
15. Shiller RJ (2005) Irrational exuberance, 2nd edn. Princeton University Press, Princeton
16. Taylor JB (1993), Macroeconomic policy in a world economy: from econometric design to practical operation. W.W. Norton, New York
17. Taylor JB (2007) Housing and monetary policy. NBER Working Paper Series 13682

Reconstructing Macroeconomics Based on Statistical Physics

Masanao Aoki and Hiroshi Yoshikawa

Abstract We believe that time has come to integrate the new approach based on statistical physics or econophysics into macroeconomics. Toward this goal, there must be more dialogues between physicists and economists. In this paper, we argue that there is no reason why the methods of statistical physics so successful in many fields of natural sciences cannot be usefully applied to macroeconomics that is meant to analyze the macroeconomy comprising a large number of economic agents. It is, in fact, weird to regard the macroeconomy as a homothetic enlargement of the representative micro agent. We trust the bright future of the new approach to macroeconomies based on statistical physics.

1 Introduction

Macroeconomics has gone astray. In the past 30 years, macroeconomics has become less relevant. Events in the world economic crisis since fall 2008 have unmistakably demonstrated this fact.

The mainstream macroeconomics today begins with optimization of the representative consumer. The "optimum" growth theory once meant to be a guide for government policy is now being taught as a theory which describes the actual working of our market economy (see, for example, a major macroeconomics textbook [7]). It is the neoclassical equilibrium theory. By construction, it broadly underlines the efficiency of market albeit with mild admission of the so-called "market failures". In reality, far from being efficient, most of the time, the economy must move on a bumpy road. It is simply misleading and wrong to analyze

M. Aoki
Department of Economics, University of California, 403 Hilgard Avenue, Los Angeles, CA 90095-1477, USA
e-mail: aoki@econ.ucla.edu

H. Yoshikawa (✉)
Faculty of Economics, The University of Tokyo, 7-3-1 Hongo, Bunkyo-ku, Tokyo 113-0033, Japan
e-mail: yoshikawa@e.u-tokyo.ac.jp

M. Takayasu et al. (eds.), *Econophysics Approaches to Large-Scale Business Data and Financial Crisis*, DOI 10.1007/978-4-431-53853-0_9, © Springer 2010

such problems as business cycles, unemployment, deflation, and financial turmoil – the subject matters of macroeconomics – with the neoclassical equilibrium theory.

Nevertheless, many economists still believe that the first principle of economics is the optimization of economic agents such as household and firm. This principle and the notion of equilibrium, namely equality of supply and demand, constitute the core of the neoclassical theory. To some, this is the only respectable economic theory on earth. For example, a Nobel laureate Lucas [19] concluded his Yrjo Jahnsson Lectures as follows:

> The most interesting recent developments in macroeconomic theory seem to me describable as the reincorporation of aggregative problems such as inflation and the business cycle within the general framework of 'microeconomic' theory. If these developments succeed, the term 'macroeconomic' will simply disappear from use and the modifier 'micro' will become superfluous. We will simply speak, as did Smith, Ricardo, Marshall and Warlras, of *economic* theory. If we are honest, we will have to face the fact that at any given time there will be phenomena that are well-understood from the point of view of the economic theory we have, and other phenomena that are not. We will be tempt, I am sure, to relieve the discomfort induced by discrepancies between theory and facts by saying that the ill-understood facts are the province of some other, different kind of economic theory. Keynesian 'macroeconomics' was, I think, a surrender (under great duress) to this temptation. It led to the abandonment, for a class of problems of great importance, of the use of the only 'engine for the discovery of truth' that we have in economics. Now we are once again putting this engine of Marshall's to work on the problems of aggregate dynamics. [19, pp. 107–108]

Thus, over the last 30 years, economics has attempted, in one way or another, to build maximizing microeconomic agents into macroeconomic models. To incorporate these agents into the models, the assumption of *the representative agent* is usually made. By and large, these exercises lead us to the neoclassical macroeconomics. The real business cycle (RBC) theory (e.g. [18]) praised so highly by Lucas [19] is the foremost example.

We maintain that this "micro-founded" macroeconomics represented by RBC is misguided, and that a fundamentally different approach is necessary to analyze *the macroeconomy*. Such an approach is based on the method of *statistical physics*. It is commonly used in physics, biology and other natural sciences when one studies a system consisting of a large number of entities.

The basic idea of statistical physics is explained in every textbook. Here is an example:

> Many systems in nature are far too complex to analyze directly. Solving for the behavior of all the atoms in a block of ice, or boulders in an earthquake fault, or the nodes on the Internet, is simply infeasible. Despite this, such systems often show simple, striking behavior. Statistical mechanics explains the simple behavior of complex system. [33, p. 1]

Thus, statistical physics begins by giving up the pursuit of the precise behavior of individual units, and grasps the system as a whole by statistical methods. This approach, which is nothing but common sense in natural sciences, is indeed in stark contrast to the modern *micro-founded* macroeconomics. The latter analyzes the precise behavior of the representative micro agent, and regards the macroeconomy as a homothetic enlargement of such a micro unit. We will shortly argue that there is no fundamental reason why the method so successful in natural sciences cannot be applied to economics. Contrary to Lucas' assertion, to study the macroeconomy, we *do* need "some other, different kind of economic theory".

A new approach to macroeconomics based on statistical physics has gradually emerged. Meanwhile, a closely related research area has come to be broadly dubbed *econophysics*. We can safely acknowledge that econophysics has established itself in finance. (See, for example, [24], [25], and [37].) However, we must recognize that there is still a significant gap between econophysics-based finance and the traditional finance theory.

For example, following a seminal work by Mandelbrot [22],[1] econophysicists appear to have established a stylized fact that probability distributions of (changes in) most asset prices are power laws. There is also a good consensus among economists that distributions in question are not Gaussian, and that they have heavy tails. However, traditional economists are not quite happy with the statement that the distributions of changes in asset prices are power laws. In fact, in an extreme case where the second moment or variance does not exist, the mean-variance analysis, arguably the most important framework in finance would not make sense. Also to the extent that real economic activity like consumption follows a different probability distribution from asset prices, the standard consumption-based capital asset pricing model (CAPM) breaks down (see [3]). We will later discuss the problems in finance in detail.

In contrast to finance, the research in the areas of economics is still in its infancy. To be sure, econophysicists have made important empirical findings that just as financial variables such as stock returns, many real economic variables such as personal incomes and the size of firms also obey the Pareto distribution, or the *power-laws* in their tails. The size distribution of firms and cities has, in fact, a long research history in economics [16]. Let alone Pareto [31], personal incomes have also been studied by many researchers [10].

Despite their importance, so far the impact of these empirical studies on economics has been rather limited, to say the least; they were often ignored by a majority of economists. The reason is that their relation to the mainstream economic *theory* is by no means clear. Discussing the Gibrat Law, a particular type of power law allegedly found for many economic variables, Sutton [38] argues as follows:

> It seems to have been widely felt that these models might fit well, but were "only stochastic". The aim was to move instead to a program of introducing stochastic elements into conventional maximizing models. [38, p. 45]

This trend is certainly in accordance with the motto of modern micro-founded macroeconomics. However, as the system under investigation becomes larger and more complicated, the importance of stochastic elements increases. In fact, it is almost the definition of macro-system to which the basic approach based on statistical physics can be usefully applied.

[1] Mandelbrot's related works are collected in [23].

2 Is the Statistical Approach Applicable to Economics?

The fundamental method based on statistical physics has been extremely successful in natural sciences ranging from physics to biology. Because the macroeconomy consists of a large number of economic agents, typically of the order of 10^6 to 10^7, we can expect that this method should show the same analytical power in macroeconomics as in natural sciences.

A common argument to the contrary is, however, that natural science analyzes system comprising inorganic particles such as atoms or molecules whereas economics analyzes the economy in which agents with brains purposefully pursue their respective goals. This understandable skepticism on the applicability of the method based on statistical physics to economics is actually not warranted. The truth is the method can be fruitfully applied to the analysis of system comprising a large number of micro units regardless of the nature of micro unit. A good example is analysis of traffic jams on turnpike. Here, micro unit is a driver, a purposeful human being with brains! And yet, traffic jams have been beautifully analyzed by the method based on statistical physics (see, for example, [27]).

It is not essential for studying macro system whether micro units comprising the macro system under investigation are human beings with brains or inorganic particles. The point is that because the number of micro units is large, it is impossible and meaningless to pursue precise behavior of each micro unit. Every economist knows that the economic agent who does intertemporal optimization maximizes the Hamiltonian. Likewise, every physicist knows that the inorganic particle in Newtonian motion also minimizes the Hamiltonian. Thus, in this respect, sophisticated human being pursuing intertemporal optimization and inorganic particle are on an equal footing. The issue is not whether micro unit is human utility/profit maximizer or not. Rather it is whether the method based on the representative micro unit makes sense or not. We believe that it is incorrect to analyze macro-system by the method based on the representative micro unit. That is what natural sciences have demonstrated time and again. The main-stream micro-founded macroeconomics, however, clings to utility/profit maximization of the representative agent. It is useful to consider concrete examples.

3 The Standard Approach Based on the Representative Agent: Some Examples

It is well recognized that stock market is volatile. It is indeed one of the major findings econophysics has made. Rather than accepting that stock prices are too volatile to be consistent with the standard capital asset pricing model (CAPM), a majority of economists have attempted to reconcile the alleged volatility with efficiency or "rationality" of market.[2]

A seminal work of Shiller [34] and subsequent studies have shown us that volatility of stock prices cannot be properly explained only by volatility of dividends or profits. One way to explain volatility of stock prices is to allow significant changes in the discount rate or the required return on stocks. In fact, in the neoclassical macroeconomic theory, the following relationship between the rate of change in consumption, C, and the return on capital, r must hold in equilibrium (see, for example, [7]):

$$-\left[\frac{u''(C)C}{u'(C)}\right]\left(\frac{\dot{C}}{C}\right) = \frac{1}{\eta(C)}\left(\frac{\dot{C}}{C}\right) = r - \delta. \tag{1}$$

Here, the elasticity of intertemporal substitution η is defined as

$$\frac{1}{\eta(C)} = \frac{u''(C)C}{u'(C)} > 0.$$

In general, η depends on the level of consumption, C Equation (1) says that the rate of change in consumption over time is determined by η and the difference between the rate of return on capital, r and the consumer's subjective discount rate, δ This equation, called *the Euler equation*, is derived as the necessary condition of the representative consumer's maximization of the Ramsey utility sum.

The return on capital, r in (1) is the return on capital equity or stocks, which consists of the expected capital gains/losses and dividends. Thus, according to the neoclassical macroeconomics, the return on stocks must be consistent with the rate of change in consumption over time in such a way that (1) holds.

To explain volatility of stock prices, consumption must be volatile enough to make the return on stock, r volatile. However, we know that in reality, consumption C is not volatile. If anything, it is *less* volatile than dividends or profits. Thus, given equation (1), the volatility of stock prices or their rate of return r must be explained ultimately by sizable fluctuations of the elasticity of intertemporal substitution η which depends on consumption. Consequently, on the representative agent assumption, researchers focus on the "shape" of the utility function in accounting for the volatility of stock prices [13]. It is not an easy task, however, to reconcile the theory with the observed data if we make a simple assumption for the elasticity of intertemporal substitution, η; η must change a lot despite of the fact that changes in consumption are small.

A slightly different assumption favored by theorists in this game is that the utility, and therefore, this elasticity η depend not on the current level of consumption C_t but on its deviation from the "habit" level, $\hat{C}_t$, namely, $C_t - \hat{C}_t$. By assumption, the habit $\hat{C}_t$ changes much more slowly than consumption C_t itself so that at each moment in time, $\hat{C}_t$ is almost constant. The trick of this alternative assumption is that

[2] For extensions of CAPM in the field of finance, see [21]. Here, we focus on problems in macroeconomics.

although C_t does not fall close to zero, $C_t - \hat{C}_t$ can do so as to make the elasticity of intertemporal substitution η, now redefined as

$$\frac{1}{\eta} = -\frac{u''(C - \hat{C})(C - \hat{C})}{u'(C - \hat{C})} > 0, \tag{2}$$

quite volatile. Campbell and Cochrane [8] is a primary example of such an approach. Though ingenious, the assumption is not entirely persuasive. Why does the consumer's utility become minimal when the level of consumption is equal to the habit level even if it is extremely high? In any case, this is the kind of end point we are led to as long as we keep the representative agent assumption in accounting for the volatility of stock prices.

The above analysis is typical of micro-founded macroeconomics in that it attempts to explain macro phenomenon – volatility of stock prices in relation to consumption – by way of the behavior of the representative consumer. In the case of consumption based CAPM, researchers attribute volatility of stock prices to a particular type of preference of the representative consumer.

Mehra and Prescott [26] is another example of the representative agent model for asset prices. They considered a simple stochastic Arrow–Debreu model. The model has two assets, one the equity share for which dividends are stochastic, and the other the riskless security. Again, on the representative agent assumption, the "shape" of the utility function and the volatility of consumption play the central role for prices of or returns on two assets. For the reasonable values of η, which may be more appropriately called the relative risk aversion in this stochastic model, and the US historical standard deviation of consumption growth, Mehra and Prescott calculated the theoretical values of the returns on two assets. The risk premium, namely the difference between the return on the equity share and the return on the riskless security implied by their model, turns out to be mere 0.4% In fact, the actual risk premium for the US stock (the Standard and Poor 500 Index, 1889–1978) against the short-term security such as the Treasure bill, is 6% Thus, the standard model with the representative consumer fails to account for such high risk premium that is actually observed. Mehra and Prescott posed this result as a puzzle. Since then, a number of authors have attempted to explain this puzzle: see [8].

The "puzzles" we have seen are, of course, puzzles conditional on the assumption of the representative-agent. Indeed, Deaton [11] laughs away the so-called "puzzles" as follows:

> There is something seriously amiss with the model, sufficiently so that it is quite unsafe to make any inference about intertemporal substitution from representative agent models ...
>
> The main puzzle is not why these representative agent models do not account for the evidence, but why anyone ever thought that they might, given the absurdity of the aggregation assumptions that they require. While not all of the data can necessarily be reconciled with the microeconomic theory, many of the puzzles evaporate once the representative agent is discarded. [11, pp. 67, 70]

We second Deaton's criticism. Having said that, here, we note that the standard analyses all focus on the *variance* or the *second moment* of asset prices or returns; see [9], for example, and the literature cited therein. Econophysicists recognize that the variance or standard deviation may *not* be a good measure of *risk*.

To repeat, micro-founded macroeconomics attempts to explain *macro phenomena* – volatilities of consumption and stock price in this particular case – in terms of the micro behavior of the representative agent. According to micro-founded macroeconomics, the key of understanding macro phenomena lies in behavior of micro agent. To grasp precise behavior of intertemporally maximizing economic agent, economist must make a crucial assumption of the representative agent whose behavior mimics macro dynamics. Econophysicists may not be so interested in the games economists play, and may simply laugh them away. However, to make progress in macroeconomics, both physicists and economists must make every effort to understand each other.

4 The Approach Based on Statistical Physics

The approach based on statistical physics is in stark contrast to micro-founded macroeconomics. In this approach, on purpose, we give up pursuit of detailed behavior of micro unit. Instead, we make as simple an assumption as possible of behavior of micro unit. In this respect, we follow the spirit underlying the model of ideal gas where behavior of molecule is assumed to be extremely simple to the extent that taken literally it is actually unrealistic. This assumption is justified because proper statistical aggregation is much more important for the purpose of understanding macro system than detailed behavior of micro unit. In their discussion of stock prices and financial risks, Malevergne and Sornette [21] state as follows:

> The understanding of the large-scale organization as well as the sudden macroscopic changes of organization due to small variation of a control parameter has led to powerful concepts such as "emergence": the macroscopic organization has many properties not shared by its constituents. For the markets, this suggests that its overall properties can only be understood through the study of the transformation from the microscopic level of individual agents to the macroscopic level of the global market. [21, p. 22]

It is our view that the same approach is called for not only in finance but also in economics. Perhaps, we can regard Slutzky's [35] early work on business cycles entitled "the summation of random causes as the sources of cyclic processes" as such an attempt.[3]

To persuade economists that the new approach is necessary in macroeconomics, we need good examples. Here, we take up Yamada et al. [40] as an example. It is a market model meant to explain power laws observed for asset prices. The model

[3] For business cycles, see [1] and Chap. 6 of [4]. Nirei [29] is a good example of the combination of "statistical" and traditional approaches to business cycles.

consists of two traders whose reservation prices for "buy" and "sell" are assumed to be random walks. When the "buy price segment" of one trader and the "sell price segment" of the other overlap each other, one unit of asset is traded, and at the same time, the market price is determined as the mid point of reservation prices of buyer and seller. Obviously, this assumption on micro behavior is extremely simple, perhaps, too simple if it is literally taken. However, it is *not* a defect of the model, but is actually its merit because such a simple model with a little modification (model 3 in the paper) can beautifully account for the power law distribution of price changes as well as the distribution of transaction interval.

The results obtained in such a simple model strongly suggest that proper statistical aggregation is indeed much more important for the purpose of understanding macro system – the probability distribution of price changes in this case – than detailed behavior of micro agent, and that the approach based on statistical physics is very promising in macroeconomics. Thus, the challenge for economics is *not* to create sophisticated models of micro optimization, but rather to pin down a particular stochastic process for the problem under investigation (see [15], for example).

5 Productivity Dispersion

One of the major insights statistical physics provides to economics is that *equilibrium* is not a point in space but a *distribution*. Instead, the general equilibrium theory, the hard core of the main stream neoclassical economics, regards equilibrium as a solution of simultaneous equations or a fixed point of mapping. Foley [12] is a seminal work which challenges the standard notion of equilibrium by means of the method of statistical physics. It advances new notions of equilibrium and efficiency.

Here, we explain *productivity dispersion* in relation to Keynesian economics [17].[4] In the Walrasian general equilibrium theory, the marginal products of production factor such as labor are equal in all the sectors and firms. This is required for the Pareto efficiency, and constitutes the concept of the equilibrium. The equality of productivities across sectors/firms is the fundamental reason why demand does not matter in determining the aggregate output. Demand affects only composition of outputs. Conversely, the existence of *underemployment*, which is equivalent to inequality of *value* marginal product across sectors, means that total output (GDP) can be increased by changing demand.

Traditionally, the presence of (involuntary) unemployment has been taken by many economists as *the* essential condition for Keynesian economics to make sense. However, (involuntary) unemployment is merely a particular form of *underemployment*, and that underemployment defined as differences in productivity of production factors is more essential for the relevance of demand in the determination

[4] For Keynesian economics and the development of macroeconomics, see [42].

of total output than unemployment. To repeat, for demand to play an essential role in the determination of aggregate output, the inequality of productivity across sectors/activities, or underemployment is the generic condition [41].

In the real economy, labor productivity, in fact, differs across firms and industries [6, 28].[5] Why and how does the inequality of productivity persist? One might think that differences in productivity across sectors imply unexploited profit opportunities, and therefore, contradict equilibrium. However, heterogeneous economic agents actually have different thresholds for a change in their strategies. Therefore, we cannot suppose that all the economic agents and production factors instantaneously move to the sector with the highest productivity. We must describe their behavior by the transition rate in the jump Markov process. Consequently, at each moment in time, we have a distribution in productivity. Aggregate demand affects the aggregate output because it affects the transition rate of production factors across sectors/firms with different productivities.

To the extent that the economy is incessantly subject to sectoral demand shocks, and that demand affects productivity, differences in productivity across sectors necessarily persist. Tobin [39, p. 9] proposes a notion of "stochastic macro-equilibrium". He argues that it is "stochastic, because random intersectoral shocks keep individual labor markets in diverse states of disequilibrium; macro-equilibrium, because the perpetual flux of particular markets produces fairly definite aggregate outcomes." The concept of equilibrium which we propose is very similar to what Tobin called "a theory of stochastic macro-equilibrium". By way of affecting the transition rate of production factors, the aggregate demand affects "stochastic macro-equilibrium" or, more precisely, distribution of productivity in the economy, and, therefore, the level of total output.

What would be a distribution of productivity in the economy? And how does it depend on the aggregate demand? Here come the methods of statistical physics with which physicists are familiar.

Suppose that there are K different levels of productivity in the economy, $x_1 < x_2 < \cdots < x_K$. The number of production factors or simply workers working in the sector with productivity x_i is n_i. The total number of workers N is given

$$\sum_i^K n_i = N. \tag{3}$$

Given the aggregate demand D, the endowment N, and a vector of productivities $(x_1, \ldots, x_K)$, we must have

$$\sum_i^K x_i n_i = D. \tag{4}$$

[5] Houthakker [14] and Sato [32] explore how productivity dispersion across firms can be aggregated to macro production function. We believe that econophysics can shed new lights on this old problem.

Then, following the general principal in physics, Yoshikawa [41] conjectures that given a large number of economic units/production factors and randomness, the distribution of productivity is *the Boltzmann distribution*:

$$P(x_i) = \exp(-x_i/kD) \Big/ \sum_i \exp(-x_i/kD) \quad (k > 0),$$

where $P(x_i)$ is the probability or share of workers working with productivity x_i $(i = 1 \cdots K)$, namely n_i/N, and D is the aggregate demand.

Let us explain it briefly. First, we observe that the possible number of a particular configuration $(n_1, \ldots, n_K)$ under constraint (3) is

$$\frac{N!}{\prod_{i=1}^{K} n_i!}.$$

Because the total number of configurations is K^N, the probability of occurrence of such a configuration, $P(n)$, on the assumption of equal probabilities for all configurations, is given by

$$P(n) = \frac{1}{K^N} \frac{N!}{\prod_{i=1}^{S} n_i!}. \tag{5}$$

Following the fundamental principle of statistical physics we postulate that the configuration $\{n_1, n_2, \ldots, n_s\}$ that maximizes $P(n)$ under two constraints (3) and (4) is realized in equilibrium. This postulate is similar to the method of maximum likelihood in statistics or econometrics. Then we obtain the Boltzmann distribution.[6]

Note that the aggregate demand D (relative to N) in economics corresponds to temperature in physics. When D rises, this distribution becomes flatter. That is, when the aggregate demand is high, economic units/production factors are mobilized to higher productivity. Okun [30] makes a similar point saying that workers climb up a 'ladder' of productivity in a 'high pressure economy'. Yoshikawa [41] and Aoki and Yoshikawa [4] argue that this is the proper microeconomic foundation for Keynes' principle of effective demand. This theory also provides precise definition of Tobin's [39] "stochastic macro-equilibrium".

Meanwhile, Aoyama et al. [6] demonstrates that the empirical distribution of labor productivity is actually not the Gibbs (exponential) distribution but the power distribution. Their paper not only explores the empirical distribution, but also suggests a theoretical framework for understanding the obscured power-law. The framework is called *superstatistics* in which the level of aggregate demand is allowed to fluctuate rather than is simply assumed to be constant.

---

[6] Unlike particle in physics, production factor in economics differs in size. However, we can suppose that apparently 'large' production factor actually consists of a large number of the 'basic' units all of a size, and, therefore, we can basically apply the same method of physics to economics.

Because we observe productivity dispersion, plainly, the Walrasian equilibrium theory cannot be literally applied to the real economy. Search theory allegedly fills this gap by encompassing apparent "disequilibrium" phenomena such as productivity dispersion in the neoclassical equilibrium framework. Aoki and Yoshikawa [5] take up Lucas and Prescott [20] as an example, and argue that the observed productivity dispersion cannot be explained within the framework of the standard equilibrium search theory. Our argument rests on the concept of non-self-averaging.

6 Non-self-averaging: A Key Concept for Macroeconomics

Non-self-averaging is little known to economists. It is an important concept that undermines micro-founded macroeconomics. Consider a sequence or a group of random variables $X_n (n = 1, 2, \ldots)$. If the coefficient of variations, namely the standard deviation of X_n divided by its mean, approaches zero as n goes to infinity, then X_n is said to be self-averaging. If not, X_n is *non-self-averaging*.

The Gausian normal distribution and the Poisson process so commonly assumed in economic analysis are self-averaging. However, these cases of self-averaging cannot be actually taken as the norms. Put it differently, non-self averaging is not pathological case to be left only for mathematical curiosity, but rather quite naturally emerges and is generically present in nature (see, for example, [36, p. 369]).

Suppose, for example, that n random variables form K_n clusters. Clusters may be the subsets of firms or sectors/industries to be distinguished from each other by their respective characteristics. Now, in a two-parameter model in which one parameter, θ, affects the probability existing clusters grow while the other parameter, α, influences the probability a new cluster (of initial size one) is born. This model is known as the Poisson–Dirichlet two-parameter process, PD (α, θ). In this model, the number of clusters, K_n, is self-averaging with $\alpha = 0$, but is non-self-averaging for $\alpha > 0$; see [2]. In other words, if the number of clusters is given, we obtain self-averaging, but when the number of clusters stochastically grows, we obtain non-self-averaging. Note that the standard Poisson process is nothing but a special case of $\alpha = 0$. Aoki and Yoshikawa [5] demonstrate that non-self-averaging quite naturally emerges in a simple model of economic growth.

Now, non-self-averaging means that we cannot ignore the fluctuations around the mean even if n (typically, the number of agents in the model) becomes large. This, in turn, means that optimization exercises in the standard micro-founded macroeconomics do not make much sense because such analyses are meant to capture the dynamics of the means.

7 Conclusion

We believe that time has come to integrate the new approach based on statistical physics or econophysics into macroeconomics. Toward this goal, there must be more dialogues between physicists and economists. On one hand, economists must show physicists where the crucial problems lie in macroeconomics, and how relevant or, in certain cases, irrelevant the findings made by econophysicists are to economics. Physicists, on the other, should pay more attention to economic theory. For example, the mathematical models which generate power-laws have been much studied (see [36] and references cited therein). The basic question is then how we interpret such models as economic models (see [3], for example). To this end, we cannot do without the knowledge of economic theory.

In this paper, we argued that there is no reason why the methods of statistical physics so successful in many fields of natural sciences cannot be usefully applied to macroeconomics that is meant to analyze the macroeconomy comprising a large number of economic agents. It is, in fact, weird to regard the macroeconomy as a homothetic enlargement of the representative micro agent. We trust the bright future of the new approach to macroeconomies based on statistical physics.

References

1. Aoki M (1998) Simple model of asymmetrical business cycles: interactive dynamics of a large number of agents with discrete choices. Macroecon Dyn 2:427–442
2. Aoki M (2006) Patterns of non-exponential growth of macroeconomic models: two-parameter Poisson–Dirichlet model. CIRJE-F-449, Faculty of Economics Discussion Paper. University of Tokyo, Tokyo
3. Aoki M, Yoshikawa H (2006) Stock prices and the real economy: power law versus exponential distributions. J Econ Interact Coord 1:45–73
4. Aoki M, Yoshikawa H (2007) Reconstructing macroeconomics: a perspective from statistical physics and combinatorial stochastic processes. Cambridge University Press, Cambridge, MA
5. Aoki M, Yoshikawa H (2008) The nature of equilibrium in macroeconomics: a critique of equilibrium search theory. Economics E-journal Discussion Paper, No 2008-37. http://www.economics-ejournal.org
6. Aoyama H, Yoshikawa H, Iyetomi H, Fujiwara Y (2009) Labour productivity superstatistics. Progress of theoretical physics supplement 179:80–92
7. Blanchard O, Fischer S (1989) Lectures on macroeconomics. MIT Press, Cambridge, MA
8. Campbell J, Cochrane J (1999) By force of habit: a consumption-based explanation of aggregate stock market behavior. J Polit Econ 107:205–251
9. Cecchetti S, Lam P, Mark N (2000) Asset pricing with distorted beliefs: are equity returns too good to be true? Am Econ Rev 90(4):787–805
10. Champernowne D (1953) A model of income distribution. Econ J 83:318–351
11. Deaton A (1992) Understanding consumption. Oxford University Press, Oxford
12. Foley D (1994) A statistical equilibrium theory of markets. J Econ Theory 62:321–345
13. Grossman S, Shiller R (1981) The determinants of the variability of stock market prices. Am Econ Rev 71:222–227
14. Houthakker H (1955) The Pareto distribution and the Cobb–Douglas production function in activity analysis. Rev Econ Stud 23(1):27–31

15. Ijiri Y Simon HA (1975) Some distributions associated with Bose–Einstein statistics. Proc Natl Acad Sci USA 72(5):1654–1657
16. Ijiri Y, Simon HA (1979) Skew distributions and the sizes of business firms North-Holland, Amsterdam
17. Keynes J (1936) The general theory of employment, interest, and money. Macmillan, London
18. Kydland F, Prescott E (1982) Time to build and aggregate fluctuation. Econometrica 50(6):1345–1370
19. Lucas RE (1987) Models of business cycles. Blackwell, Oxford
20. Lucas RE, Prescott E (1974) Equilibrium search and unemployment. J Econ Theory 77:721–754
21. Malevergne Y, Sornette D (2006) Extreme financial risks. Springer, Berlin
22. Mandelbrot B (1963) The variation of certain speculative prices. J Bus 36:394–419
23. Mandelbrot B (1997) Fractals and scaling in finance. Springer, New York
24. Mantegna R, Stanley HE (2000) An introduction to econophysics: correlations and complexity in finance. Cambridge University Press, Cambridge
25. McCauley J (2004) Dynamics of markets: econophysics and finance. Cambridge University Press, Cambridge
26. Mehra R, Prescott E (1985) The equity premium. J Monet Econ 15:145–161
27. Montroll E (1987) On the dynamics and evolution of some socio-technical systems. Bull Am Math Soc 16(1):1–46
28. Mortensen DT (2003) Wage dispersion. MIT Press, Cambridge, MA
29. Nirei M (2006) Threshold behavior and aggregate fluctuation. J Econ Theory 127(1):309–322
30. Okun A (1973) Upward mobility in a high-pressure economy. Brooking Papers on Economic Activity 1:207–261
31. Pareto V (1897) Cour d'Economie Politiquc. Lausanne, Rouge
32. Sato K (1974) Production functions and aggregation. North-Holland, Amsterdam
33. Sethna J (2006) Statistical mechanics: entropy, order parameters, and complexity. Oxford University Press, Oxford
34. Shiller R (1981) Do stock prices move too much to be justified by subsequent changes in dividends? Am Econ Rev 71(3):421–436
35. Slutzky E (1937) The summation of random causes as the sources of cyclic processes. Econometrica 5:105–146
36. Sornette D (2000) Critical phenomena in natural sciences. Springer, Berlin
37. Stanley HE, Gopikrishnan P, Plerou V (2006) Statistical physics and economic fluctuations. In: Gallegati M et al (eds) The complex dynamics of economic interaction. Springer, New York
38. Sutton J (1997) Gibrat's legacy J Econ Lit 35:40–59
39. Tobin J (1972) Inflation and unemployment. Am Econ Rev 85:150–167
40. Yamada K, Takayasu H, Ito T, Takayasu M (2009) Solvable stochastic dealer models for financial markets. Phys Rev E 79:051120
41. Yoshikawa H (2003) The role of demand in macroeconomics. Japanese Econ Rev 54(1):1–27
42. Yoshikawa H (2009) The general theory: toward the concept of stochastic macro-equilibrium. In: Bateman BW, Hirai T, Marcuzzo MC (eds) The return of Keynes; Keynes and Keynesian policies in the new millennium. Harvard University Press, Cambridge, MA

How to Avoid Fragility of Financial Systems: Lessons from the Financial Crisis and St. Petersburg Paradox

Hideki Takayasu

Abstract Firstly, I point out that the financial crisis occurred in 2008 has many analogous points with a physical phenomenon of brittle fracture in the sense that it is a highly irreversible phenomenon caused by concentration of stress to the weakest point. Then, I discuss distribution of gain–loss of continuous transactions of options which can be regarded as a source of stress among financial companies. The historical problem of Saint Petersburg paradox is reviewed and it is argued that the paradox is solved by decomposing the process into a combination of a fair gamble and an accompanied financial option. By generalizing this fair gamble it is shown that the gain–loss distribution in this problem is closely related to the distribution of gain–loss of business firms in the real world. Finally, I pose a serious question to the ordinary way of financing money to business firms with compound interest rates. Instead we introduce a new way of financing business firms without applying prefixed interest rates, in which financial stress is shared by all involved firms. This method is expected to reduce the risk of both financial firms and business firms, and is applicable even in the non-growing society.

1 Financial Crisis Viewed as Brittle Fracture

A bankrupt occurs when a firm can not pay its debt by the due date. This situation occurs when the sum of firm's cash and bank balance is smaller than the debt. This rule applies independent of the size of the firm, and the information of bankruptcy is immediately announced to all financial companies. As a result all bank accounts related to this firm are shut down immediately after the breach of contract, so even a big firm having huge amount of assets and huge number of employees can be bankrupted instantly. This phenomenon is characterized by a highly nonlinear irreversible dynamics as shown in Fig. 1a. The activity of the firm does not change

H. Takayasu (✉)
Fundamental Research Group, Sony Computer Science Laboratories, 3-14-13 Higashigotanda, Shinagawa-ku, Tokyo 141-0022, Japan
e-mail: takayasu@csl.sony.co.jp

M. Takayasu et al. (eds.), *Econophysics Approaches to Large-Scale Business Data and Financial Crisis*, DOI 10.1007/978-4-431-53853-0_10, © Springer 2010

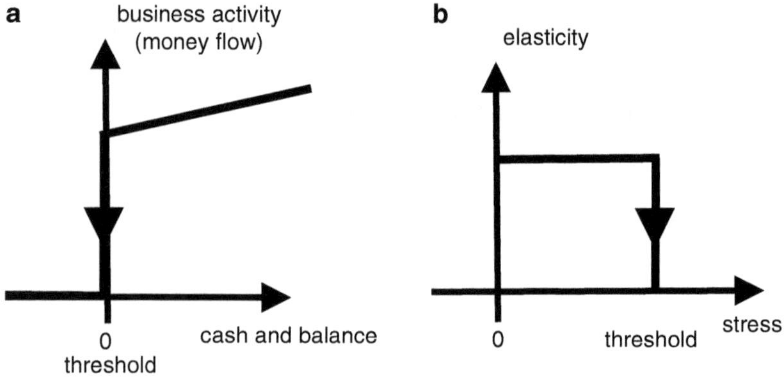

Fig. 1 (**a**) Threshold dynamics characterizing bankruptcy (*left*). (**b**) Threshold dynamics characterizing brittle fracture (*right*)

much when the sum of cash and balance is not negative, however, once this sum goes into a negative region by trying to pay a debt, the activity of the firm is forced to be 0, meaning the death of the firm.

It is interesting that a purely physical phenomenon of brittle fracture is also described by a threshold dynamics. Assume the situation that we bend a brittle material such as glass bar by gradually adding force. The elasticity of the bar is nearly constant when the stress of the bar stays below a threshold value [1]. When the stress goes beyond the threshold then fracture occurs suddenly and the elasticity becomes 0 in an irreversible manner.

Similarity between bankrupt and brittle fracture is not limited to this elementary process. In the case of brittle fracture of non-uniform materials such as a rock, the first fracture can cause successive breakdowns as the accumulated stress is re-distributed to the neighbors resulting excess of the fracture threshold at some new points. In the case of financial breakdown debts play the role of stress and they are re-distributed when a firm is bankrupted in the form that the scheduled money flows are cancelled and the debts remain at the lenders. By this re-distribution all the lenders' balances are lowered by the amount of the debts, so there is a possibility that some of the lenders fall into the situation that the sum of cash and balance becomes negative resulting successive bankrupts.

Another similarity between the brittle fracture and financial bankrupt is the tendency of accumulation of stress to the weakest point. In the case of brittle fracture stress accumulates nearly uniformly when the applied external force is gentle, however, as the force becomes stronger there is a tendency that stress concentrates on a weakest point where deformation is largest. In the case of finance there is a tendency that interest rates of the debts are higher for weaker firms. This is realized as a result of competition: For a strong firm whose financial situation is good and the risk of bankrupt is small, many banks will offer lending money with lower interest rates so the firm can choose the lowest interest rate. On the other hand for a weak firm which has a big risk of bankrupt, no bank will offer lending money with low interest rate, as a result the firm has to accept a loan with high interest rate. By this effect financial stress tends to concentrate on weak firms.

In September 2008 a big and long-lived financial company Lehman Brothers bankrupted suddenly due to the increase of financial stress caused by the decay of value of financial products related to the subprime loan. By this breakdown financial stresses of debts are distributed to those firms which lent money to this firm. As the information of borrowing and lending money between financial firms is not open to public, so the risk of bankrupt seemed to be increased for all financial firms which might have business relation to Lehman Brothers. By this increase of potential risk of bankrupt the interest rates between financial firms jumped up in the short-term money market. This practically means that money flows among financial firms almost stopped and it became difficult for all financial firms to make ends meet. This situation was quite analogous to the physical situation that wind blew a window glass and a crack appeared at a point, and the whole glass was almost broken into pieces.

Facing with this crisis of financial systems the governments of major countries immediately supplied money to money markets to lower the interest rates and to keep ordinary money flow among financial firms. This prompt action was effective and the worst scenario of successive failure of financial firms which actually occurred in 1929 had been avoided. Using the above analogy of window glass, the government treatment was similar to an emergency treatment of taping the cracked glass.

2 Diverging Loss in Continuous Trading of Options

Options are financial products for markets originally designed to reduce the risk of market price fluctuations. For example, a European put-option for a stock market is the right of selling stocks at price K on T days in the future. Assuming that the market price at the T-th day is y, then an investor who already has a stock can surely get money of amount not less than K at T-th day by buying one-unit of put-option with premium S: If $K \leq y$ he can sell the stock in the market, and if $K > y$ he can use the option and can get K. From the viewpoint of a financial firm selling this option the income is the premium U, and it pays K to the option holder if $K - y$ is positive. In this case the stock can be sold immediately in the market with the price y, so the loss of the seller is $K - y$.

The fair value of premium of this option, which is sold at time t when the market price is x, can be estimated by the expectation value of loss of the seller. As the payment of the seller appears only when y is less than K and the loss is $K - y$, the fair premium is given by the following equation:

$$U(x,t;K,t+T) \equiv \int_0^K (K-y)p(x,t;y,t+T)dy, \tag{1}$$

where $p(x,t;y,t+T)$ denotes the market price probability density to take the value y at time $t+T$. By assuming the time evolution rule of market price the option price can be determined either theoretically or numerically by estimating the future price probability.

Nowadays options are quite popular because they can be used both as insurance and as speculation. However, here, I point out that continuous transaction of options increase the risk of financial firms severely by the following reasons.

Firstly, as often mentioned the pricing formulation such as Black–Scholes equation neglects the effect of so-called long-tails [2]. This means that if an event with large deviation occurs, then gain or loss will become extremely large. This risk is already noticed and many researchers are working on revising the estimation of the future price density.

Secondly, it is less pointed out but there is a severe problem on duration time of payment for option. In the above notation the option seller must pay $K - y$ when $K > y$. The time interval which satisfies $K > y$ for a fixed value of K is given by the time interval of level-set as schematically shown in Fig. 2, in this case the option seller with the contract price K should pay during the period $[t, t + s]$.

In an idealistic case that $y(t + T)$ follows a Brownian motion, the time interval of such level set is known to follow a power law with an exponent $-1/2$ [3].

$$W(> s) \propto s^{-1/2}, \tag{2}$$

where $W(> s)$ denotes the cumulative distribution that an interval is longer than s. It should be noted that the mean value of this period is infinite, and there is a possibility that such losing period continues even an infinite time.

This phenomenon can be characterized also by an autocorrelation. From the viewpoint of option seller, let us assign $+1$ for the moment of winning, that is, $K \leq y(t + T)$, and assign -1 for the other case of losing situation, then the autocorrelation of this signal becomes

$$C(T) \propto T^{-1/2}. \tag{3}$$

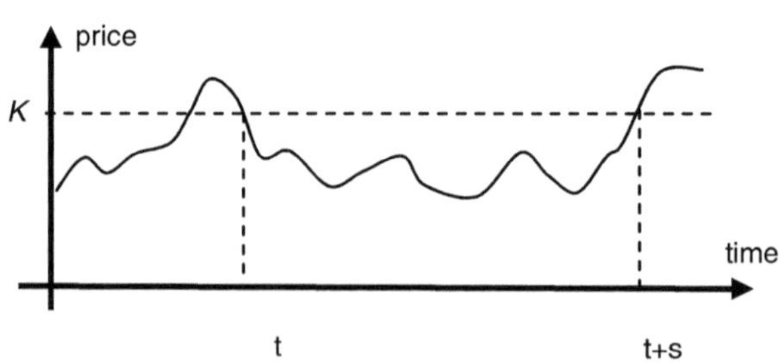

Fig. 2 A schematic graph showing the period that the option seller must pay $K - y(t)$

Namely, winning or losing in this option trade is strongly auto-correlated. In the case of ordinary gambles in which win or lose occurs independently at every time step, the whole risk will decrease by repeating such gamble continuously due to the central limit theorem. However, in the case of strongly correlated gambles like the case of this option, the risk increases for more deals.

In order to understand this situation more clearly, let us consider the gross loss during one interval of $K > y$ which is proportional to the area surrounded by the dotted line and the price curve in Fig. 2. The cumulative distribution of this area is known to follow the next power law [4].

$$P(> m) \propto m^{-1/3}. \tag{4}$$

From the value of the exponent, $-1/3$, it is easy to show that both the variance and average diverge in this gross loss distribution. This fact implies that even if we can estimate the right option price of (1) for each time step, the cumulative loss in each interval of $K > y$, $m' \equiv m - sU$, also follows the same distribution as (2), so the loss of a financial firm can become extraordinarily large. It should be noted that this accumulation of risk can not be lowered by a pair-trading of selling call-options simultaneously, it simply doubles the risk.

The third risk of option dealing is caused by non-stationarity. Option pricing is always based on the assumption of stationary statistics, so if there occurs an unusual event that changes the basic statistics of the market, then, the option sellers might suffer a huge risk. Also, in the process of option pricing a trend term is usually neglected theoretically, however, in the real markets trends can be found in various scales, which enhance the risk of option sellers. Volatility is the square of price changes and is known to have a long autocorrelation. Long correlation in volatility is practically quite similar to the case of non-stationary statistics, so the risk of estimation of option pricing is always very large.

As a summary of this section it is mentioned that the more option deals the more risk financial firms will have. A single deal of an option can be insurance, however, continuous deals of options make a dangerous gamble.

3 The Saint Petersburg Paradox: Another Solution by Options

In 1738 Daniel Bernoulli wrote a paper on risk of a lottery posing a serious question on a diverging mean value [5]. He considered a simple coin-toss gamble: "You can toss a coin repeatedly until a tail appears. In the case the number of heads is n, you can get 2^n dollars as reward. Then, what is the fair entrance fee of this gamble?" The probability of getting 2^n dollars reward is 2^{-n-1}, so the normal answer to this question is given by the following expectation value of reward.

$$E = \sum_{n=0}^{\infty} 2^n 2^{-n-1} = \frac{1}{2}\sum_{n=0}^{\infty} 1 = \infty. \tag{5}$$

This divergence makes a paradox, as this result sounds ridiculous: The reward of this gamble is not so big, 1 dollar with probability 0.5, 2 dollars with probability 0.25, 4 dollars with probability 0.125, ..., so, intuitively no one will pay entry fee higher than 100 dollars.

Bernoulli's answer to this paradox was to introduce the concept of utility. He considered that human impression of economical value is a nonlinear function of the quantity of money such as a logarithmic function, so the above expectation should be replaced by the utility which takes a reasonable finite value. Bernoulli's idea attracted economists' interest 200 years later [6] and the concept of utility became one of the key concepts in economics [7]. Also, he discussed about probability weighting, that is, human tendency to neglect rare events with very low but finite possibility. An extension of this idea developed as the work on prospect theory by Kahneman and Tversky [8].

There are other ways of solving this paradox. Samuelson discussed that such lottery with infinite expectation will not be held in general simply because the host of the lottery will have infinite loss [9]. Also it is easy to consider finiteness of the host's finance and re-formulate using with only finite numbers [10]. It should be noted that even if the host introduces the upper limit of the reward as the whole GDP of the world, the fair entry fee becomes only 24 dollars (http://en.wikipedia.org/wiki/St._Petersburg_paradox). This means that any entry fee above this value is practically non-sense.

Here, I show a new solution of this paradox by applying a kind of option to this lottery. Let us start with a simple fare gamble: A gambler bets money and tosses a coin, if it is head then the reward is the doubled value of bet, and if it is tail then he loses everything. He repeats this gamble betting all the assets. The time evolution of this gambler's asset at the t-th trial, $x(t)$, is described by the following simple probabilistic rule,

$$x(t+1) = \begin{cases} 2x(t) & \text{prob } 1/2, \\ 0 & \text{prob } 1/2. \end{cases} \tag{6}$$

It is obvious that this gambler will lose everything sooner or later, if he does not stop at some time, as known generally by the name of gambler's ruin. However, it is easy to show that the mathematical value of expectation is a constant for any t.

$$< x(t) >= x(0). \tag{7}$$

It should be noted that a kind of serious deviation between expectation and intuition appears in this simplest version of gamble, which may be also called a paradox.

As this simplest gamble is too dangerous for gamblers, the host can introduce a kind of insurance. It is an option, $Op(t_0, t_1)$, the right to reverse one time step valid during the time steps from t_0 to t_1. For example, in the case that a gambler had a successive 10 heads and a tail at the 11-th trial, he can get the maximum asset, 2^{10} dollars, instead of 0, if he had bought an option which is valid at the 11th time step. As known from this example, the Saint Petersburg lottery is the special case that $x(0) = 1$ and the option is automatically included for any number of t, $Op(0, \infty)$.

Now, the problem of the Saint Petersburg paradox becomes a simpler question for evaluation of the price of this option.

The fair price of this option can be calculated easily by estimating the expected increase of asset by this reversion of one time step. In the above simplest gamble described by (6), the asset at time step t is either $x(0) \times 2^t$ or 0 with the probability of occurrence 2^{-t} or $1 - 2^{-t}$, respectively. The case that $Op(t, t)$ works is the situation that $x(t-1) = x(0) \times 2^{t-1}$ and $x(t) = 0$, which realizes with probability 2^{-t}. Therefore, the expectation of gain by this option is $x(0)/2$. Generalizing this we have the fair price of option,

$$Op(t_0, t_1) = (t_1 - t_0 + 1) \times x(0)/2. \tag{8}$$

In the case of the Saint Petersburg lottery this option price actually diverges, so the entrance fee of this lottery diverges. Now, the reason of divergence becomes apparent, the lottery has too much insurance. We can consider the cases with reasonable insurance. For examples, if the gambler buys $Op(1, 20)$, which is 10 dollars, then he can start the gamble by betting 1 dollar. The maximum asset value during the gamble does not vanish up to $t = 20$ protected by this option, namely, he can get at most $2^{20} \approx 1$ million dollars in the lucky case.

The gambler can choose any option he likes at the start of this gamble with a fair price. Also, from the viewpoint of host of this gamble, it is required not to sell too much insurance which the host can not cover. It is now clear that no paradox remains in this problem from both sides, the gambler and the host.

4 A New Way of Financing Firms: An Alternative Proposal for Financial Firms

The simple fair gamble introduced in the preceding section captures an essential property. Here, we consider the situation that a gambler is repeating a generalized gamble adding a dollar at every trial without using the option. The time evolution of the asset in this situation is described by the following stochastic equation.

$$x(t+1) = b(t)x(t) + f(t), \tag{9}$$

where $b(t)$ is a random variable which takes either 2 or 0 with equal probabilities, and $f(t) = 1$. This process can be solved exactly and the distribution of $x(t)$ is obtained as,

$$P(\geq x) = 2/(1+x). \tag{10}$$

As known from this functional form the mean value and the standard deviation are both diverging in this distribution which belongs to the so-called Zipf's law.

Stochastic processes described in the form of (9) with general $b(t)$ and $f(t)$ are called the random multiplicative processes and the basic statistical properties are

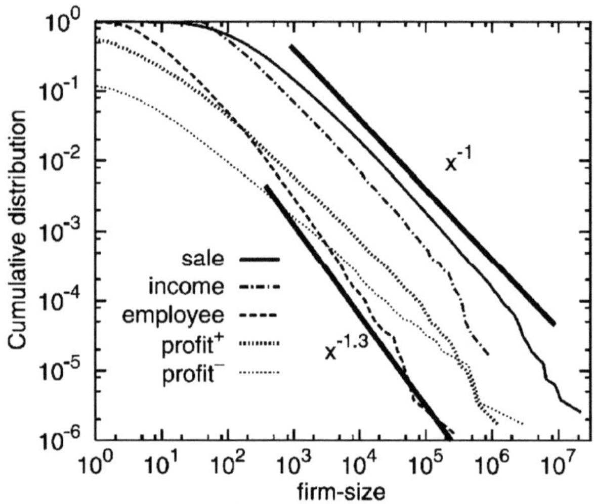

Fig. 3 Log-log plot of distributions of firm sizes measured by sale, income, employee and profit (positive and negative) [14]

studied [11]. It is known that Zipf's law realizes whenever the following equation holds independent of $f(t)$,

$$< b(t) >= 1. \tag{11}$$

For examples, in the cases that the distribution of $b(t)$ is symmetric around $b(t) = 1$ we always have a Zipf's law, such as the case of $b(t) = 0.9$ or 1.1 with probabilities both 0.2, and $b(t) = 1$ otherwise. This case, 10% growth or decay, is not the level of gambles, it is closer to business firms' annual growth rate fluctuation.

As for real firms' statistics it is known that firms' sales and incomes in the real world generally follow a Zipf's law or a similar power law with the exponent close to -1 [12], and time evolutions of sales or incomes are modeled by a kind of multiplicative processes [13]. Figure 3 shows the distributions of sales and income for about 800,000 Japanese firms in 2005 which are based on an exhaustive database provided by Tokyo Shoko Research covering practically all active firms in Japan [14]. From the lines in this figure we can confirm that Zipf's laws are actually valid for sales, incomes and profits.

Growth rates of firms have been also intensive recent studies. It is found that growth rates are fairly independent of the past and other quantities, and the distribution of annual growth rate of sales can not be approximated by a normal distribution but the tail parts are approximated by power laws [15]. Figure 4 shows distributions of real growth rates of firms for sales and incomes using the same data as Fig. 3 [14]. As known from this plot the growth rates are concentrated sharply around 1 and the distributions are nearly symmetric, namely, (10) and (11) can be used for real business firms' growth phenomenon. In this plot a normal distribution is given by a parabolic curve, and we can confirm that the real firm growth rates scatter much wider than the normal distribution. And, this wide distribution is the target of proposing new financing method.

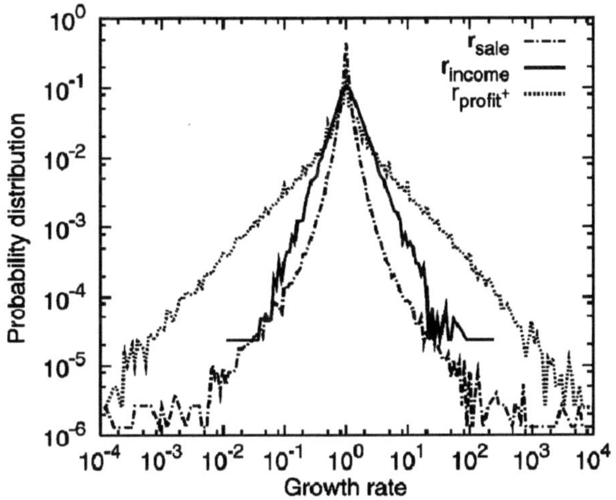

Fig. 4 Log-log plot of probability densities of growth rates measured by sale, income and profit [14]

In general financing business firms is the most important role of financial firms. Now it is a commonsense to lend money with a prefixed interest rate under free competition, namely, a borrower asks interest rate for several banks and chooses the bank which proposes the lowest interest rate. However, as mentioned in the first section interest rates produce a kind of continuous stress among business firms, and free competition makes the whole economic system fragile. Here, we propose a new financing method that does not use interest rate.

Instead of fixing an interest rate before financing, we consider a new loan system in which the amount of repayment is determined after the financing period by evaluating the growth of the firms. The policy is to share the luck, that is, highly grown firms repay larger amount and non-growing firms repay just the amount they borrowed.

In order to realize this loan system it is necessary to make a contract which determines the relationship between growth rate of sales during the loan period and the amount of repayment. Let the rate of repayment at growth rate b be $h(b)$, and assume a function which satisfies $h(b) = 1$ for $b < 1$, meaning that the borrower should return just the borrowed amount if the firm unfortunately could not make its sales bigger at the end of loan period compared with the amount of sales of the starting point. The expected profit rate of gross income of the lender J is given as,

$$J = \int_0^{\infty} h(b)j(b)db - Q - 1, \tag{12}$$

where $j(b)$ is the probability density of growth rate of sales of a firm at the end of loan period, and Q is the probability of bankruptcy of borrowers implying direct loss of the lender. Here, we assume the normalization, $1 = \int_0^{\infty} j(b)db + Q$.

As a typical example we can assume the following type of function:

$$h(b) = \begin{cases} 1, & 1 > b \geq 0; \\ 1.02, & 1.1 \geq b \geq 1; \\ 1.1, & 2 \geq b \geq 1.1; \\ 1.5, & b \geq 2. \end{cases} \tag{13}$$

In this situation the value of J is estimated from the distribution of $j(b)$ from the empirical results given in Fig. 4 for r_{sale} as

$$J = 0.036 - Q. \tag{14}$$

Namely, the lender can get a positive profit even if the bankrupt rate is about 3% which is higher than ordinary bankrupt rate which is less than 2%. This is just an example, and we can design other profiles for $h(b)$, for example, by a continuous function.

The profit of lender by this new method is not so high compared with ordinary interest rate system, however, the merit is that this system can be managed in a non-growing society where the GDP keeps a constant value.

5 Summary and Discussions

In the first section we showed that the present financial system has a fragility that is quite similar to fragility of glass against external force. The corresponding force in the financial system is caused by the mismatch of interest rate and actual growth rate. The stress accumulates to weaker firms which failed growing, and once a breakdown or a bankrupt occurs successive failure tends to continue to firms having business connections. In the second section we showed that financial derivatives such as options are enhancing the risk of financial firms, especially, it is proved that the bankers with diverging mean value if the firm is continuously selling options. In the third section we solved the classical problem of Saint Petersburg paradox by showing that the paradox is a composite of a fair gamble of no average growth and a kind of option of which price diverges if it covers the infinite number of trials. We generalize this fair gamble and consider the situation that artificial firms are repeating fair gambles adding unit of money at every time step. It is shown that the distribution of asset of such firms follows the so-called Zipf's law, which is also the distribution of sales or incomes of real firms. In the fourth section we introduced a new financing method which does not assume a pre-fixed interest rate, instead the amount of repayment is determined by the growth rates at the end of loan period.

The new financing method may not sound so attractive for those people who are keen about earning money with high interest rate under competition, however, the world with competitive interest rates is considered to have fragility and stresses are accumulated by the difference of interest rates and the real growth rate. So, the present financial system, if it goes without any reformation, will meet another financial crisis like that of Lehman shock in 2008 in the near future.

In the proposing new financing system only successful firms return money more than loaned amount, so stresses do not accumulate to unlucky firms. It is shown that the lender can get positive profit even the whole society is non-growing and the effect of bankrupt is taken into account.

This new financing system is expected to be compatible with the present financing method. However, in that case the lender should check carefully the possibility of large growth of the borrower. If there is no possibility of making the sales two times bigger at the end of loan period, the lender should not lend money to the candidate firm. Because this new system is sustained by firms with high growth rate whose occurrence probability is rather small, for example about 2% in real data.

Another point that the lender should pay much attention is to check the correctness of the treasurer's report of the borrower. There are cases that firm's accounting is not honestly calculated, in such a case the repayment could be lowered. As the system is similar to progressive taxation, honest report is strictly requested.

References

1. Takayasu H (1985) Prog Theor Phys 74:1343–1351
2. Mandelbrot BB, Hudson RL (2004) The (mis)behavior of markets. Basic Books, New York
3. Feller W (1950) An introduction to probability theory and its applications. Wiley, New York
4. Takayasu H, Nishikawa I, Tasaki H (1989) Phys Rev Lett 63:2563–2566
5. Bernoulli D (1954) Econometrica 22(1):22–36 (originally published in 1738, translated by Dr. Lousie Sommer)
6. Menger K (1934) Zeitschrift für Nationalökonomie 5(4):459–485
7. Arrow KJ (1974) Q J Econ 88(1):136–138
8. Kahneman D, Tversky A (1979) Econometrica 47:263–291
9. Samuelson P (1960) Int Econ Rev 1(1):31–37
10. Martin R (2004) The Stanford encyclopedia of philosophy, fall 2004 edn. Stanford University, Stanford, http://plato.stanford.edu/archives/fall2004/entries/paradox-stpetersburg/
11. Takayasu H, Sato A-H, Takayasu M (1997) Phys Rev Lett 79:966–969
12. Takayasu H, Okuyama K (1998) Fractals 6:67–79
13. Mizuno T, Takayasu M, Takayasu H (2004) Physica A 332:403–411
14. Ohnishi T, Takayasu H, Takayasu M (2009) Prog Theor Phys 179:157–166
15. Riccaboni M et al (2008) Proc Natl Acad Sci 105:19595–19600

Part 3
General Methods and Social Phenomena

Data Centric Science for Information Society

Genshiro Kitagawa

Abstract Due to rapid development of information and communication technologies, the methodology of scientific research and the society itself are changing. The present grand challenge is the development of the cyber-enabled methodology for scientific researches to create knowledge based on large scale massive data. To realize this, it is necessary to develop a method of integrating various types of information. Thus the Bayes modeling becomes the key technology. In the latter half of the paper, we focus on time series and present general state-space model and related recursive filtering algorithms. Several examples are presented to show the usefulness of the general state-space model.

1 Change of Society and Scientific Research

By the progress of information and communication technologies (ICT), large-scale massive heterogeneous data have accumulated in various fields of scientific researches and society. As examples, we may consider the microarray data in life science, POS data in marketing, high-frequency data in finance, all-sky CCD image in astronomy, and various data obtained in environmental science and earth science, etc.

These rapid developments changed the society and the research methodologies in science and technology. In the information society, the information became as worthy as the substances and the energy, and the quantity of information was the crucial factor for the success in the society. However, in this twenty-first century, the so-called ubiquitous society is becoming widespread, where everybody can access to huge amount of information anywhere and anytime. If such ubiquitous society is actually realized, the value of information itself will be depreciated, because

G. Kitagawa (✉)
The Institute of Statistical Mathematics, Research Organization of Information and Systems, 10-3 Midori-cho, Tachikawa, Tokyo 190-8562, Japan
e-mail: kitagawa@ism.ac.jp

M. Takayasu et al. (eds.), *Econophysics Approaches to Large-Scale Business Data and Financial Crisis*, DOI 10.1007/978-4-431-53853-0_11, © Springer 2010

everybody can share most information in common. Therefore, the interest in the development of the methods and technologies for information extraction and knowledge creation has grown, because the success and failure in the ubiquitous society depends on whether one can extract essential information from massive data.

2 Data Centric Science: A Cyber-Enabled Methodology

2.1 Expansion of Research Object and Change in Scientific Methodology Based on ICT

The scientific research until the nineteenth century has developed basically under Newton–Descartes paradigm based on a mechanic view of the world. In this deductive approach, i.e., in theoretical sciences, mathematics played an important role as the language of science.

However, the evolutionism advocated by C. Darwin in mid-nineteenth century concluded that every creature in real world evolves and changes with time. Motivated by such changes of view on the real world, K. Pearson declared in 1891 that everything in the real world can be an object of scientific research, and advocated *the grammar of science* [15]. The mathematical statistics has been developed as a tool to realize the grammar of science. By the establishment of the method of experimental sciences, not only biology but also many stochastic phenomena in real world such as economy and psychology, became the objects of scientific research.

In the latter half of the twentieth century, by the progress of the computers, computing science has been developed, and numerical computation and Monte Carlo method were applied to the nonlinear dynamics, complex systems, and intertwined high degree of freedom systems that have been difficult to handle by the conventional analytic approach based on theoretical science.

The development of the information technology resulted in accumulation of large-scale massive data in various fields of scientific researches and society, and a huge cyber-world is being created. It is not possible to talk about future development of the science and the technology without establishing methods of effective use of large-scale data. In this article, the cyber-enabled methodology based on the large-scale data set will be called the *data centric science*.

The computing science and the data centric science are newly establishing cyber-enabled deductive and inductive methods while the conventional methodologies, theoretical science and experimental science, relies on the researcher's expertise and experiences. Now having been developed the computing science, it is indispensable to promote this data centric science strategically to realize well-balanced scientific researches in the information era (Fig. 1).

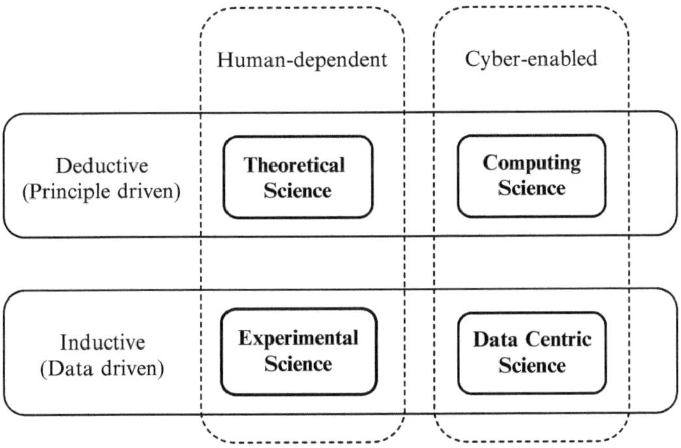

Fig. 1 Four methodologies that drive scientific researches

2.2 *Active Modeling*

In the area of statistical science, the role of a model is changing along with the change of the scientific methodology and the image of the knowledge. In the conventional setting of the mathematical statistics, by assuming that the data is obtained from the true distribution, we aimed at performing an objective inference concerning the true structure. However, in the statistical modeling for information extraction or knowledge creation, it is rather natural to consider that the model is not true or close replica of the truth but is a "tool" for information extraction.

Once the statistical model is considered like this, a flexible standpoint of modeling is obtained. Namely, in statistical modeling we should use not only the present data but also the knowledge on that subject, empirical knowledge, and any other data that have been obtained so far, and even the objective of the modeling. Once the model is obtained, the information extraction, knowledge discovery, prediction, simulation, control, and management, etc., can be achieved straightforwardly or deductively (Fig. 2). Needless to say, the result of knowledge acquisition using the model leads to refine modeling. In this article, such a process will be called active modeling.

To enhance and promote the data centric approach, the technologies for the knowledge integration and for the personalization are needed. However, in the modeling for personalization, ultimate conditioning is required, and the difficult problem called "new NP problem" arises, in which the number of variables is much more than the number of observations.

Anyways, the technology that becomes a key to achieve information integration and an ultimate conditioning is the Bayes modeling. It is because various prior information from knowledge, experiences and data can be integrated by the use of the Bayes model. Although the Bayes' theorem was discovered in the mid-eighteenth century, and the superiority of the inference based on the Bayes' theorem was

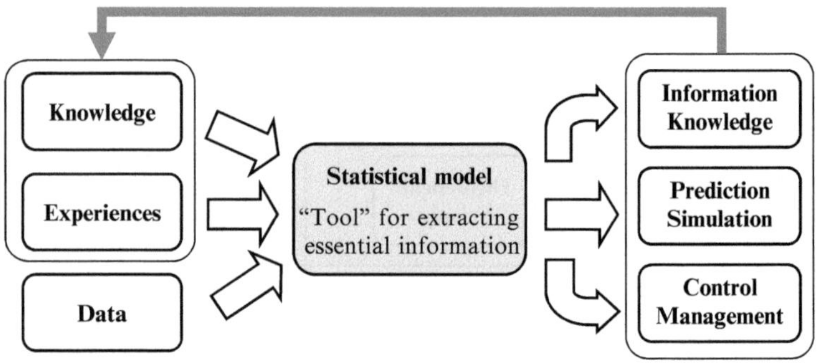

Fig. 2 Active modeling and the use of identified model

well-known, application to real problems was rather rare, due to philosophical controversy, difficulty in determining the prior distributions, and the difficulty in computing the posterior distribution, etc. However, owing to the development of statistical science such as the change in the viewpoint of modeling, the development of model evaluation criterion [1,13], and statistical computing methods such as MCMC and sequential Monte Carlo methods [4,9], now the Bayes method becomes the main tool in information extraction, information integration, and information retrieval, etc. [14].

Although the Bayes modeling is becoming of practical use, there still remains one difficulty in the modeling. Namely, there is no established methodology to derive appropriate class of models for particular problem. Therefore, the researcher's art is still demanded in the most important part of statistical modeling, i.e., the presentation of a good family of models. The raison d'etre of the researchers, in particular of the statisticians in a cyber world can be found here.

3 Time Series Modeling

Hereinafter, we restrict our attention to time series and consider the data centric approach in time series analysis. Here the state-space model plays an important role as a unified tool.

3.1 State-Space Model and the Kalman Filter

Consider the linear Gaussian state-space model for the time series y_n,

$$x_n = F_n x_{n-1} + G_n v_n, \tag{1}$$

$$y_n = H_n x_n + w_n, \tag{2}$$

where x_n is an unknown state vector, v_n and w_n are the system noise and the observation noise with $v_n \sim N(0, Q_n)$ and $w_n \sim N(0, R_n)$, respectively. The initial state x_0 is assumed to be distributed as $x_0 \sim N(x_{0|0}, V_{0|0})$. Equations (1) and (2) are called the system model and the observation model, respectively. The state-space model has been used in the modeling of various types of time series, such as the nonstationary time series in the mean, the variance or the covariance [5, 11].

The set of information from the observations up to time j is denoted by Y_j, namely, $Y_j \equiv \{y_1, \ldots, y_j\}$. The problem of state estimation is to evaluate $p(x_n|Y_j)$, the conditional density of x_n given the observations Y_j and the initial density $p(x_0|Y_0) = p(x_0)$. For $n > j, n = j$ and $n < j$, the problem is called the prediction, filtering and smoothing, respectively. The state estimation problems are important because many problems in time series analysis, such as increasing horizon prediction, interpolation, parameter estimation and signal extraction, can be solved using the estimated state vector.

It is well known that for linear-Gaussian state-space model, the conditional density $p(x_n|Y_m)$ also becomes Gaussian and the mean vector $x_{n|m}$ and the variance covariance matrices $V_{n|m}$ can be obtained by the Kalman filter and the fixed interval smoothing algorithms [3].

3.2 Smoothness Priors Modeling of Time Series

The state-space model provides a power tool for approaching to the so-called new NP problems, where the number of unknown parameters is equal to or larger than that of the observations. To exemplify the problem, consider the problem of fitting a curve $\{f_n\}$ to the data $\{y_n\}$

$$y_n = t_n + w_n, \quad n = 1, \ldots, N, \tag{3}$$

where w_n is the noise at time n. Here, if we assume a parametric function for f_n, then we can estimate it by the least squares method or the maximum likelihood method. However, in the modeling of massive data, parametric models are sometime too rigid and cannot extract useful information from data. In this situation, if we consider f_n as an unknown parameter, then we can get a very flexible method of expressing arbitrary function. However, in estimating the unknown parameters, we face to the simplest version of the new NP problem, since the number of unknown parameters f_n is the same as the number of observations.

Even in this situation, it is well-known that we can get a reasonable estimates of f_n by the penalized least squares method that minimizes

$$\sum_{n=1}^{N} (y_n - f_n)^2 + \lambda^2 \sum_{n=2}^{N} (\delta^k f_n)^2. \tag{4}$$

The crucial problem here was the selection of the trade-off parameter λ^2. Akaike [2] gave a solution from a Bayesian interpretation of the problem and proposed the ABIC criterion.

It is very interesting that, in time series context, the use of the penalized method is equivalent to assume the following two simple time series models

$$\begin{aligned} t_n &= t_{n-1} + v_n, \\ y_n &= t_n + w_n. \end{aligned} \tag{5}$$

We notice that these models are a special case of the state-space model and that the trade-off parameter λ^2 is naturally determined as the signal-to-noise ratio, i.e., $\lambda^2 = \sigma^2/\tau^2$. This means the so-called trend estimation problem can be easily solved by the state-space model. Further, since the above model is the simplest case of the state-space model, it suggests that by using the state-space model, we can perform more sophisticated nonstationary time series modeling. Actually, for example, seasonal adjustment and various signal extraction problems can be performed by using the state-space model [11].

As an example of the use of the state-space model for time-varying structure modeling, we shall briefly show estimation of time-varying AR model and changing spectrum. Let y_n be nonstationary time series with time-varying spectrum. Since the spectrum of a stationary time series can be reasonably approximated by an AR model with constant coefficients, if the characteristics of the series changes with time, it is natural to consider an AR model with time-varying parameters

$$y_n = \sum_{j=1}^{m} a_{j,n} y_{n-j} + w_n, \quad w_n \sim N(0, \sigma_n^2), \tag{6}$$

where $a_{j,n}$ is an AR coefficient of order j at time n and σ_n^2 is the innovation variance at time n. Once the estimators of these parameters, $a_{j,n}$ and σ_n^2, are obtained from data, instantaneous spectrum of nonstationary time series is obtained by substituting the estimates to the formula of AR spectrum [7]. This time-varying AR model of order m has $n \times m$ AR coefficients for n observations and thus estimation of this model is a typical example of a new NP problem. The estimation can be realized by introducing a proper model for time evolution of the AR coefficients such as the k-th order random walk model for the AR coefficients with respect to time n,

$$\Delta^k a_{j,n} = v_{j,n}, \quad v_{j,n} \sim N(0, \tau^2). \tag{7}$$

Then the AR model (6) and the model for the evolution of the coefficients (7) can be combined into a state-space model form. Therefore, by the Kalman filter and the smoothing algorithm, we can estimate the time-varying coefficients $a_{j,n}$ [11].

4 General State-Space Modeling

Although the above mentioned state-space model is very useful in time series modeling, there are many situations where the linear-Gaussian state-space model is inadequate. For such situations, generalization of the state-space model has been proposed. Consider a nonlinear non-Gaussian state-space model for time series y_n,

$$x_n = F_n(x_{n-1}, v_n), \tag{8}$$

$$y_n = H_n(x_n, w_n), \tag{9}$$

where x_n is an unknown state vector, v_n and w_n are the system noise and the observation noise with densities $q_n(v)$ and $r_n(w)$, respectively. The initial state x_0 is assumed to be distributed according to the density $p_0(x)$. $F_n(x, v)$ and $H_n(x, w)$ are possibly nonlinear functions of the state and the noise inputs. Obviously, this model is a generalization of the linear-Gaussian state-space model considered in the previous section.

The above nonlinear non-Gaussian state-space model specifies the conditional density of the state given the previous state, $q(x_n|x_{n-1})$, and that of the observation given the state, $r(y_n|x_n)$. These are the essential features of the state-space model, and it is sometimes convenient to express the model in the general form based on the conditional distributions

$$x_n \sim Q_n(\,\cdot\,|x_{n-1}), \tag{10}$$

$$y_n \sim R_n(\,\cdot\,|x_n). \tag{11}$$

Note that, with this model, it is possible to treat the discrete process as well.

4.1 Non-Gaussian Filter and Smoother

For general state-space models, the conditional distributions become non-Gaussian and their distributions cannot be completely specified by the mean vectors and the variance covariance matrices. Therefore, in general, the Kalman filter cannot be applied to the general state-space model and various algorithms have been proposed so far for state estimation [3], e.g., the extended Kalman filter and the Gaussian-sum filter.

In particular, the following non-Gaussian filter and smoother [8] can yield an arbitrarily precise posterior densities.

[Non-Gaussian Filter]

$$p(x_n|Y_{n-1}) = \int p(x_n|x_{n-1})p(x_{n-1}|Y_{n-1})dx_{n-1},$$

$$p(x_n|Y_n) = \frac{p(y_n|x_n)p(x_n|Y_{n-1})}{p(y_n|Y_{n-1})}, \tag{12}$$

where $p(y_n|Y_{n-1}) = \int p(y_n|x_n)p(x_n|Y_{n-1})dx_n$.

[Non-Gaussian Smoother]

$$p(x_n|Y_N) = p(x_n|Y_n)\int \frac{p(x_{n+1}|x_n)p(x_{n+1}|Y_N)}{p(x_{n+1}|Y_n)}dx_{n+1}. \tag{13}$$

However, the direct implementation of these formulas require computationally very costly numerical integration and can be applied only to lower dimensional state-space models.

4.2 Monte Carlo Filter and Smoother

To mitigate the computational burden, Monte Carlo (particle) methods have been developed. In the Monte Carlo filtering [4,9], we approximate each density function by many particles which can be considered as realizations from that distribution. Specifically, assume that each distribution is expressed by using m ($m = 10{,}000$, say) particles as $\left\{p_n^{(1)}, \ldots, p_n^{(m)}\right\} \sim p(x_n|Y_{n-1})$, $\left\{f_n^{(1)}, \ldots, f_n^{(m)}\right\} \sim p(x_n|Y_n)$ and $\left\{s_{n|N}^{(1)}, \ldots, s_{n|N}^{(m)}\right\} \sim p(x_n|Y_N)$. Then it can be shown that a set of realizations expressing the one-step-ahead predictor $p(x_n|Y_{n-1})$ and the filter $p(x_n|Y_n)$ can be obtained recursively as follows ([4,9], Fig. 3):

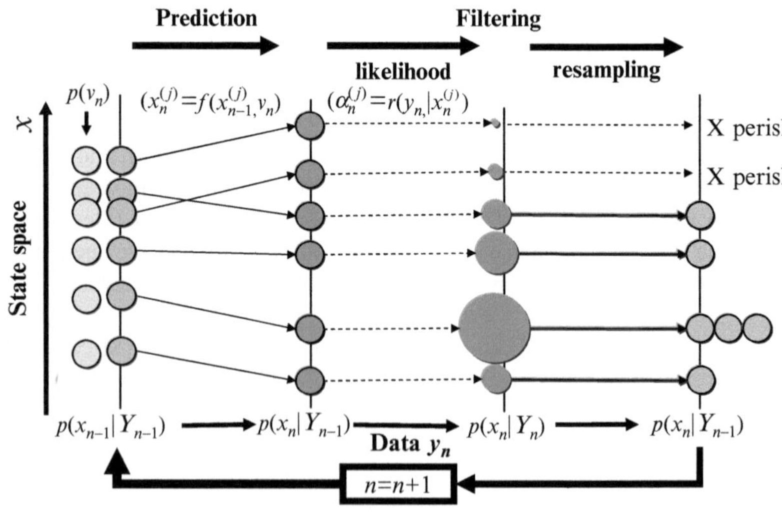

Fig. 3 One step of the Monte Carlo filter algorithm

[Monte Carlo Filter]

1. *Generate a random number* $f_0^{(j)} \sim p_0(x)$ *for* $j = 1, \ldots, m$.
2. *Repeat the following steps for* $n = 1, \ldots, N$:

   (a) *Generate a random number* $v_n^{(j)} \sim q(v)$ *for* $j = 1, \ldots, m$.
   (b) *Compute* $p_n^{(j)} = F\left(f_{n-1}^{(j)}, v_n^{(j)}\right)$, $j = 1, \ldots, m$.
   (c) *Compute* $\alpha_n^{(j)} = p\left(y_n | p_n^{(j)}\right)$ *for* $j = 1, \ldots, m$.
   (d) *Generate* $f_n^{(j)}$ *for* $j = 1, \ldots, m$ *by the resampling of* $p_n^{(1)}, \ldots, p_n^{(m)}$.

An algorithm for smoothing is obtained by replacing the Step 2(d) of the algorithm for filtering by

(d-L) *For fixed* L, *generate* $\left\{\left(s_{n-L|n}^{(j)}, \cdots, s_{n-1|n}^{(j)}, s_{n|n}^{(j)}\right)^T, j = 1, \ldots, m\right\}$ *by the resampling of* $\left\{\left(s_{n-L|n-1}^{(j)}, \cdots, s_{n-1|n-1}^{(j)}, p_n^{(j)}\right)^T, j = 1, \ldots, m\right\}$ *with* $f_n^{(j)} = s_{n|n}^{(j)}$.

This is equivalent to applying the L-lag fixed lag smoother. The increase of lag, L, will improve the accuracy of the $p(x_n|Y_{n+L})$ as an approximation to $p(x_n|Y_N)$, while it is very likely to decrease the accuracy of $\left\{s_{n|N}^{(1)}, \cdots, s_{n|N}^{(m)}\right\}$ as representatives of $p(x_n|Y_{n+L})$. Since $p(x_n|Y_{n+L})$ usually converges rather quickly to $p(x_n|Y_N)$, it is recommended to take L not so large, e.g., $L = 30$.

5 Applications of General State-Space Modeling

5.1 Automatic Change Point Detection

As a simple example of the general state-space modeling, we consider the trend estimation with a state-space model with non-Gaussian system noise [11]:

$$\begin{aligned} t_n &= t_{n-1} + v_n, \\ y_n &= t_n + w_n, \end{aligned} \tag{14}$$

where w_n is a Gaussian white noise $w_n \sim N(0, \sigma^2)$ but v_n is not necessarily Gaussian. Specifically, we consider the following two cases, $v_n \sim N(0, \tau^2)$ or $C(0, \tau^2)$. The parameters of the model, σ^2 and τ^2 can be estimated by the maximum likelihood method.

Figure 4 shows the marginal posterior distributions of the smoother with Gaussian noise (left) and Cauchy noise (right) models, respectively [11]. The estimates by the Gaussian model is wiggly and cannot clearly detect the jumps. On the other hand, by the Cauchy noise model, the jumps of the trend are clearly detected and

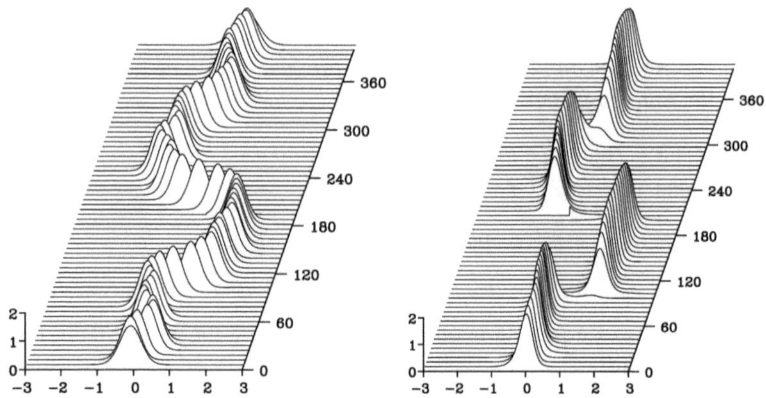

Fig. 4 Posterior distributions of the trend component. *Left*: Gaussian model, *Right*: Cauchy model

the estimated trend between the jumps is very smooth. This example shows that by properly selecting the noise distribution, the jump of the trend, and structural changes in general, can be detected automatically.

5.2 Nonlinear Filtering and Smoothing

The Monte Carlo filtering and smoothing algorithm can be used for filtering and smoothing for nonlinear state-space model. A test data was generated by the nonlinear state-space model [11]

$$\begin{aligned} x_n &= \frac{1}{2}x_{n-1} + \frac{25x_{n-1}}{x_{n-1}^2+1} + 8\cos(1.2n) + v_n, \\ y_n &= \frac{x_n^2}{10} + w_n, \end{aligned} \tag{15}$$

where y_n is the observation, x_n is unobservable state and $v_n \sim N(0,1)$, $w_n \sim N(0,10)$, $v_0 \sim N(0,5)$. Here, we consider the problem of estimating the unknown state x_n based on 100 observations, $y_n, n = 1, \dots, 100$. Because of the nonlinearity and the sinusoidal input in the system model in (15), the state x_n occasionally shift between the positive and negative regions. However, since in the observation model, the state x_n is squared, and the x_n^2 contaminated with an observation noise w_n is observed, it is quite difficult to discriminate between positivity and negativity of the true state.

It is well-known that the extended Kalman filter occasionally diverges for this model. Figure 5 shows the posterior distributions $p(x_n|Y_m)$ for $m = n-1, n, n+1$ and $n+2$, obtained by the Monte Carlo filter and smoother. Two cases for $n = 30$ and 43 are shown. It can be seen that by the nonlinear filtering, the predictive

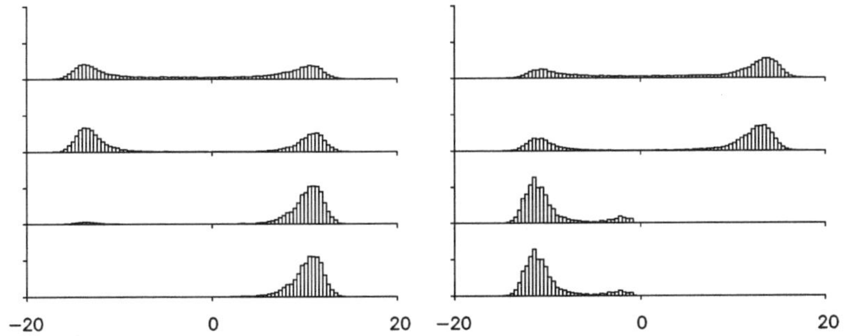

Fig. 5 Fixed point smoothing for $n = 30$ (*left*) $n = 43$ (*right*). *From top to bottom*: predictive distributions, filter distributions, 1-lag smoothers and 2-lag smoothers (from [12])

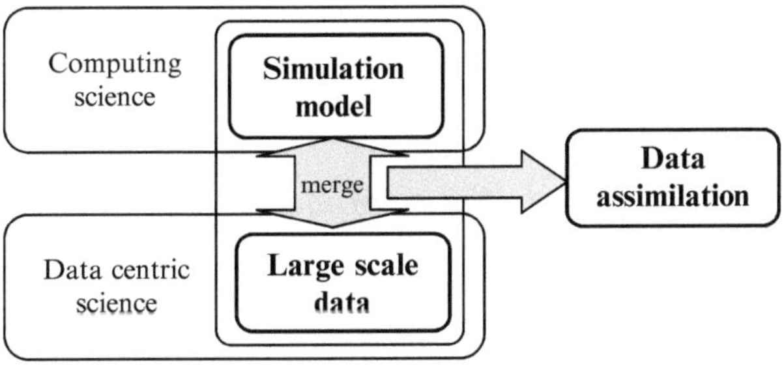

Fig. 6 Data assimilation

distributions and filter distributions are usually bimodal, and that as the observations increases in smoothing, the distributions converge to either positive or negative side. On the other hand, the extended Kalman filter approximates the distribution by a single Gaussian distribution, and thus cannot yield the sudden jump of the distributions.

5.3 Data Assimilation

Data assimilation emerged in the field of global simulation in meteorology and oceanography. In ordinary global simulation, "simulation model" is solved numerically under properly specified initial condition and boundary condition. However, in actuality, the model, the initial condition, the boundary condition and even the phenomena themselves contain various uncertainty. On the other hand, recently because of the development of ICT, huge amount of data are obtained from various sensors such as satellite. The data assimilation is a methodology to integrate simulation model and observed data. By the integration of two types of information, or two types of methodologies, we expect that we will be able to improve the model and the simulation (Fig. 6). The data assimilation is becoming popular not only in earth science, but also in the areas of space science and life science, etc.

Typically the data assimilation can be realized by using the general state-space model with nonlinear system model and linear observation model

$$\begin{aligned} x_n &= F(x_{n-1}, v_n), \\ y_n &= H x_n + w_n. \end{aligned} \tag{16}$$

Therefore, the data assimilation problem can be handled within the framework of the general state-space model. Significant difference from other problems is the dimension of the state and the observation. In a typical problem, the dimensions of the state x_n and the observation y_n are $10^4 \sim 10^6$ and $10^2 \sim 10^5$, respectively. So the computational difficulty occurs in the application of the filter and the smoother and various modified algorithms such as the ensemble Kalman filter and the merging particle filter are developed [14].

5.4 *Self-Organizing State-Space Modeling*

In this subsection, we consider a method of simultaneous estimation of the state x_n and the unknown parameter θ of the state-space model. To realize the simultaneous estimation, we define an augmented state vector as

$$z_n = \begin{bmatrix} x_n \\ \theta \end{bmatrix}. \tag{17}$$

Then the state-space model for this augmented state vector z_n is given by

$$\begin{aligned} z_n &= F^*(z_{n-1}, v_n), \\ y_n &= H^*(z_n, w_n), \end{aligned} \tag{18}$$

where the nonlinear functions $F^*(z, v)$ and $H^*(z, w)$ are defined by $F^*(z, v) = [F(x, v), \theta]^T$, $H^*(z, w) = H(x, w)$.

Assume that we obtain the posterior distribution $p(z_n|Y_N)$ given the entire observations $Y_N = \{y_1, \cdots, y_N\}$. Since the original state vector x_n and the parameter vector θ are included in the augmented state vector z_n, it immediately yields the marginal posterior densities of the parameter and of the original state. It was a common understanding that "Although this extended Kalman filter approach appears perfectly straightforward, experiences has shown that with the usual state-space model, it does not work well in practice" [3, p. 284]. However, by using precise nonlinear filter and smoother, we can actually perform simultaneous estimation of the state and the parameters [10].

This method of Bayesian simultaneous estimation of the parameter of the state-space model can be easily extended to a time-varying parameter situation where the parameter $\theta = \theta_n$ evolves with time n. Also, if the self-organizing state-space model is implemented by the Monte Carlo filter, the original formulation of the self-organizing state-space model without system noise for the parameter does not work well in practice. In this case, inclusion of the system noise enables application of the Monte Carlo filter/smoother [10].

5.5 Semi-Markov Switching Model

In the analysis of economic or financial time series, we sometimes notice the trend switches from upward to downward and vice versa. Obviously for such series, the ordinary stochastic trend model such as the random walk model is inappropriate. Markov switching model has been investigated extensively for such cases [6]. However, in many cases, once the switching occurs it usually stay in the regime for a while. This phenomenon cannot be properly expressed by a simple Markov chain. The semi-Markov switching slope model is developed to cope with this problem.

Consider a model for time series y_n, $y_n = t_n + w_n$, where t_n is an unknown trend component and w_n is a Gaussian observation noise with mean 0 and the variance σ^2. We assume that the trend component can be expressed as $t_n = t_{n-1} + \Delta t_n$. The distribution of the slope Δt_n $(= t_n - t_{n-1})$ depends on its previous value Δt_{n-1} and a bivariate Markov chain S_n. In the semi-Markov switching model, the switching (transition) occurs according to a probability distribution P_k. Here k is the sojourn (duration) time in the new regime. If K is the realization of the distribution P_k at time n, then we have $S_n = S_{n+1} = \cdots = S_{n+K}$ and

$$p(S_{n+k} = i | S_{n+k-1} = \cdots = S_n = j) = p_{ij}. \tag{19}$$

The semi-Markov chain does not possess the Markovian property but the semi-Markov switching model can be expressed in general state-space model form by defining the state vector as $x_n = (t_n, \Delta t_n, S_n, R_n)^T$, where R_n denotes the time that the semi-Markov chain has stayed in the current regime since the previous switching. The transition of R_n is given by

$$R_n = \begin{cases} 0 & \text{with probability } \beta^{-1} P_{R_{n-1}+1}, \\ R_{n-1} + 1 & \text{with probability } 1 - \beta^{-1} P_{R_{n-1}+1}, \end{cases} \tag{20}$$

where $\beta = \sum_{j=R_{n-1}+1}^{\infty} P_j$.

As a parametric model for the sojourn time probability, we may, for example, use the negative binomial distribution (Fig. 7)

$$P_{k+\ell}(\ell, p) = \binom{k + \ell - 1}{k} p^k (1 - p)^{\ell}. \tag{21}$$

Fig. 7 Distribution of switching time by the Markov model and semi-Markov model

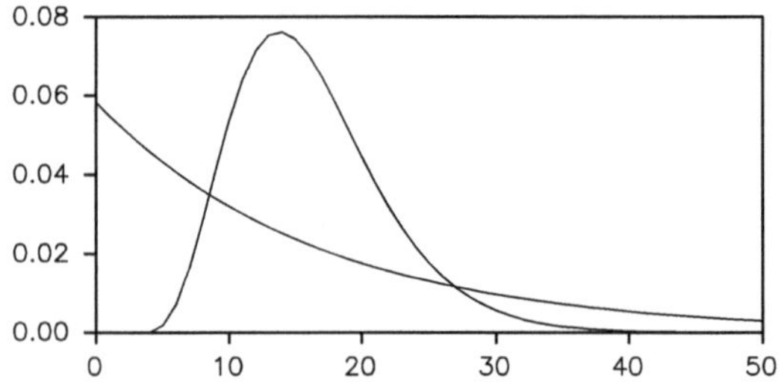

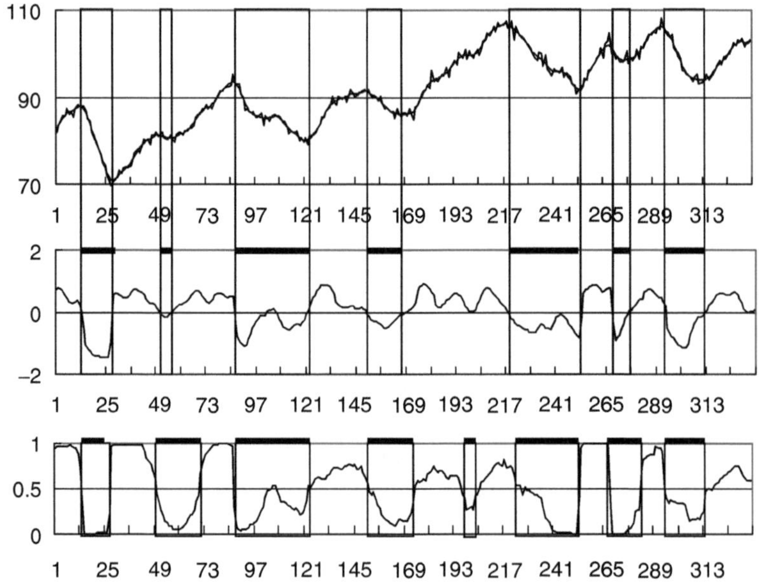

Fig. 8 Regime switching estimated by the semi-Markov switching model: *from top to bottom*, observed IIP data, estimated slope and the posterior probability of the upward and downward regimes

Under these assumptions, we can define the joint conditional distribution of the state x_n given x_{n-1}, and therefore define a general state-space model.

Note that $P_k(\ell, p) = 0$ for $k = 0, \ldots, \ell$. In the ordinary Markov switching model, the probability of sojourn time k is given by $p_{jj}^k(1 - p_{jj})$ if $S_0 = j$. Therefore the probability of sojourn time is a monotone decreasing function of time k. On the other hand, that of the semi-Markov process with the negative binomial distribution attains its maximum at $k = P_0^{-1}(\ell - 1)$.

Figure 8 shows the original IIP (index of industrial production) series and the posterior mean of the trend, $E[t_{n|N}]$, (top plot), the posterior mean of the slope, $E[\Delta t_{n|N}]$, (middle plot), and the posterior mean of the semi-Markov chain, $E[S_{n|N}]$, respectively. Vertical lines show the switching times detected from the estimated slope and the mean of the semi-Markov chain, respectively. The gray lines indicate the detected depression periods.

References

1. Akaike H (1973) Information theory and an extension of the maximum likelihood principle. In: Petrov BN, Csaki F (eds) Proc. 2nd international symposium on information theory. Akademiai Kiado, Budapest, 267–281
2. Akaike H (1980) Likelihood and the Bayes procedure. In: Bernardo JM, DeGroot MH, Lindley DV, Smith AFM (eds) Bayesian statistics. University Press, Valencia, 143–166
3. Anderson BDO, Moore JB (1979) Optimal filtering. Prentice-Hall, New Jersey
4. Doucet A, Freitas F, Gordon N (2001) Sequential Monte Carlo methods in practice. Springer, New York
5. Harrison PJ, Stevens CF (1976) Bayesian forecasting. J R Stat Soc B 38:205–247
6. Kim CJ, Nelson CR (1999) State-space models with regime switching. MIT Press, Cambridge, MA
7. Kitagawa G (1983) Changing spectrum estimation. J Sound Vibration 89(3):433–445
8. Kitagawa G (1987) Non-Gaussian state-space modeling of nonstationary time series (with discussion). J Am Stat Assoc 82:1032–1063
9. Kitagawa G (1996) Monte Carlo filter and smoother for non-Gaussian nonlinear state space model. J Comput Graph Stat 5:1–25
10. Kitagawa G (1998) Self-organizing state space model. J Am Stat Assoc 93:1203–1215
11. Kitagawa G, Gersch W (1996) Smoothness priors analysis of time series. Springer, New York
12. Kitagawa G, Sato S (2001) Monte Carlo smoothing and self-organizing state-space model. In: Doucet A, de Freitas N, Gordon N (eds) Sequential Monte Carlo methods in practice. Springer, New York
13. Konishi S, Kitagawa G (2008) Information criteria and statistical modeling. Springer, New York
14. Nakano S, Ueno G, Higuchi T (2007) Merging particle filter for sequential data assimilation. Nonlinear Process Geophys 14:395–408
15. Tsubaki H (2002) Statistical science aspects of business. Proc Japan Soc Appl Sci 16:26–30 (in Japanese)

Symbolic Shadowing and the Computation of Entropy for Observed Time Series

Diana A. Mendes, Vivaldo M. Mendes, Nuno Ferreira, and Rui Menezes

Abstract Order, disorder and recurrence are common features observed in complex time series that can be encountered in many fields, like finance, economics, biology and physiology. These phenomena can be modelled by chaotic dynamical systems and one way to undertake a rigorous analysis is via symbolic dynamics, a mathematical-statistical technique that allows the detection of the underlying topological and metrical structures in the time series. Symbolic dynamics is a powerful tool initially developed for the investigation of discrete dynamical systems. The main idea consists in constructing a partition, that is, a finite collection of disjoint subsets whose union is the state space. By identifying each subset with a distinct symbol, we obtain sequences of symbols that correspond to each trajectory of the original system. One of the major problems in defining a "good" symbolic description of the corresponding time series is to obtain a generating partition, that is, the assignment of symbolic sequences to trajectories that is unique, up to a set of measure zero. Unfortunately, this is not a trivial task, and, moreover, for observed time series the notion of a generating partition is no longer well defined in the presence of noise. In this paper we apply symbolic shadowing, a deterministic algorithm using tessellations, in order to estimate a generating partition for a financial time series (PSI20) and consequently to compute its entropy. This algorithm allows producing partitions such that the symbolic sequences uniquely encode all periodic points up to some order. We compare these results with those obtained by considering the Pesin's identity, that is, the metric entropy is equal to the sum of positive Lyapunov exponents. To obtain the Lyapunov exponents, we reconstruct the state space of the PSI20 data by applying an embedding process and estimate them by using the Wolf et al. algorithm.

D.A. Mendes (✉), N. Ferreira, and R. Menezes
Department of Quantitative Methods, ISCTE-IUL and UNIDE, Avenida Forças Armadas, 1649-026 Lisbon, Portugal
e-mail: diana.mendes@iscte.pt; nuno.ferreira@iscte.pt; rui.menezes@iscte.pt

V.M. Mendes
Department of Economics, ISCTE-IUL and UNIDE, Lisbon, Portugal
e-mail: vivaldo.mendes@iscte.pt

M. Takayasu et al. (eds.), *Econophysics Approaches to Large-Scale Business Data and Financial Crisis*, DOI 10.1007/978-4-431-53853-0_12, © Springer 2010

1 Introduction

There are several phenomena in the surrounding world that simply we cannot explain, or then we are able explain only parts of its observed behavior. This is because nature is neither regular nor easily predictable due to its extreme complexity.

The basic idea of classical complexity was to differentiate between regular, chaotic and stochastic behavior. Nowadays, there are several techniques and algorithms to distinguish between the main characteristics of these dynamic processes. Chaos is a technical word which intends to describe an extremely complex and irregular structure in time and space, in a fully deterministic dynamic framework. The complexity of the dynamics in a chaotic model arises from within the very internal structure we are modelling. On the contrary, in a stochastic process, its dynamics and the associated potential complexity has nothing to do with its internal structure but with the external random part that is added to the model, and therefore the deterministic component of such models has an irrelevant role to play in the underlying dynamics.

There are few similarities between these two kind of processes and the most obvious and tricky one is to be found in the geometry of the time series associated with each process. However, by placing stochastic and deterministic processes under the same analytical framework, the researchers have established a connection between the two scientific areas, which were usually regarded as mutually exclusive and, as a consequence, they may obtain more powerful models to analyze reality.

In the fields of finance and economics, there is widespread evidence that nonlinear dynamics should play a significant role in generating the data that we are able to observe. For example, why there are systematic financial bubbles despite the lack of any credible big shock that would explain those bubbles in a fully stochastic framework? Or as Cochrane [9] argues in a classic paper "we haven't found large, identifiable, exogenous shocks to account for the bulk of output fluctuations" (p. 296).

In the last years, the application of tools from nonlinear analysis, in particular from chaos theory, to the study of complex economic and financial time series, seems to be quite relevant and has generated extensive research programs (see [1,6,7,23,30,34,37–39], among others). In particular, chaotic dynamics can provide appealing techniques for the analysis of financial time series, not only because both systems exhibit irregular and apparently random behavior, but also because financial data is largely available even in extremely high frequencies. However, before examining the data with the purpose of searching for deterministic nonlinear dynamics, one should note that while chaotic systems are typically low dimensional, financial markets may be characterized as high dimensional structures. With the main aim focused on understanding the complexity associated with financial markets, it is important to carefully analyze the basic structures of the data and to have significant contextual knowledge of the evolution of the markets over time, in order to take into account possible complementary information (e.g., a major shock that has occurred in a particular point in time).

It is well known that topological and metric invariants like entropy, Lyapunov exponents, information and correlation dimensions, are fundamental in the local and global characterization of the behavior of a nonlinear dynamical system. Many of these invariants are related to each other. For example, the Lyapunov exponents are related to the Kolmogorov–Sinai entropy via the Pesin identity, and the information dimension and Lyapunov exponents seem to be related (at least under some conditions) through the Kaplan–Yorke conjecture [25]. On the other hand, the entropy which is considered one of the most powerful invariant was already successfully applied to quantify information and uncertainty in financial data (see, e.g. [11,12,31]).

The relationship between Shannon and Renyi entropies and the generalized correlation integral has also been pointed out by a number of authors [28,36]. It was demonstrated that, by using this relationship, more accurate estimates of many information theoretic statistics can be obtained, as compared to conventional box-counting methods.

Time-delay embedding theorem [40] is a remarkable and very useful result, that can be applied to time series, in order to obtain information about the underlaying dynamics and to estimate some invariants. The former works very well when data has a significant deterministic component. When data is noisy these measures and invariants are often difficult to compute. The method of surrogate data [24,38] has been proposed as a sanity check on these estimates. Surrogate data may be applied to confirm that the estimates of the dynamic invariants obtained from the data are distinct from what one would obtain from linear noise processes. By doing this, we can be sure that the data is not linear noise, and also that the obtained value for the correlation dimension (for example) is an accurate measure of the deterministic nonlinearity exhibited by the data.

More recently, the techniques and ideas of symbolic dynamics [29] have also found significant applications. The theory of symbolic dynamics arose as an attempt to study systems by means of discretization of space and time. The basic idea is to divide the set of possible states into a number of pieces. Each piece is associated with a symbol, and in this way the evolution of the system is described by a sequence of symbols. This leads to a symbolic dynamical system that helps us to understand the dynamical behavior of the original system.

The paper of Milnor and Thurston [33] sets up an effective method for describing the qualitative behavior of the successive iterates of a piecewise monotonic map and has been applied with significant success to the study of one-dimensional difference equations. A properly constructed symbolic dynamics, being a coarse-grained description, provides a powerful tool to capture global and topological aspects of the dynamics. The success of symbolic dynamics depends on how the coarse-graining is performed on the partition of the phase space and significant efforts were concentrated on the question of choosing an adequate partition [4,13,17,21,26].

When we already have a generating partition and in consequence a trustful symbolic dynamics, by using the ordering of the symbolic sequences, we can easily locate all allowed periodic orbits up to a certain order. Over the last few years, several papers have been devoted to provide some generalization of these techniques to two-dimensional dynamics with some success [10,17,32]. This extension from one-dimensional to two-dimensional maps is by no means trivial, and there are several open problems related to this subject. Nowadays, symbolic dynamics represents a rapidly growing and essential part of dynamical systems theory, and can be also encountered in nonlinear data analysis, in particular reflected in several applications in scientific areas, such as, physiology, engineering, economy and biology (see [3, 22, 23, 37, 39, 41], among others).

The concept of shadowing [27, 35] is of great importance when one wants to verify, in a rigorous mathematical sense, the existence of complicated behavior detected via numerical experiments. The shadowing problem is that of finding a deterministic orbit as close as possible to a given noisy orbit. Farmer and Sidorowich [15] presented an optimal solution to this problem in the sense of least-mean-squares, which also provides an effective and convenient numerical method for noise reduction for data generated by a dynamical system. More recently, Hirata et al. [21] have combined symbolic dynamics and shadowing in order to obtain a very powerful algorithm to detect a generating partition from a time series.

In the remain of the paper we study the Portuguese Stock Index (PSI20) data with the purpose of detecting whether determinism can be accepted as a fundamental ingredient of the time series. We begin by performing some descriptive statistics of the data and then we embed the scalar time series in order to obtain the attractor of the underlying dynamics. We employ the Hirata et al. [21] algorithm to search for a good partition of the reconstructed phase space to be used when defining the symbolic dynamics for the data. Finally, we numerically compute the entropy, as the sum of the Lyapunov exponents and as defined by the frequency of patterns in the symbolic sequences.

2 Symbolic Dynamics and Shadowing

In what follows, suppose that we observe a noisy time series $\{y_t\}$, $t = 1, 2, \ldots, N$, that can be decomposed as

$$y_t = x_t + n_t,$$

where x_t is the desired (or deterministic) part of the series and n_t is the noise. The main objective is to get rid off the noise; that is, instead of considering the noisy signal y_t, we wish simply to highlight the pure signal x_t. Furthermore, we assume that the time series $\{x_t\}$ is a trajectory of a known (or unknown) invertible dynamical system f, i.e.,

$$x_{t+1} = f(x_t),$$

and the noise $\{n_t\}$ might also be a trajectory of a different dynamical system g. We have now that the noisy orbit y_t evolves according to a mixture of known (or unknown) deterministic dynamics and noise, with a dynamic equation defined by

$$y_t = f(x_t) + \sigma n_t.$$

In this case we call the noise as dynamical noise, since it is coupled to the dynamics. The noise due to measurement errors that are independent of the dynamics are called observational noise. We intend to find a deterministic orbit $x_{t+1} = f(x_t)$ that stays close to y_t, or in other words, a deterministic orbit that shadows the noisy orbit.

The problem of shadowing and noise reduction are closely related. On one hand, shadowing is not a noise reduction problem *de per se*, since the shadowing orbit x is an artificial construction. On the other hand, if once we have found the former we can trivially write

$$y_t = x_t + \tilde{n}_t,$$

where $\tilde{n}_t = y_t - x_t$ is the effective noise. Conversely, if we intend that observational noise is dynamical noise, then the true orbit x is automatically a shadowing orbit.

The shadowing property of dynamical orbits (periodic and chaotic) was investigated by many authors (see [15, 19, 20, 27, 35] and references therein) but was initially proved by Anosov [2] and Bowen [5]. Assume that f is everywhere hyperbolic, and the noise is bounded, i.e., $\|n_t\| < \varepsilon$, where $\varepsilon > 0$. Anosov and Bowen demonstrated that for every $\delta > 0$ there is an ε such that every noisy orbit y_t of f is shadowed by an orbit x_t with $\|y_t - x_t\| < \delta$ for all t. The Anosov–Bowen construction depends on the fact that stable and unstable manifolds are never parallel where they intersect. The distance between two trajectory segments x_t and y_t is defined as

$$D(x, y) = \sqrt{\frac{1}{N} \sum_{t=1}^{N} \|y_t - x_t\|^2},$$

where $\|y_t - x_t\|$ is the Euclidean distance between the vectors x_t and y_t.

Recently, some generic results about of the periodic orbital shadowing property in the space of discrete dynamical systems of a compact topological manifold of dimension at least 2 have been presented [27], and so we may give a valuable guarantee for the applicability of this method to the embedded data. The shadowing property is of great importance to the problem of verifying, in a rigorous mathematical sense, the existence of complicated behavior detected via numerical experiments.

In most practical applications, f is unknown and must be learned directly from the data. We begin by embedding the data in a state space and then attempting to approximate the graph of a function that maps present states into future states. We use an iterative procedure: first we approximate f and apply the noise reduction algorithm, and then approximate f again using the smoothed data, repeating this until the results stop to improve. If we use a sufficiently large approximation neighborhood, the learned representation of f averages over the behavior of neighboring points, providing an initial noise reduction. Application of the noise reduction algorithm averages together points from other regions as well, including those from distant parts of the state space, thus amplifying the initial noise reduction (for more technical details see for example [15]). The ability to reduce noise is ultimately limited by the accuracy of the approximation to the true dynamics of f. Nowadays, performing software and trustful algorithms in discrete dynamical system are mainly based on the shadowing property, when searching for orbits in models and data.

The theory of symbolic dynamics [29] is a powerful tool for the investigation of discrete dynamical systems and time series, more recently see e.g. [3, 4, 6, 41]. Symbolic time series analysis involves the transformation of the original data into a sequence of discretized symbols that are processed to extract useful information about the state of the system that generates the process. The application of symbolic time analysis requires a partition of the signal, that is, a finite collection $\mathcal{P}$ of disjoint subsets whose union is the state space. The choice of a partition is the starting point to symbolize the original time series data. If we identify each $A \in \mathcal{P}$ with a unique symbol, then, we have a sequence of symbols that correspond to each trajectory of the original system – the sequence is produced as the evolving state visits different regions of the partition. This idea is extremely powerful when the partition is chosen to be a generating partition, that is, when the assignment of symbol sequences to trajectories is unique, up to a set of measure zero.

As might be expected, to find a generating partitions is a very difficult task. When the dynamical system is defined by a known map, there are methods to find generating partitions by using primary tangencies and also by considering the unstable periodic orbits [10, 13, 17, 18, 25]. In cases where the dynamical system is not fully known, for example, in cases where only a time series of observations is available, there appears to be no general methods for obtaining a generating partition. A sub-optimal partition induces improper projections or degeneracies [4], where a single observed symbolic orbit may correspond to more than one topologically distinct state space orbit. This leads to finding the wrong topological entropy, because some distinct transitions are improperly merged. In short, under these circumstances there is no satisfactory general theory saying how to find a generating partition, except for one dimensional maps where the partition is defined by the critical points (minima, maxima, or discontinuities).

However, Kennel and Buhl [26] have recently proposed a method for finding a generating partition from a time series which uses symbolic false nearest neighbors to localize the region specified by a finite block of symbols and Hirata et al. [21] presented a method for estimating a generating partition from a time series were they approximate the generating partition by tessellations of state space and use certain fundamental properties of generating partitions observed by Eckmann and Ruelle [13].

In what follows, we describe the symbolic shadowing process presented by Hirata et al. [21]. Let $f : X \to X$ be a continuous map defined in $X \subset \mathbb{R}^n$. For any initial condition $x_0 \in X$, the iterations of the map $f(x_0)$ give way to a unique discrete trajectory

$$\ldots, x_{-1}, x_0, x_1, \ldots \in X,$$

satisfying $x_{t+1} = f(x_t)$. We suppose that $\mathcal{P}$ is a partition of the phase space X and $\mathcal{A}$ is the alphabet associated with the partition $\mathcal{P}$, where $\mathcal{A}$ consists of n symbols if the partition has n subsets. To any numerical trajectory (time series) $\ldots, x_{-1}, x_0, x_1, \ldots \in X$ we can associate a symbolic sequence $\ldots s_{-1}s_0.s_1s_2\ldots$, where $s_t = \phi(x_t)$ and $\phi : X \to \mathcal{A}$. Moreover, since the initial point determines the symbolic sequence, we can define another map $\Phi : X \to \mathcal{A}^Z$, which assigns to x_0 the symbolic sequence $\Phi(x_0) = \ldots s_{-1}s_0.s_1s_2\ldots$. We define the fullshift

$$\mathcal{A}^{Z} = \Sigma_n = \{\ldots s_{-1}s_0.s_1s_2\ldots \text{ where } s_i \in \mathcal{A}\}$$

to be the set of all possible infinite symbolic strings of symbols from the alphabet $\mathcal{A}$. Then, any given infinite symbolic sequence is a singleton in the fullshift space. The map $\sigma : \mathcal{A}^{Z} \to \mathcal{A}^{Z}$ is called the shift map and is defined by

$$\sigma(\ldots s_{-1}s_0.s_1s_2\ldots) = \ldots s_{-1}s_0s_1.s_2\ldots.$$

In general, not all symbolic sequences correspond to the trajectory of an initial condition x_0. Restricting the shift map to a subset of Σ_n consisting of all the itineraries that are realizable, yields the subshift $\Sigma \subset \Sigma_n$. It can be observed that $\Phi(f(x_0)) = \sigma(\Phi(x_0))$, and we say that $\mathcal{P}$ is a generating partition if Φ is one-to-one, that is, the original dynamics and the symbolic dynamics are conjugate. So, the key is to choose a partition where neighbors in the symbolic representation are neighbors in continuous state space. To do this we need to define what we mean by neighbors in the symbolic space and then optimize the partition to satisfy this criterion as best as possible.

The Hirata et al. [21] algorithm defined to find a generating partition is presented below (for further details we refer to the same paper):

(1) Prepare an initial partition and find unstable periodic points from a time series. Assign to each unstable periodic point a substring of length l of type $S_{[-m,n]} = S_{-m}S_{-m+1}\ldots S_{-1}.S_0S_1\ldots S_n$, $m = [l/2]$ and $n = [(l-1)/2]$, over alphabet $\mathcal{A}$ so that the unstable periodic points are encoded uniquely. Let each unstable periodic point be the representative[1] $r_{S_{[-m,n]}}$ of the substring $S_{[-m,n]} \in \mathcal{A}^l$.

(2) For each observed point x_t, find its closest representative $r_{S_{-m}S_{-m+1}\ldots S_{-1}\cdot S_0\ldots S_n}$. Then make s_t to be S_0.

(3) Classify x_t depending on its substring $s_{[t-m,t+n]}$, and let

$$C_S = \left\{x_t : s_{[t-m,t+n]} = S,\ m+1 \le t \le N-n\right\}.$$

Set C_S as the set of points whose currently allocated substring is S.

(4) For each substring $S \in \mathcal{A}^l$, update its representative:

$$r_S = \sum_{y \in C_S} \frac{y}{|C_S|}.$$

---

[1] The representatives can be chosen so that

$$\sup_{x\in X} \left\| x - r_{\Phi_{[-k,k]}(x)} \right\| \to 0,\ k \to \infty,$$

where $\Phi_{[-k,k]}(x)$ is the set of all possible subsequences of length $2k+1$ and Φ is localizing, that is

$$\sup_z diam\left(\Phi^{-1}_{[-k,k]}(z)\right) \to 0,\ k \to \infty.$$

Following Eckmann and Ruelle [13], localizing is a sufficient condition for $\mathcal{P}$ being a generating partition.

(5) Return to step (2) until the set of representatives and the symbolic sequence are no longer subject to change, or until they cycle.

(6) Increase the length of the substrings by $l \to l + 1$, $m \to [l/2]$ and $n \to [(l-1)/2]$. Return to step (3) until a stopping criterion is achieved.[2]

The partition is approximated by tessellating state space with representatives, that is, points in state space, each of which has a distinct substring of a certain length. Using this scheme, Hirata el al. [21] stated the problem of finding a generating partition as finding the minimum discrepancy between a series of points in the data and one specified by a symbol sequence and representatives. By solving this minimization problem approximately, using an iterative algorithm, an estimate for a generating partition can be found.

If the dimension of embedding is high, it is advisable to keep some doubts about the geometry of the generating partition. In this case, and in other inconclusive cases, we can symbolize the data directly from the original time series, choosing the partition in the state space of the given signal by trial and error. Several approaches have been used to symbolize the time series (see [6, 41]), and the most common approach is to divide the time series in two bins and assign the value 1 or 0 to each bin according to its occurrence.

Given a time series x_t the approach to transform the time series into symbols may be:

$$s_t = \begin{cases} 1 & \text{if } x_t \geq \text{threshold}, \\ 0 & \text{if } x_t < \text{threshold}, \end{cases}$$

or may be defined like

$$s_t = \begin{cases} 1 & \text{if } |x_{t+1} - x_t| \geq \text{threshold}, \\ 0 & \text{if } |x_{t+1} - x_t| < \text{threshold}, \end{cases}$$

or by many other ways, depending on data, partition and context (see, e.g. [41] and [3]).

After symbolization, the next step is the construction of symbolic strings from the series by collecting groups of symbols together in temporal order. This process typically involves a definition of a finite-length template that can be moved along symbolic series one step at a time, each step revealing a new sequence. If each possible sequence is represented in terms of a unique identifier, the end result will be a new time series often referred to as a code series.

[2] The algorithm can be stopped when the length of substrings reaches a certain length, or when the discrepancy is sufficiently small.

The discrepancy is defined by

$$\sum_{i=m+1}^{N-n} \left\| x_i - r_{s_{[i-m,i+n]}} \right\|^2 / (N - m - n).$$

However, for the analysis of financial time series a more pragmatic approach is required. A main problem is to find an appropriate transformation into symbols. It has to be chosen in context-dependent and in this way some information about the system is lost, however, the coarse graining of the behavior can be analyzed. For this reason, some measures of complexity can be developed n the basis of such context-dependent transformations, which have a close association to economic phenomena and are relatively easy to interpret.

Symbolic sequence analysis depends on quantitative measures of symbolic string frequencies. These measures can be based on information theory. The information theoretic measures of the symbolic sequence analysis include Shannon and generalized Renyi entropy of order q and are defined, respectively, as

$$H = -\sum p_i \log_2 p_i \quad \text{and} \quad H^q = -\frac{1}{1-q} \log_2 \sum p_i^q,$$

where p is the histogram of the symbolic sequence frequencies and $q \neq 1$ is a real number. Usually we can estimate information theoretic measures by using empirical probabilities, but these estimates are affected by random error in numbers and also by a systematic error or bias.

Let $x = \{x_i, i = 1, \ldots, N\}$ be a given time series that can be transformed into a symbolic sequence

$$s^\alpha = \{s_i^\alpha, i = 1, \ldots, N\}$$

having a fixed number of α values, labelled from 0 to $\alpha - 1$. By using quantization level 2 (symbols 0 and 1) the original time series can be transformed into a symbolic sequences defined by

$$s^\alpha = \begin{cases} 1 & |x_i - \bar{x}| \geq \theta, \\ 0 & |x_i - \overline{x}| < \theta, \end{cases}$$

where θ is the threshold and $\bar{x}$ is the mean value of the time series. The symbolic sequence can be divided to make word sequences (blocks) of length L of three or more symbols

$$s_{L,i}^\alpha = \left\{s_i^\alpha, s_{i-1}^\alpha, \ldots, s_{i-L+1}^\alpha\right\}.$$

Finally, the code series is generated as

$$w_i = \left\{s_i^\alpha \alpha^{L-1}, s_{i-1}^\alpha \alpha^{L-2}, \ldots, s_{i-L+1}^\alpha \alpha^0\right\}.$$

For a symbolic sequence of length L the number of all possible words is α^L, where α is the quantization level.

Shannon entropy of order L is defined by

$$SE(L, \alpha) = -\sum p\left(s_L^\alpha\right) \log_2 p\left(s_L^\alpha\right),$$

where $p\left(s_L^\alpha\right)$ is the probability of s_L^α being the pattern. The Corrected Shannon Entropy [14] can be obtained as

$$CSE\left(L,\alpha\right)=SE\left(L,\alpha\right)+\frac{C_R-1}{2M\ln 2},$$

where M is the total number of words and C_R is the number of occurring words among all possible words. The value of Corrected Shannon Entropy will be maximum for a certain word length L and quantization level α when all M words occur with a uniform distribution in a data series. Hence, it will be given by

$$CSE^{\max}\left(L,\alpha\right)=-\log_2\left(\frac{1}{M}\right)+\frac{M-1}{2M\ln 2}.$$

Since it is not possible to compare two values of the Corrected Shannon Entropy for two different word length at same threshold level, one can introduce the Normalized Corrected Shannon Entropy for such purpose, that is:

$$NCSE\left(L,\alpha\right)=\frac{CSE\left(L,\alpha\right)}{\left(-\log_2\left(\frac{1}{M}\right)+\left(\frac{M-1}{2M\ln 2}\right)\right)}.$$

The value of Normalized Corrected Shannon Entropy will vary between 0 to 1 for any word length and quantization level.

The word length L is important in unfolding the dynamics of the system in the underlying time series. By quantizing the same time series and dividing the former into a sequence of symbols of a certain word length, one can get a coarse grained description of the system. Using word length L is similar to embed the time series in L dimensions. As the word length is increased, one can obtain a larger degree of complexity based on the larger size of patterns in the time series. In order to get the best estimate of complexity, the size of word length should be optimized. But the selection of the optimal word length is constrained by the number of data points available in the data, and, as a consequence, the size of word length L should be much smaller than the total number of data points N in the time series.

3 Data

The time series considered in this paper is given by the Portuguese Stock Market Index (PSI20) and consist of 4,789 daily observations; 5 days per week (see Fig. 1), from February 1990 to December 2008. The data was taken from DataStream and it is clearly non-stationary, with high skewness and kurtosis, and non-normal distribution, as can be observed from Fig. 2. By considering the log-difference data, the time series became stationary and its shape is also illustrated in Fig. 1.

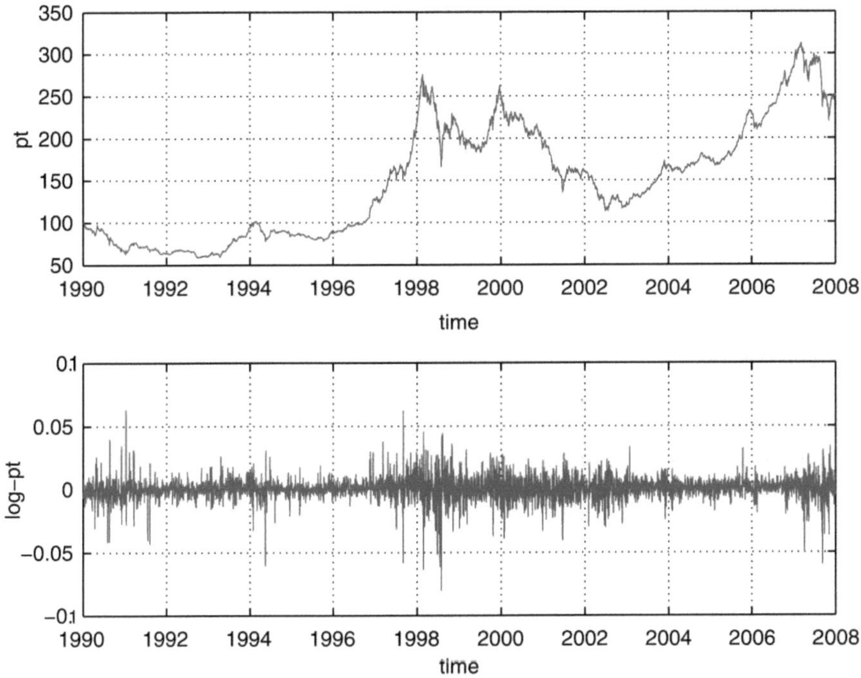

Fig. 1 The time series corresponding to the Portuguese Stock Market Index and to the log-difference data

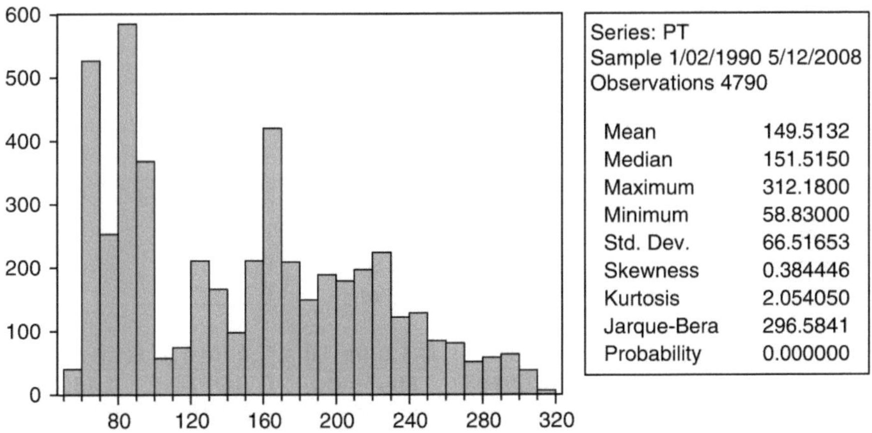

Fig. 2 Descriptive statistics of the Portugal Stock Market Index time series

The BDS test was first defined by Brock, Dechert and Scheinkman in 1987 [8] and is a powerful tool for detecting serial dependence in time series. The BDS statistics tests the null hypothesis of independent and identical distribution (i.i.d.) in the data against an unspecified departure from i.i.d. A rejection of the i.i.d. null hypothesis in the BDS test is consistent with some type of dependence in the data, which could result from a linear stochastic system, a nonlinear stochastic system, or a nonlinear deterministic system. Under the null hypothesis, the BDS test asymptotically converges to a standard normal distribution. The BDS statistics cannot test chaos directly, only nonlinearity, provided that any linear dependence has been removed from the data.

For our time series, the BDS statistics clearly shows that there is no linear dependence in the data (see Fig. 3). We applied the BDS test to our data for embedding dimensions of $m = 2, 3, 4, 5$ and 6. We use the quantiles from the small sample simulations reported by Brock et al. [8] as approximations to the finite-sample critical values of our BDS statistics. The i.i.d. null hypothesis is rejected in all cases for yield changes.

All testes and numerical analysis were initially done for the original non-stationary data and later extended to the log-difference data. We would like to stress that all the initial parameters and the algorithms implemented are the same for the original and for the log-difference data.

BDS Test for PT
Date: 02/24/09 Time: 15:12
Sample: 1/02/1990 5/12/2008
Included observations: 4790

Dimension	BDS Statistic	Std. Error	z-Statistic	Prob.
2	0.204947	0.000698	293.7345	0.0000
3	0.348908	0.001103	316.2462	0.0000
4	0.449787	0.001306	344.2839	0.0000
5	0.520359	0.001354	384.3905	0.0000
6	0.569610	0.001298	438.9545	0.0000

Raw epsilon	103.0320		
Pairs within epsilon	16144790	V-Statistic	0.703658
Triples within epsilon	5.71E+10	V-Statistic	0.519276

Dimension	C(m,n)	c(m,n)	C(1,n-(m-1))	c(1,n-(m-1))	c(1,n-(m-1))^k
2	8026628.	0.700107	8067551.	0.703676	0.495160
3	7992952.	0.697461	8065114.	0.703757	0.348553
4	7963700.	0.695198	8062685.	0.703839	0.245411
5	7937382.	0.693191	8060263.	0.703922	0.172831
6	7913042.	0.691354	8057816.	0.704003	0.121743

Fig. 3 BDS statistics for Portuguese Stock Market time series

4 Numerical Results

The study of a irregular signal measured experimentally represents in most cases a serious challenge to the analyst. These signals are more difficult to model than regular ones, and, if a prior model exists, we have to be very careful when comparing it to the data. Moreover, we should not expect that the output of the model should necessarily conform exactly with the observed data. Usually, for observed time series, instead of proceeding to a direct comparison, we should reconstruct the underlying dynamics from the scalar data, and later, compare the obtained output to the model if this is available.

The first step, before trying to find a partition, is to choose a proper embedding space for the time series. If the embedding dimension d_e is not correctly chosen, then a generating partition may not be found. For example, if d_e is chosen too low, then we will have a large number of false nearest neighbors in the state space (further details in [1]). These false neighbors are part of different regions of the attractor and thus should have completely different symbolic sequences. Therefore, the map defined from the state space to the symbolic space is no longer a homeomorphism and the methods for finding a good partitions will fail.

Let $\{x_t\}_{t=1,2,\ldots,N}$ be a time series of scalar observations. According to the Takens embedding theorem [40], we can determine an embedding dimension d_e and reconstruct the vector time series $\{z_t\}_{t=1,2,\ldots,N}$. of N points with

$$z_t = \left(x_t, x_{t-1}, x_{t-2}, \ldots, x_{t-(d_e-1)\tau}\right),$$

where τ is the time lag (delay). The time lag was introduced with the purpose of observing, for example, the noise level in the time series and to avoid the perturbations in the dynamics. In ideal conditions, it has been shown that for d_e and τ sufficiently large, the evolution of z_t is topologically equivalent to the underlying dynamic flow [1,40]. A flow itself is not topologically conjugate to a symbolic dynamics; rather it is some embedded map, such as a Poincaré map or a constant time-step map, that is topologically conjugate to a symbolic dynamics.

In order to capture the nonlinear correlations, and in particular the optimal time lag, τ, we use the first local minimum of the mutual information function [16] $I(T)$, which is defined by

$$I(T) = \sum P(x_t, x_{t+\tau}) \log 2 \frac{P(x_t, x_{t+\tau})}{P(x_t)P(x_{t+\tau})},$$

and where, $P(x_t, x_{t+\tau})$ is the probability of observing x_t and $x_{t+\tau}$, and $P(x_t)$ is the probability of observing x_t.

Further, we use the method of false nearest neighbors to obtain the embedding dimension. This method is based on the idea that the vectors can be very close if a small embedding dimension is considered. Thus, the embedding dimension is estimated as the lowest value of d_e for which the number of neighbors remains constant [1]. This criterion gives a correct value of d_e if the data follows deterministic chaotic dynamics, but if the time series correspond to a stochastic process then the size of the embedding can be very large unless the system presents low correlations.

We can also determine the minimum embedding dimension by measuring some invariants on the attractor. This alternative method consists in increasing the embedding dimension, d_e, used for the computation until we note that the value of the invariant stop changing.

Figures 4 and 5 illustrate the embedding process and the reconstructed attractor for the original data and for the stationary log-difference time series. The embedding dimension is given by the first kink in the graphic (Cao method), namely $d_e = 3$ in both cases (we also analyzed the case $d_e = 2$ and 4 for the two time series). The optimal time delay, estimated as the first local minimum of the mutual information function, is $\tau = 10$ for the first time series and $\tau = 7$ for the second. Therefore, one can easily conclude that the data in our case obeys to some of the fundamental characteristics of nonlinear deterministic processes, since the embedding dimension of the reconstructed space is low and the attractors are limited in a small region of the phase space.

Two other major features which have emerged as classifiers of deterministic dynamical systems are the fractal dimensions (information and correlation

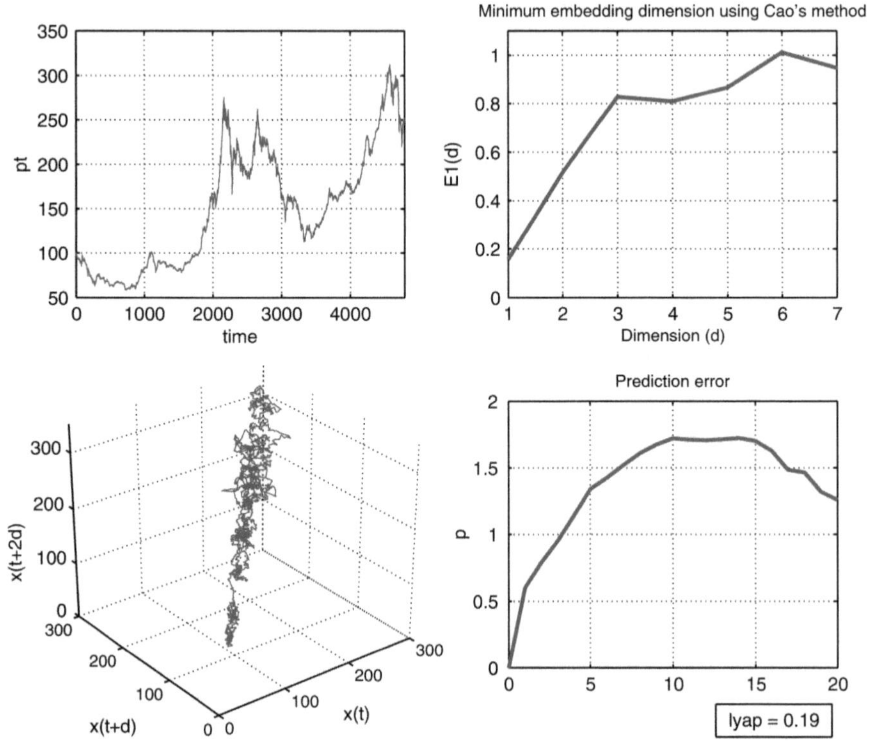

Fig. 4 Embedding dimension, reconstructed attractor and Lyapunov exponents for the original time series

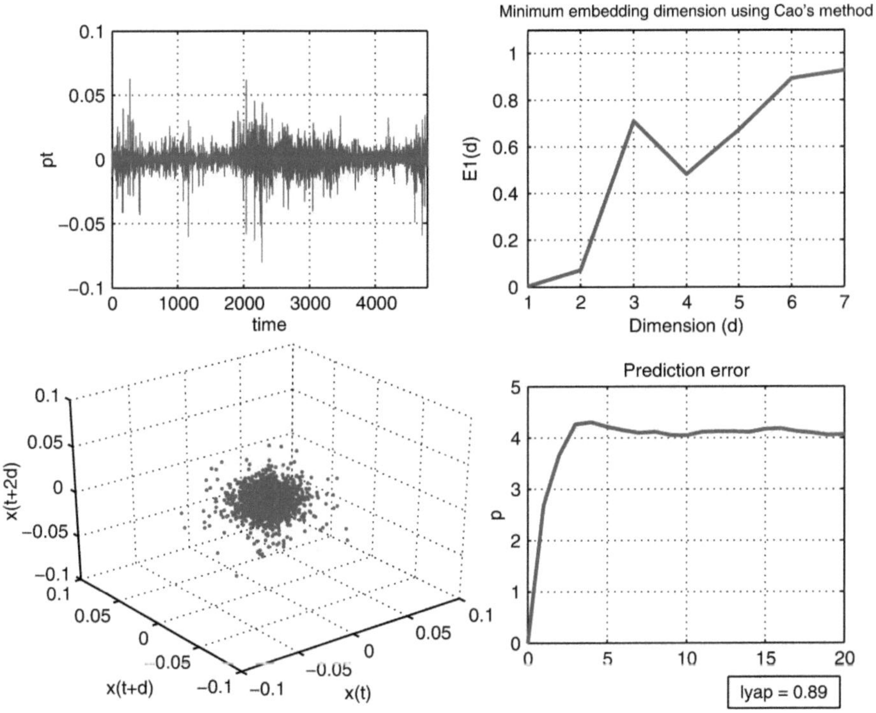

Fig. 5 Embedding dimension, reconstructed attractor and Lyapunov exponents for the log-difference stationary time series

dimensions) and the Lyapunov exponents. Fractal dimensions are representative for the geometric shape of the attractor and are related to the way the points on the attractor are distributed in the d_e-dimensional space. Lyapunov exponents show how orbits on the attractor move apart (or together) under the evolution of the dynamics and can be numerically approximated by using, for example, the Wolf et al. algorithm [42].

In our case, the largest positive Lyapunov exponent was estimated as the slope of the curve in the lower right-hand panel of Figs. 4, 5 and was given by 0.19 for the original data and 0.89 for the log-difference data. This suggests that we are in the presence of sufficient conditions to accept (under some confidence) the evidence of chaotic motion in the underlying dynamics of the financial time series under discussion [1, 42].

The time series exhibits a complex structure and the reconstructed attractor is limited, and looks as strange enough. For given r, we obtain the correlation integral value based on the following formula

$$C(r,m) = \frac{1}{N(N-1)} \sum_{i,j=1(i \neq j)}^{N} \theta\left(r - \left\|z_i - z_j\right\|\right),$$

where θ is the Heaviside function

$$\theta(y) = \begin{cases} 0 & \text{if } y < 0, \\ 1 & \text{if } y > 0, \end{cases}$$

z_i is a phase point in the d_e-dimensional embedding space and r is a critical distance between two phase points. The correlation integral measures the fractions of pairs of points that are closer than ε. Then, the correlation dimension is given by

$$D_2 = \left| \frac{\ln(C_2(r, d_e))}{\ln(r)} \right|$$

and captures the power-law relation between the correlation integral of an attractor and the neighborhood radius of the analyzed hyper-sphere.

The embedding dimension d_e is not increased until d_e is up to d_{ec} (the saturation embedding dimension), and

$$D_2(d_{ec}) = D_2(d_{ec} + 1) = D_2(d_{ec} + 2) = \cdots.$$

Hence, we have that $D_2 = D_2(d_{ec})$ and then the value $D_2(d_{ec})$ is the corresponding fractal dimension of the attractor. If d_{ec} does not reach the saturation level, then $D_2(d_{ec}) \to \infty$, which means that no attractor exists, and then the time series should be considered as produced by a stochastic process. If $D_2(d_{ec})$ is up to saturation, then we would be dealing with a chaotic system. Moreover, a non-integer correlation dimension is a necessary condition for chaos.

The scaling of the correlation dimension is represented in Fig. 6. By using the Takens estimator, we obtain the following fractional correlation dimension (correlation sum approach) for the original and log-difference time series: $D_2 = 1.6977$ and $D_2 = 2.8103$ respectively. In order to check the dimension of the embedding, the Takens estimator about the correlation dimension (D_2) was computed. The choice of the minimum embedding of the signal occurs when the fractal dimension saturates for increasing the embedding. The values of d_e estimated by both methods are very similar (we recall that we initially found $d_e = 3$). Moreover, Fig. 6 reveals derivative correlation sum curves very close up to medium-larger scales ($\log r > -3.0$ and $\log r < 0$), and slightly different for low scales ($\log r < -3.0$).

The algorithm proposed by Hirata et al. [21] gives useful results for noisy time series. Comparing different kinds of symbolic transformations, we found that the strategy of two symbols (0 and 1), as explained above, is most appropriate for the analysis of the financial time series. Instead of detecting the unstable periodic points, we initialized the algorithm in the following way. We start with a generic partition for the time series state space; we label each piece of the partition by a symbol and convert the original time series into a symbolic sequence. A threshold value (mean, median, or other) θ is defined. We experiment the algorithm with several threshold values, and the best choices was $\theta \simeq 60$. For a quantization level of 2 (symbols 0 and 1), all the data values of mean subtracted absolute time series above

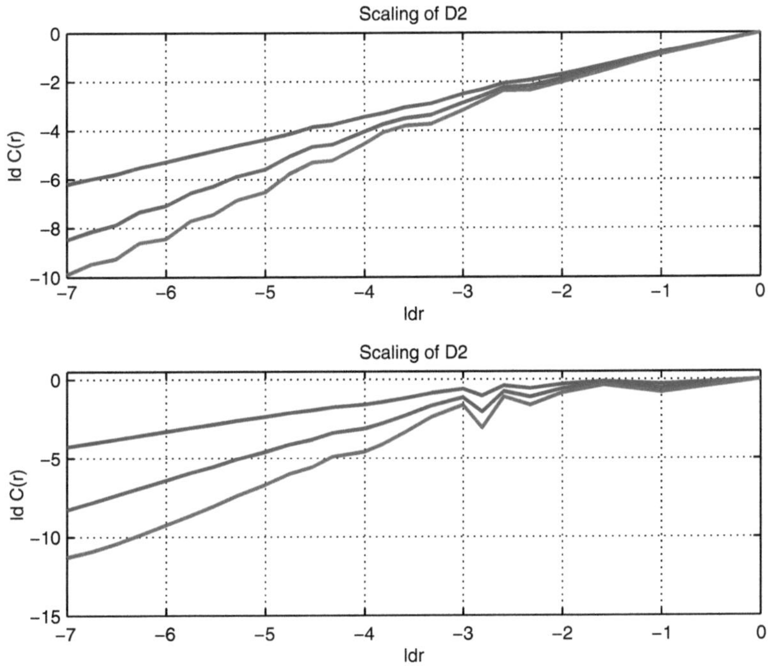

Fig. 6 Correlation dimensions for original data and for log-difference data

the threshold value are converted to 1 and the rest to 0. This process will generate the symbolic series that is illustrated in Fig. 7. After defining a word length $L = 10$ (since the embedding dimension was considered up to 2), the symbolic series is converted into code series. We start now the algorithm from step (3) by classifying points of the time series using their substrings of length 2. We stopped the algorithm when it converged with substrings of length 10. The histogram generated from this code series can be used to calculate the Normalized Corrected Shannon entropy. Moreover, if we introduce a metric distance between two symbolic time series we can derive a hierarchical organization. Figure 8 shows the output of the numerical approximation of a two-symbol generating partition for the PSI20 time series.

The metric entropy can be compared with the sum of the positive Lyapunov exponents under the assumption that Pesin's identity holds. So, we compute the sum of the Lyapunov exponents and estimate the entropy from the obtained symbolic structure of the data. The numerical value for the sum of positive Lyapunov exponent of the time series is known to be $\lambda \simeq 0.19$, for $d_e = 3$, as shown earlier, and $\lambda \simeq 0.21$ for $d_e = 2$. For the topological entropy (which is the supremum of the metric entropies), we obtained the theoretical value using the numbers of unstable periodic points, that is, by considering that $N(p)$ is the number of fixed points for the p-time map, the topological entropy is then approximated by $1/p \,\log N(p)$. Averaging over those values obtained for periods between 1 and 10, we get the following estimate for the topological entropy: $h \simeq 0.2931$. Of course, if we consider a larger

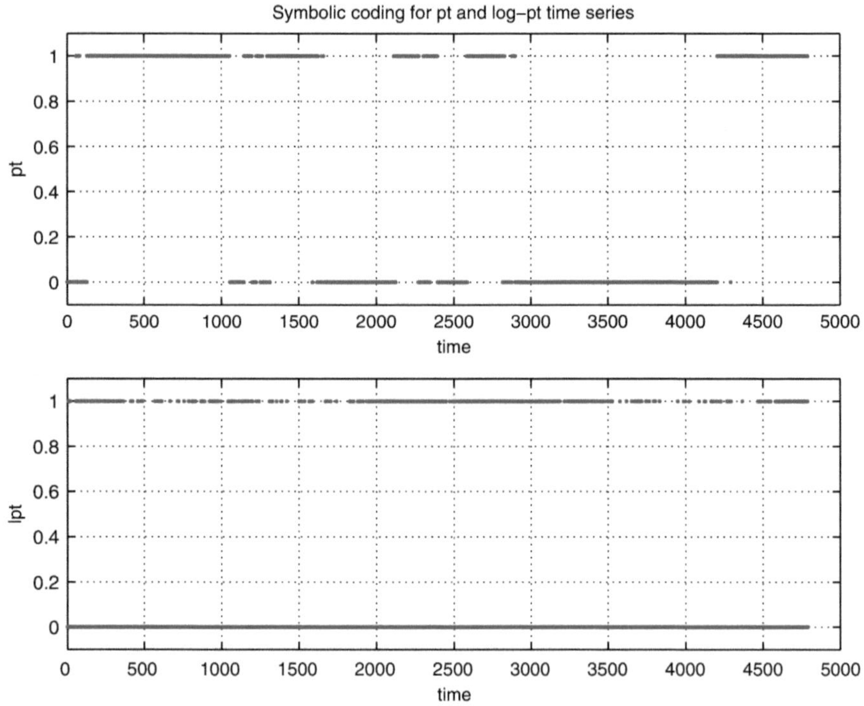

Fig. 7 Symbolic coding for the time series

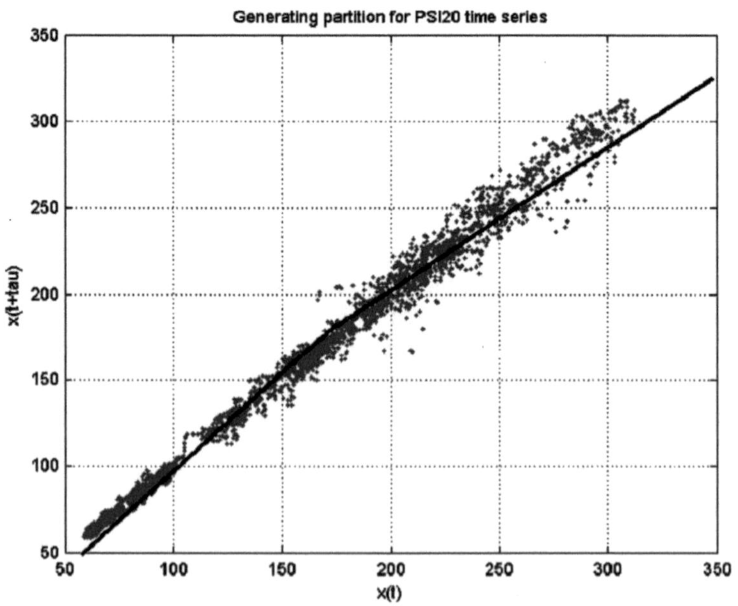

Fig. 8 Approximate generating partition for PSI20 time series

number of periods, we will obtain a more appropriate entropy estimation. These similarities between the values of the estimated entropy and the sum of Lyapunov exponents lead us to suggest that determinism is present in the characterization of the financial time series under consideration; or at least to guarantee that linear dependences and high amount of noise are not sufficient to characterize the data.

Acknowledgements Financial support from the Fundação Ciência e Tecnologia, Lisbon, is grateful acknowledged by the authors, under the contracts No PTDC/GES/73418/2006 and No PTDC/GES/70529/2006.

References

1. Abarbanel H (1996) Analysis of observed chaotic data. Springer, New York
2. Anosov DV (1967) Geodesic flows and closed Riemannian manifolds with negative curvature. Proc Steklov Inst Math 90:1–235
3. Aziz W, Arif M (2006) Complexity analysis of stride interval time series by threshold dependent symbolic entropy. Eur J Appl Physiol 98:30–40
4. Bolt EM, Stanford T, Lai Y-C, Zyczkowski K (2001) What symbolic dynamics do we get with a misplaced partition? On the validity of threshold crossings analysis of chaotic time-series. Physica D 154(3–4):259–286
5. Bowen R (1975) ω-Limit sets for axiom A diffeomorphisms. J Diff Eqs 18:333–339
6. Brida JG, Gómez DM, Risso WA (2009) Symbolic hierarchical analysis in currency markets: an application to contagion in currency crises. Expert Syst Appl 36:7721–7728
7. Brock WA (1986) Distinguishing random and deterministic systems: abridged version. In: Grandmont J-M (ed) Nonlinear economic dynamics. Academic, New York, pp 168–195
8. Brock WA, Dechert W, Scheinkman J (1987) A test for independence based on the correlation dimension. Working paper, University of Winconsin at Madison, University of Houston, and University of Chicago
9. Cochrane J (1994) Shocks. Carnegie-Rochester Conf Ser Public Policy 41:295–364
10. Cvitanovic P, Gunaratne GH, Procaccia I (1988) Topological and metric properties of Hénon-type strange attractors. Phys Rev A 38(3):1503–1520
11. Darbellay G (1998) Predictability, an information-theoretic perspective. In: Prochazka A, Uhlır J, Rayner PJW, Kingsbury NG (eds) Signal analysis and prediction. Birkhauser, Boston, pp 249–262
12. Dionísio A, Menezes R, Mendes DA (2006) Entropy-based independence test. Nonlinear Dyn 44(1–4):351–357
13. Eckmann J-P, Ruelle D (1985) Ergodic theory of chaos and strange attractors. Rev Mod Phys 57:617–656
14. Eguia MC, Rabinovich MI, Abarbanel HD (2000) Information transmission and recovery in neural communications channels. Phys Rev E 62(5B):7111–7122
15. Doyne Farmer J, Sidorowich JJ (1991) Optimal shadowing and noise reduction. Physica D 47:373–392
16. Fraser AM, Swinney HL (1986) Independent coordinates for strange attractors from mutual information. Phys Rev A 33:1134–1140
17. Grassberger P, Kantz H (1985) Generating partitions for the dissipative Henon map. Phys Lett A 113(5):235–238
18. Grassberger P, Procaccia I (1983) Characterization of strange attractors. Phys Rev Lett 50:346–349
19. Gu R (2008) On ergodicity of systems with the asymptotic average shadowing property. Comput Math Appl 55:1137–1141

20. Hammel SM, Yorke JA, Grebogi C (1987) Do numerical orbits of chaotic processes represent true orbits? J Complexity 3:136–145
21. Hirata Y, Judd K, Kilminster D (2004) Estimating a generating partition from observed time series: symbolic shadowing. Phys Rev E 70:016215
22. Hirata Y, Judd K (2005) Constructing dynamical systems with specified symbolic dynamics. Chaos 15:033102
23. Iseri M, Caglar H, Caglar N (2008) A model proposal for the chaotic structure of Istanbul stock exchange. Chaos Solitons Fractals 36:1392–1398
24. Kantz H, Schreiber TH (1997) Nonlinear time series analysis. Cambridge University Press, Cambridge
25. Katok A, Hasselblat B (1999) An introduction to the modern theory of dynamical systems. Cambridge University Press, Cambridge
26. Kennel MB, Buhl M (2003) Estimating good discrete partitions from observed data: symbolic false nearest neighbors. Phys Rev Lett 91:084102
27. Koscielniak P, Mazur M (2007) Chaos and the shadowing property. Topol Appl 154:2553–2557
28. Liebert W, Schuster HG (1988) Proper choice of the time delay for the analysis of chaotic time series. Phys Lett A, 142:107–111
29. Lind D, Marcus B (1995) An introduction to symbolic dynamics and coding. Cambridge University Press, Cambridge
30. Mantegna RN, Stanley HE (1999) An introduction to econophysics: correlations and complexity in finance. Cambridge University Press, Cambridge
31. Maasoumi E, Racine J (2002) Entropy and predictability of stock market returns. J Econom 107:291–312
32. Mendes DA, Sousa Ramos J (2004) Kneading theory for triangular maps. Int J Pure Appl Math 10(4):421–450
33. Milnor J, Thurston W (1988) On iterated maps of the interval. In: Alexander J (ed) Dynamical systems, Proceedings of Special Year at the University of Maryland. Lecture Notes in Mathematics, vol 1342. Springer, Berlin, pp 465–563
34. Nakamura T, Small M (2006) Nonlinear dynamical system identification with dynamic noise and observational noise. Physica D 223:54–68
35. Pearson DW (2001) Shadowing and prediction of dynamical systems. Math Comput Model 34:813–820
36. Pompe B (1993) Measuring statistical dependences in a time series. J Stat Phys 73:587–610
37. Serletis A, Gogas P (1997) Chaos in East European black market exchange rates. Res Econ 51:359–385
38. Small M, Tse CK (2003) Evidence for deterministic nonlinear dynamics in financial time series data. CIFEr 2003, Hong Kong
39. Small M, Tse CK (2003) Determinism in financial time series. Stud Nonlinear Dyn Econom 7(3):5
40. Takens F (1981) Detecting strange attractors in turbulence. In: Rand D, Young L (eds) Dynamical sysytems and turbulence. Springer, Berlin, pp 366–381
41. Wessel N, Schwarz U, Saparin PI, Kurths J (2007) Symbolic dynamics for medical data analysis. http://citeseerx.ist.psu.edu/viewdoc/summary?doi=10.1.1.15.7153
42. Wolf A, Swift J, Swinney H, Vastano J (1985) Determining Lyapunov exponents from a time series. Physica D 16:285–292

What Can Be Learned from Inverse Statistics?

Peter Toke Heden Ahlgren, Henrik Dahl, Mogens Høgh Jensen, and Ingve Simonsen

Abstract One stylized fact of financial markets is an asymmetry between the most likely time to profit and to loss. This gain–loss asymmetry is revealed by inverse statistics, a method closely related to empirically finding first passage times. Many papers have presented evidence about the asymmetry, where it appears and where it does not. Also, various interpretations and explanations for the results have been suggested.

In this chapter, we review the published results and explanations. We also examine the results and show that some are at best fragile. Similarly, we discuss the suggested explanations and propose a new model based on Gaussian mixtures. Apart from explaining the gain–loss asymmetry, this model also has the potential to explain other stylized facts such as volatility clustering, fat tails, and power law behavior of returns.

1 Introduction

For years, time series analysis has been a central research topic in many disciplines within physics, economics and finance. In finance the time series are directly observable from stock indices, individual stock quotes, interest rates, exchanges rates, etc. In physics, on the other hand, time series typically appear as a result of delicate experimental measurements. One notable physical system where such measurements have been of importance in more than a century is in the motion of fluids, like boiling

P.T.H. Ahlgren (✉) and H. Dahl
Nykredit Asset Management, Otto Mønsteds Plads 9, 1780 Copenhagen, Denmark
e-mail: ahl@nykredit.dk; heda@nykredit.dk

M.H. Jensen
Niels Bohr Institute, Blegdamsvej 17, 2100 Copenhagen, Denmark
e-mail: mhjensen@nbi.dk

I. Simonsen
Department of Physics, Norwegian University of Science and Technology (NTNU), 7491 Trondheim, Norway
e-mail: ingve.simonsen@ntnu.no

M. Takayasu et al. (eds.), *Econophysics Approaches to Large-Scale Business Data and Financial Crisis*, DOI 10.1007/978-4-431-53853-0_13, © Springer 2010

water. Here the fluid (water) molecules move among each other in an apparently random fashion, very often showing short, violent "spikes" in their motion. This phenomenon is called turbulence and represents one of the oldest unsolved problems in physics [1]. To date, there is only one possible, although incomplete, theory for turbulence proposed by Kolmogorov in 1941. In this theory, the relative velocity difference between two neighboring fluid elements is estimated and averaged over time and space, for a given distance between the fluid elements. This analysis results in so-called structure functions, that is how the averaged velocity difference scales as the distance between the fluid elements is varied, and these functions have been the central element of turbulence research over half a century [1].

In our terminology, we call this kind of analysis forward statistics, and such statistics have been thoroughly applied to financial time series: Choosing a given time scale (i.e. an investment time span) what is the typical, average return over that period? With a varied time scale, distributions of returns have been obtained which show non-normal behavior with "fat tails" of high probability [2, 3]. This is completely analogous to the situation observed in turbulence where the statistics of the velocity differences described above, for a given distance between fluid elements, also show fat tails, and it is well known that these tails are related to the spiky nature of the fluid motion, also termed intermittency. Spikyness is also very well known in financial time series and also here results in fat tails. Even though the fat-tail distributions are well established facts both in finance and turbulence, there is basically no theoretical frameworks for such behavior; it is simply difficult to integrate fat-tail distributions as compared to normal distributions. A decade ago Jensen therefore suggested that in turbulence one could ask an "inverse" question: For a given velocity difference between two fluid molecules, what is the typical, averaged distance where such velocity difference is obtained for the first time [4]? Performing the statistics and studying the averaged distance versus the velocity differences leads to the so-called inverse structure functions. Whereas the forward structure functions typically measure the spikes, the inverse structure functions measure the quiet, laminar states and thus provide alternative information about the temporal behavior. This type of inverse structure function were subsequently related to exit time statistics [5, 6].

The inverse statistics idea was taken over and applied to financial data in a series of papers [7–10]. The idea is precisely to ask the inverse question: For a given return of an investment, what is the typical, average time span needed to obtain this return (which could be a gain or a loss)? This question is obviously related to the question of first passage time. Starting at a fixed value, when is a given threshold passed for the first time? The subtle difference to first passage time statistics is that for inverse statistics any given starting point in a financial time series is considered, i.e. many different starting levels are used in performing an average over every point in the series. Applying this technique in particular to the Dow Jones Industrial Average (DJIA) index, distributions were obtained which exhibited a sharp maximum[1] followed by a long "fat" (i.e. power law) tail [7]. Again this is very similar to what is

[1] The maximum of these distributions was given the name *optimal investment horizon*.

observed in turbulence [4]. For financial indices it was furthermore found that while the maximum of the inverse statistics for a given positive return occurs at a specific time, the maximum of the inverse statistics for the same negative return appears much earlier. This is the origin of the *Gain–Loss Asymmetry* (GLA) of financial indices, which is the main topic of the present chapter.

The chapter is organized as follows: In Sect. 2 we introduce the concept of inverse statistics and discuss the GLA phenomenon. Moreover, in this section the existing literature on inverse statistics is also reviewed. In Sect. 3 we put most of these earlier findings to the test and in Sect. 4 we go further into the properties of financial time series that may cause the GLA. We present an alternative and in our opinion very credible explanation of the phenomenon in Sect. 5. Finally we conclude in Sect. 6 by outlining how the proposed explanation also has consequences for the world of financial statistics we use daily.

2 Inverse Statistics and Gain Loss Asymmetry in Financial Indices: A Review

In this section we revisit the development of inverse statistics in finance from its introduction in 2002. In particular, we will define and explain the concepts of inverse statistics and the gain–loss phenomenon. Moreover, we will review the current literature on the topics up to the present day.

2.1 Inverse Statistics

Traditionally the performance and risk of holding a certain stock is gauged from the properties of the probability distribution of returns calculated from historic data over a fixed time window, i.e. from what we refer to as the forward statistics. Let $S(t)$ denote the (dimension-less) price of a stock at time t. The logarithmic return at time t and over time window Δt can be defined as [11, 12]

$$r_{\Delta t}(t) = s(t + \Delta t) - s(t), \tag{1}$$

where $s(t) = \ln S(t)$ is the logarithmic asset price. Hence, traditional forward statistics study the historic variations of $r_{\Delta t}(t)$ over time, but for a given time window size Δt.

Contrary to the forward statistics where the length of the time window is constant, in inverse statistics one keeps the return level fixed, here called ρ, and searches for the shortest waiting time after t, $\tau_{\pm\rho}(t)$, needed to obtain a logarithmic return above (below) a predefined level $+\rho$ $(-\rho)$. Consult Fig. 1 for a schematic overview of the inverse statistics method. Note that in this notation ρ is the return *level* while a positive (negative) sign corresponds to a gain (loss). The waiting times, $\tau_{\pm\rho}(t)$, also

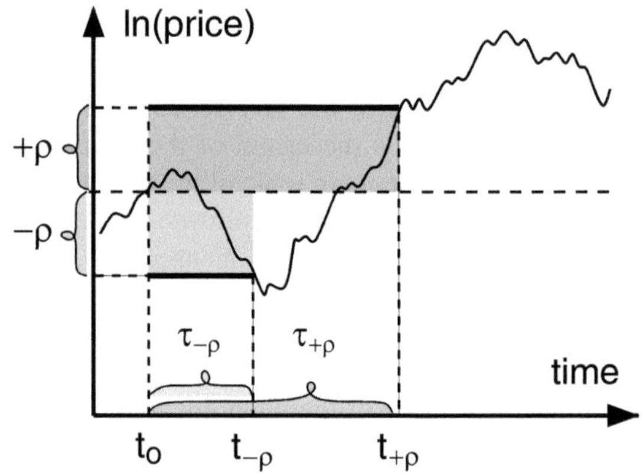

Fig. 1 The figure shows a schematic overview of the inverse statistics procedure performed on a time series (*the solid curve*). Being at t_0 one asks how long it takes to earn a return of $+\rho$ respectively $-\rho$. The analysis yields the times $\tau_{+\rho}$ and $\tau_{-\rho}$, the first moments in time following t_0 where $+\rho$, respectively $-\rho$, is reached. Figure taken from [13]

known as first passage times, can be mathematically defined as

$$\tau_{\pm\rho}(t) = \begin{cases} \inf\{\Delta t \mid r_{\Delta t}(t) \geq +\rho\}\,, \\ \inf\{\Delta t \mid r_{\Delta t}(t) \leq -\rho\}\,. \end{cases} \tag{2}$$

The inverse statistics, here denoted $p\left(\tau_{\pm\rho}\right)$, are simply the two distributions of the waiting times for gains, $\tau_{+\rho}$, and losses, $\tau_{-\rho}$, for *fixed* return level ρ. In the following we will omit the sign in front of ρ when regarding quantities that do not depend on the direction of the price movement.

2.2 Geometrical Brownian Motion Approximation to Inverse Statistics

To better understand how the inverse statistics distribution depends on the return level, ρ, it is illustrative to adapt the classic assumption that the asset price $S(t)$ constitutes a geometrical Brownian motion [14]. Under this assumption, and when using logarithmic returns,[2] we see that the inverse statistics distribution is nothing but the standard first passage probability of an unbiased[3] random walker [15–17]. The distribution of waiting times, τ_ρ, or in our language the inverse statistics, for a random walker to cross a barrier at distance ρ above (or below) its starting point is given by the density [15, 16]

$$f(\tau_\rho) = \frac{\rho}{\sqrt{4\pi D \tau_\rho^3}} \exp\left(-\frac{\rho^2}{4D\tau_\rho}\right), \tag{3}$$

[2] In this context, using logarithmic returns makes it possible to map the inverse statistics problem onto the first passage problems of diffusion.

[3] Analytic results for biased random walks are known, and the results can be found in, e.g. [16], see also Sect. 5.2.

where D denotes the diffusion constant (for the random walker) given by $D = \sigma^2/2\Delta_t$. Here σ is the standard deviation of walker's step size distribution and Δ_t is the (discrete) time interval between two consecutive jumps (here $\Delta_t = 1$). The distribution (3) is often referred to as the "first passage probability density" [16], and it belongs to the family of *inverse Gamma* and *inverse Gaussian* distributions.[4] For $\rho = 0$, (3) predicts a pure power-law decay for the inverse statistics, $p(\tau_\rho) \sim \tau_\rho^{-3/2}$, a result well known from random walk theory under the name of the *first return probability* [16, 17]. However, whenever $\rho \neq 0$ the distribution $p(\tau_\rho)$ will first go through a maximum before asymptotically reaching the power-law decay regime of exponent $-3/2$ for long waiting times (similar to that for $\rho = 0$). It is readily shown that the maximum of the distribution (3) is located at

$$\tau_\rho^* = \frac{\rho^2}{6D} = \frac{\Delta_t}{3}\frac{\rho^2}{\sigma^2}, \tag{4}$$

so that the maximum of the inverse statistics distribution should scale with the return level ρ as a power-law of exponent two. Note also that it is the barrier, ρ, renormalized by the fluctuations, σ, that enters into the expression for τ_ρ^* and $p(\tau_\rho)$.

From (3) it is important to realize that the sign of the return level does not influence the inverse statistics. In other words, under the assumption used to derive (3) the waiting times and therefore the resulting inverse statistics for *gains* or *losses* of the *same* magnitude are identical. We shall shortly see that this is not necessarily true for real financial data, which can instead show a gain–loss asymmetry.

2.3 Empirical Results

It is well documented that empirical financial fluctuations are not fully consistent with the geometrical Brownian motion assumption for price fluctuations [11,12,14]. It is therefore important to study the inverse statistics based on empirical stocks and index data.

The first study of inverse statistics within finance was conducted by Simonsen et al. [7] in 2002. These authors in particular focused on the DJIA, Fig. 2, one of the major US financial indicators. A result for the inverse statistics of return level $\rho/\sigma = 5$ is shown in the main panel of Fig. 3. Moreover, it was demonstrated in [7] that the empirical inverse statistics distributions (for the DJIA) could not be well fitted to the analytic form (3). In particular, the position of the maximum, τ_ρ^*, was shifted towards longer waiting times relative to what was predicted by (4). Moreover, it was found empirically that the scaling $\tau_\rho^* \sim \rho^\gamma$ with $\gamma=2$ (cf. (4)) did not hold true empirically. Instead, the data seemed to indicate that γ should depend on ρ in some unknown way, a dependence that seemed to become less pronounced

[4] These distributions are also known as Wald distributions.

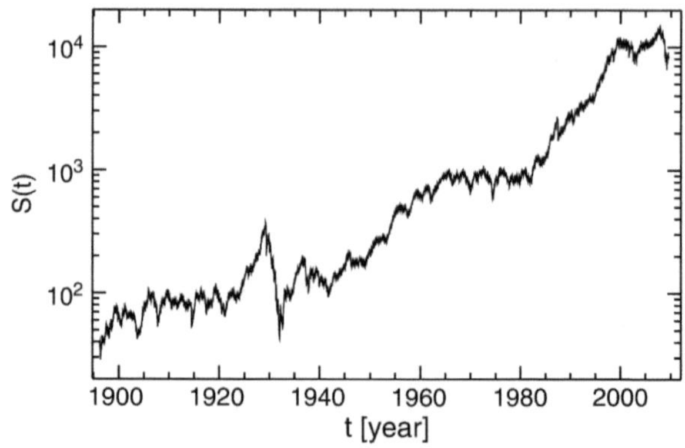

Fig. 2 The DJIA from May 1896 to July 2009 on logarithmic scale. This choice of axis corresponds to the geometrical Brownian motion assumption where one assumes the logarithm of the price process to perform a regular Brownian motion

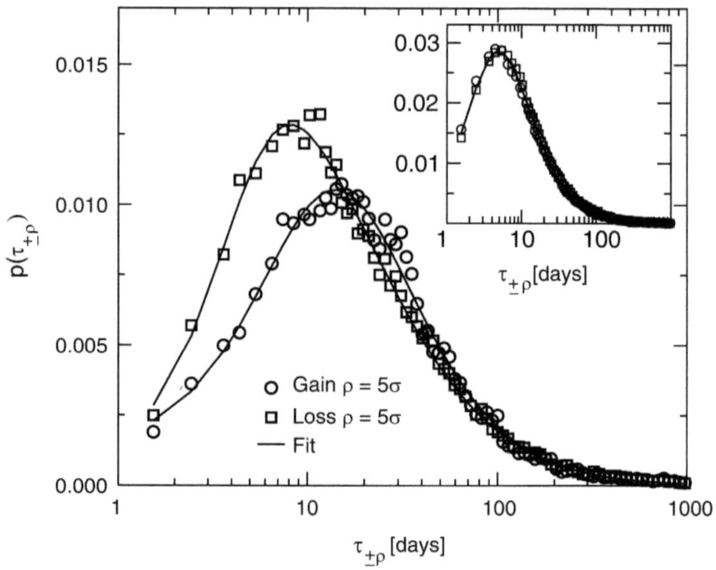

Fig. 3 The inverse statistics $p(\tau_\rho)$ *vs.* τ_ρ for the DJIA obtained on the basis of the empirical, daily close data covering its entire history. The *squared data points* describe losses while *circles* represent gains, both with a level of $\rho/\sigma = 5$. The distributions are normalized. Note the clear asymmetry between the loss and the gain statistics. The *full curves* are fits using (5). The *inset* is the distribution obtained from using the same procedure on the individual stocks of the DJIA, and subsequently averaging over the stocks. Note that the asymmetry is essentially absent for individual stocks. The plot is taken from [7]

(and maybe vanish) as ρ increased. If a scaling regime existed, the data seemed to favor $\gamma < 2$ [7, 18]. It has also been reported that the exponent γ depends on the market [18]. For these and other reasons, a shifted and more flexible version of (3) (also containing more parameters) was introduced for fitting the empirical inverse statistics distributions. The proposed *generalized inverse Gamma* distribution has the form

$$p(t) = \frac{\nu}{\Gamma\left(\frac{\alpha}{\nu}\right)} \frac{\beta^{2\alpha}}{(t+t_0)^{\alpha+1}} \exp\left\{-\left(\frac{\beta^2}{t+t_0}\right)^{\nu}\right\}, \tag{5}$$

which, since $\Gamma(1/2) = \sqrt{\pi}$, reduces to (3) in the limit of $\alpha = 1/2$, $\beta = \rho/\sqrt{4D}$, $\nu = 1$, and $t_0 = 0$. The form (5) seems to be a good approximation to the empirical inverse statistics distributions, and to the best of our knowledge it has been able to fit all such empirical distributions originating from finance well. The curves in Fig. 3 show maximum likelihood fits of the empirical data to (5), and the agreement is observed to be satisfactory.

Since the publication of [7], empirical inverse statistics studies have been performed for a number of financial systems from all over the world. They include the major US indices NASDAQ and the SP500 [9, 18], as well as stock markets from Poland [19,20], Austria (Austrian ATX index) [20], Korea (Korea Composite Stock Price Index) [21], and 40 indices from numerous countries around the world [18]. Moreover, also non stock market data have been analyzed. They include, e.g., foreign exchange data [10] and mutual funds [22].

2.3.1 Gain–Loss Asymmetry

One of the most interesting (and probably surprising) features that can be observed from the main panel of Fig. 3 is the apparent *difference* between the inverse statistics curves corresponding to gains and losses. This empirical finding contrasts what was derived in Sect. 2.2 within the geometrical Brownian motion approximation which predicted that two similar return levels of opposite signs should give rise to the same inverse statistics curve. In particular, it is empirically observed (Fig. 3) that the loss inverse statistics curve is shifted towards shorter waiting times relative to the gain curve, for an equal level of $\rho/\sigma = 5$. This difference is known as the *gain–loss asymmetry* which was first observed for the DJIA in 2003 [8]. In this publication it was also found that the difference $\tau^*_{+\rho} - \tau^*_{-\rho} > 0$ is increasing with the return level, ρ, before it seemed to saturate at a more or less constant value. Since the publication of [8], the GLA has been observed in several major stock indices including SP500 and DJIA [9]. It is consistently found that stock indices belonging to mature and liquid markets exhibit the gain–loss asymmetry with $\tau^*_{+\rho} > \tau^*_{-\rho}$. The gain–loss asymmetry has also been observed for emerging markets [19], however, here there seems to be little consensus on the sign of the asymmetry since both positive (as for mature markets) and negative values for $\tau^*_{+\rho} - \tau^*_{-\rho}$ have been reported in the literature [19].

One particularly interesting and peculiar aspect of the GLA phenomenon has been reported for the individual stocks constituting indices showing GLA. Some findings show that individual stocks do not show GLA, or if they do, this effect is substantially harder to observe empirically, as can be seen in the inset, Fig. 3. Naively one would expect that a potential asymmetry should be reduced by constructing the index since the index essentially is an (typically weighted) average of the individual stocks. It has also been speculated that the GLA is caused by the so-called leverage effect [23], but it has been shown, based on modeling, that the effect can exist with and without the leverage effect [24].

2.4 Possible Causes of the Gain–Loss Asymmetry

How can we rationalize these findings? It was already speculated in the original publication reporting the GLA [8] that the phenomenon could be due to collective effects between the individual stocks. It is well documented in financial literature that markets respond asymmetrically to positive and negative stimuli. In particular, it was speculated if negative signals, caused by, e.g. external events, could synchronize the price drops of individual assets, causing the index to show more pronounced drops.

To investigate this in more detail, a simple model termed the *fear factor model* was constructed [13]. In this model the prices of all individual assets are modeled by the standard geometrical Brownian motion for simplicity. The main idea is the introduction of a fear factor, q, so that with probability q all stocks drop simultaneously (due to fear), while with probability $1 - q$ they would move independently. Due to the synchronous drops introduced by the fear factor, the remaining moves are given a slight positive q-dependent bias in order to satisfy the assumptions regarding the individual stock price process. The interested reader is encouraged to consult [13] for additional details. From the simplified fear factor model, it has been shown that the GLA can indeed be caused by single stock synchronization. To obtain these results and to be able to compare them with empirical findings, it is important to measure the return level ρ in units of the volatility, σ, of the index and stocks [13, 25]. This is easily realized by inspection of (3), when remembering that the diffusion constant is defined as $D = \sigma^2/2\Delta_t$. Hence, the quantity of interest is the ratio ρ/σ, which is also the case for this model. Later extensions of this model has been published in [24] and [26].

Quite recently, empirical results have seemed to support the main idea behind the fear factor model [27]. In particular, it has been demonstrated by an empirical study that during market drops the correlations between individual stocks are stronger than during market rises [27].

However, other studies in the literature seem to suggest alternative mechanisms behind the gain–loss asymmetry. It is therefore fair to say that the origin underlying the GLA phenomenon is still debated in the literature, and more work is needed to clarify and settle the issue. The remainder of this chapter is devoted to this discussion.

3 Test of Earlier Results

As described in Sect. 2.3.1 there is no full consensus in the literature as to where GLA is actually found except for mature market indices where GLA is consistently found. Since there is consensus about mature markets, we shall only focus on areas where results have not been consistent. For example both inverted and regular GLA have been reported for emerging markets. In the following, we will analyze inverse statistics on a large pool of data with the intention once and for all to clarify the stylized facts of GLA.

3.1 Emerging Markets

To study the general properties of emerging markets we have chosen a major pool of the indices from MSCI-Barra (Morgan Stanley Capital International). The series are all calculated in local currency, represent total returns[5] and cover 27 countries.[6] For all countries, the daily data cover January 1999 to July 2009. Applying the inverse statistics method to each of 27 countries and constructing the average gain and loss distributions respectively give us general information on GLA in emerging markets.

As can be seen from Fig. 4, we find a not very pronounced GLA for emerging markets. The individual GLA statistics are summarized in Table 1. This reveals that approximately one-third of the indices show a gain–loss asymmetry similar to that consistently reported for mature market indices. Only two have an inverted asymmetry as has been speculated to be a general feature of emerging markets [19], while the rest, more than half, have no GLA.

The conclusion is that emerging markets in general do not possess a pronounced GLA and that earlier speculation about the existence of an inverted asymmetry is indeed not supported by data. On the contrary, if any asymmetry is present it resembles that found in mature markets.

3.2 Individual Stocks

As discussed in Sect. 2.3.1, earlier studies have found insignificant GLA in individual stock price series. The puzzle that led to the invention of the fear factor model was how single stocks possessing no significant GLA suddenly show a highly significant GLA when merely averaged in an index.

[5] Total return price series are series where dividends, splits, mergers, etc., have been taken into account and are included in the price series.

[6] The 27 countries are Argentina, Brazil, Chile, China, Colombia, Czech Republic, Egypt, Hungary, India, Indonesia, Israel, Jordan, Malaysia, Mexico, Morocco, Pakistan, Peru, Philippines, Poland, Russia, South Africa, South Korea, Sri Lanka, Taiwan, Thailand, Turkey, and Venezuela.

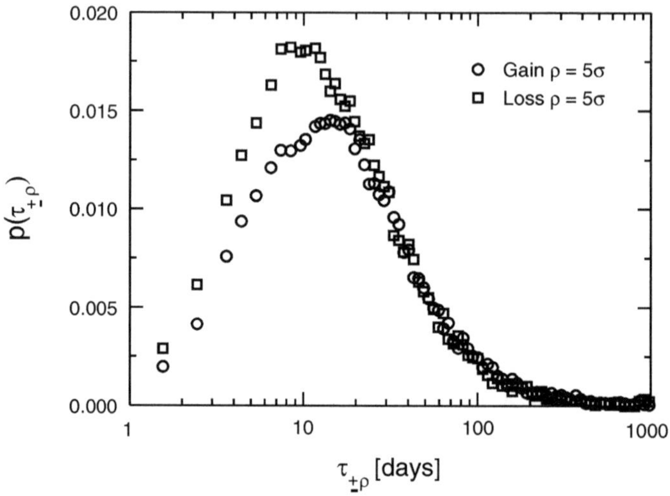

Fig. 4 Inverse statistics for MSCI emerging market indices covering 27 countries from January 1, 1999, to July 30, 2009. *Circles* show the gain distribution, while *squares* are used for the loss distribution. In both cases the chosen return level is $\rho/\sigma = 5$

Table 1 Findings of GLA in emerging market indices. Examining the sign of the difference $\tau^*_{+\rho} - \tau^*_{-\rho}$ a "+" indicates an asymmetry similar to the one found for mature markets, "−" is the inverted asymmetry while "0" symbolizes indices with no significant GLA

Country	GLA	Country	GLA	Country	GLA
Argentina	+	Indonesia	+	Poland	+
Brazil	0	Israel	0	Russia	+
Chile	0	Jordan	−	South Africa	0
China	+	Malaysia	0	South Korea	+
Colombia	0	Mexico	+	Sri Lanka	0
Czech Republic	0	Morocco	0	Taiwan	0
Egypt	0	Pakistan	0	Thailand	0
Hungary	0	Peru	+	Turkey	0
India	+	Philippines	0	Venezuela	−

Many of the explanations of GLA in indices do not necessarily disqualify as explaining the dynamics of the single stocks. Why should single stocks not have faster losses than gains? Of course synchronous stock behavior will amplify the effect but is not necessarily the only plausible explanation of the empirical GLA.

As with emerging market indices, Sect. 3.1 we observe the average of a large pool of series. We perform inverse statistics analysis for each of them and calculate the average gain and loss distributions, respectively.

We have chosen 1,123 stock time series for this analysis. The search for series has been conducted within the universe of leading regional and world indices within mature markets in order not to confuse with the mature markets-emerging markets

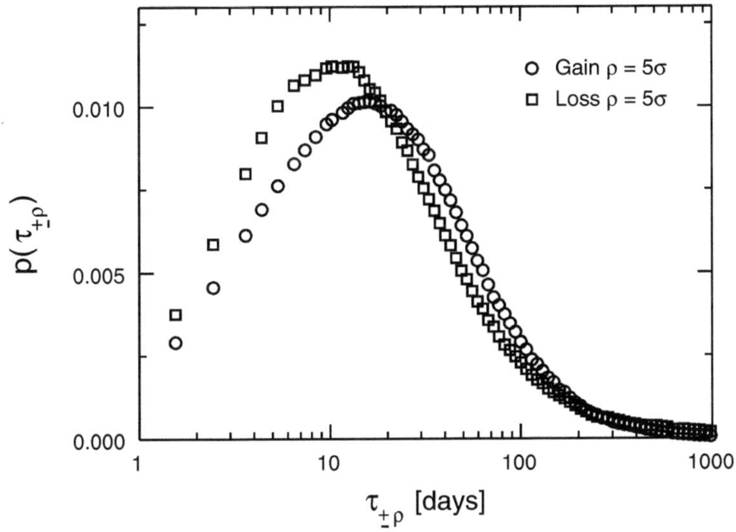

Fig. 5 The average of the inverse statistics of 1,123 different stocks each belonging to mature markets. The *graphs* represent the probability of receiving a gain or a loss of $\rho/\sigma = 5$, τ_ρ days after an investment. *Circles* indicate gains and *squares* losses. The maxima of the two distributions are found at ~10 days for the loss distribution and ~16 days for the gain. These numbers support the idea that individual stocks also have faster losses than gains, but contradict earlier reported findings [8] that single stock constituents of mature market indices do not possess any significant GLA

discussion. Only series with more than 1,000 daily observations have been used. After taking splits and mergers into account, the final list of 1,123 stocks remained. Results for a return barrier of $\rho/\sigma = 5$ are depicted in Fig. 5. The maxima of the two distributions are found after ~10 and ~16 days for the loss and gain curve respectively. This asymmetry amounts to more or less the same as what has been found for the DJIA, Fig. 3. Here the maxima are located at ~8 and ~14 days.

We conclude that individual stocks in general do possess GLA.

3.3 Bond Prices

We also find GLA in bond prices. Choosing 16 government bond total return series[7] we perform an analysis similar to the emerging markets analysis, Sect. 3.1, and the individual stock analysis, Sect. 3.2. The underlying bond indices are all calculated at 10-year maturity and range from January 1986 to July 2009. The average GLA of

[7] The chosen countries are Australia, Belgium, Canada, Denmark, Finland, France, Germany, Ireland, Italy, Japan, Netherlands, New Zealand, Spain, Sweden, United Kingdom and United States.

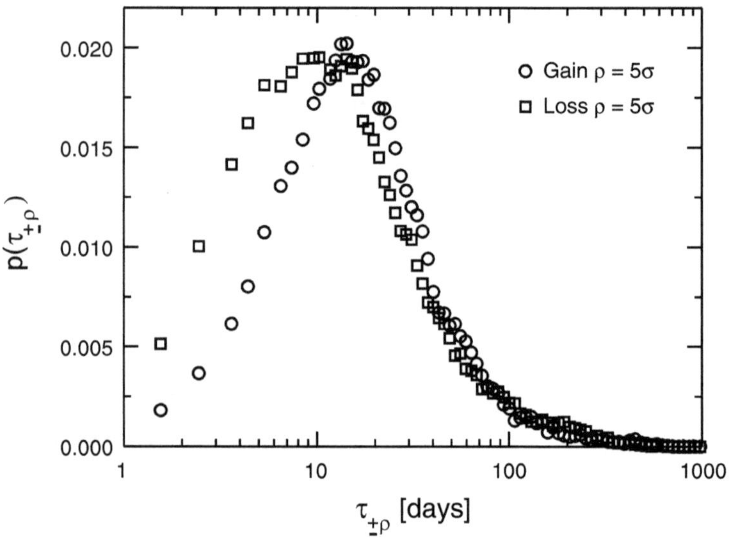

Fig. 6 GLA showing the result for an average of 16 government bond return series calculated from January 1986 to July 2009. The *curves* represent the probability of receiving a gain or loss of $\rho/\sigma = 5$, τ_ρ days after an investment. *Circles* indicate gains where as *squares* show losses

the 16 countries is shown in Fig. 6. Also here we find a difference between the most probable loss and gain waiting time of a factor about 2. The maxima are found at $\sim$10 for losses and $\sim$17 days for gains at $\rho/\sigma = 5$.

As was the case with the analysis of emerging markets and individual stocks, this is to our knowledge the first bond price analysis of its kind and size.

There is a difference between finding GLA in indices and GLA in individual stocks and bond price series. For indices there is a clear notion of what the underlying driving mechanisms are, namely the constituents. In this case it is natural to use correlations between constituents when explaining the presence of GLA. For series obtained from individual stocks or bond prices, there is no clear analogy.

4 The Anatomy of GLA

GLA is usually explained by indices moving faster down than up, an explanation that fits well with general investment experience. Mostly it takes time before an investment decision pays off, and following the usual interpretation of GLA one would further expect potential losses to arrive before a potential profit. This is a mere consequence of large market moves being mostly negative and fast in contrast to big gains that are rare and not as vivid as large losses.

4.1 A Contradiction

One might consider using GLA to attempt to make money in the stock market. Take the DJIA, Fig. 3, as an example. The density peaks at 8 days for losses and at 14 days for gains. Selling the index now, buying it back in 8 days, and selling it again in 14 days should be a successful strategy: After 8 days the stock price has gone down by 5σ whereas it has gone up by 5σ in 14 days. From 8 to 14 days, one should make a profit of 10σ.

Apart from being too good to be true, this is also a self-contradiction. First, the probability for this to actually happen is very small (c.f. the scale of the y axis). Second, the above trades amount to owning the stock for 6 days in the future. But inverse statistics make no distinction about when a 6-day period occurs. Therefore, we can look at the density of first passage times and see that 6 days is very close to the 8 days where a loss is more likely than a profit. We therefore see that a trade designed to maximize profits will actually maximize losses.

Since this trading strategy reveals nothing but a contradiction we must ask how GLA is related to the dynamics of financial time series. Whereas the principles of time varying correlations contained in the fear factor model, Sect. 2.4, may serve as a partial explanation for indices, the fact that single stocks also show GLA, c.f. Sect. 3.2, we have to go further into the emergence of the asymmetry.

4.2 Extreme Dynamics

As most other financial indices DJIA has positive but uncertain mean returns. The mean price return is 0.017% per day, or about 4.5% per year, in the period from 1928 to 2009. The standard deviation of price returns is 1.16% per day, corresponding to about 18% per year. We find a skewness of daily price returns of −0.60 and an excess kurtosis of about 25. Hence the DJIA has non-zero skewness and kurtosis. It would be reasonable to expect a connection between GLA and higher order moments, and as stated in Sect. 2, several papers focus on asymmetric behavior of extreme events as the main explanation of the phenomenon.

4.2.1 Fat Tails and Special Events

Although kurtosis is taken as a regular measure of the fat tails and therefore the degree of extremeness of a distribution, it is not a pure extreme event measure. To directly investigate whether GLA originates from extreme events we simply remove these from series initially exhibiting GLA. In the case of the DJIA, the asymmetry disappears when as few as the 2% most extreme absolute returns are removed. Since the removed events belong to many different periods in the series, one changes the small scale localized dynamics of the time series. We also change the skewness and kurtosis of the series by this procedure.

Having seen that GLA depends on the extreme events in the series, one might think that GLA is ultimately connected to higher order moments. However through simulations of t-distributions, skewed Levy distributions, etc., we find that this is not the case: GLA is not a pure result of non-zero skewness and/or kurtosis in return distributions. None of these simulations show any GLA at all.

4.2.2 Volatility Scaling of Time

A comparison of the inverse statistics of several series should be based on a scaling of the return barrier ρ with the realized volatility of each single series. This ensures that first passage times are comparable. Otherwise, a high volatility asset will be more likely to pass the barrier quickly than a low volatility asset.

However, financial markets tend to exhibit periods of relative calm development and others of high volatility. It has therefore been suggested that not only the assets but also the metric we use changes with time, for example supported by [10,28]. It is not unreasonable to say that time runs fast in highly volatile periods. If we gain ρ in a market characterized by low volatility this is less risky than in a strongly volatile market. In that sense the time it takes to reach a given level should be seen relative to the level of local volatility.

To study the dependence of GLA on volatility variations, we simply scale time with daily measures of volatility. As a simple proxy of local daily volatility, the absolute value of the daily return has proved successful [11]. We choose a new unit of time, $\widehat{\Delta t}(t)$, as a function of the original time unit $\Delta t(t)$

$$\widehat{\Delta t}(t) \sim |r_{\Delta t}(t)|. \tag{6}$$

With this definition it is easy to scale every step so that the total duration of the series is preserved. Measuring time this way we find that GLA disappears independently of the return barrier ρ.

Similar results are found when we use wavelet filtering to remove short time scale dynamics from the studied time series. When short-term effects up to 32 days are removed, there is no GLA in stock indices.

This is no mere technical observation. There are subperiods of DJIA where we find no GLA using the raw data. These subperiods can even be very long. For example, the period from 1968 to 1981 shows no significant GLA. This proves that GLA is not a constant phenomenon. Sometimes it is there, sometimes it has disappeared, but it is always found when considering the entire series. This suggests that GLA could be the result of changing market conditions on a large scale rather than being a local phenomenon.

5 Regimes

An intriguing property of financial systems is their temporal change of states. Calm, positive trending and seemingly uncorrelated markets suddenly drop in value, and correlations rise close to unity. As we have seen above, the results of performing

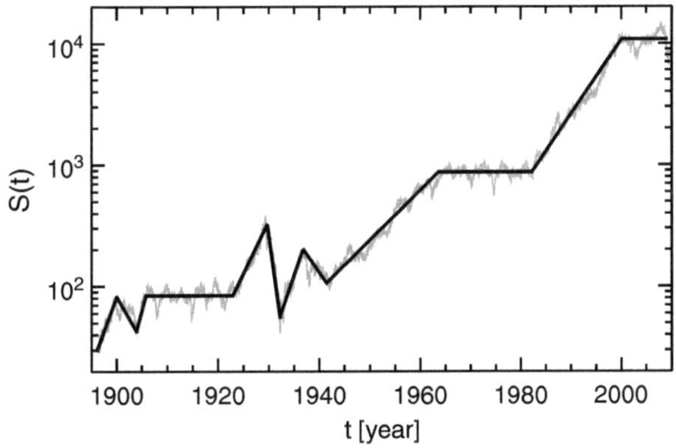

Fig. 7 The DJIA on a logarithmic price scale as in Fig. 2. In *solid black*, a set of coarsely divided regimes are shown as *straight lines*. Although some regimes can be found with negative drift, the dominant behavior of the system is either flat with no drift, or increasing with positive drift. Due to the fractal-like structure of financial time series, this observation applies irrespective of the zoom level

inverse statistics within the realm of finance touches deeply upon this feature. To better understand the importance of the temporal changes of financial systems, observe Fig. 7, depicting the DJIA on a logarithmic price scale. One easily finds flat, volatile regimes and steep but calmer regimes. Although some negative trending periods can be found, these are relatively short and not dominant over time. As a consequence of the close to fractal structure of financial time series, one could find new similar regimes when zooming in. The notion of regimes does not depend on the chosen scale.

5.1 A Simple Non-Mathematical Model

To appreciate how GLA might only represent a series as a whole and not every single time step, imagine a time series with a length of 20 years. The first 10 years have no trend while the next 10-year period shows a strong positive trend. Consider an arbitrary day from the first, flat, period. Since there is no bias, it is equally likely to get a crossing of the negative as well as positive barrier, both of absolute size ρ. Hence the distributions of waiting times are identical for gains and losses.

Now, consider the second period. With a large positive drift one could observe no crossings of the negative barrier, but many of the positive. If volatility is low enough, this period only produces observations for the positive barrier and none for the negative.

Finally, consider the entire period of 20 years. For the negative barrier, all observations stem from the first 10 years, meaning that the final GLA result only represents this period for the loss distribution. If we observe from the last part of the series, the GLA result does not represent the loss statistics.

For the positive barrier it is different. The gain statistics will be an average of observations from both periods and since both periods contribute, the gain curve represents the entire period. Although it is a very simple extreme non-stationary model, it suggests that we can get GLA from mixing the statistics of the two different states. Thus, the final result does not represent every given moment in time but only the series a whole.

The lesson is that we should be careful when analyzing data, and mixing observations from different states can lead to all sorts of misunderstanding because of confusing conditional and unconditional probabilities, and because of non-stationarity. We shall formalize this using the following mixture model.

5.2 A Mixture Model

We shall now show that a model built from nothing but the existence of regimes is capable of reproducing GLA without any relation to turbulence, small-scale effects or the like. We assume that, at any moment in time, the market can be in only one of two states: In state 1, the return generating process is flat but volatile, and in state 2, returns are generally positive and less volatile than in state 1. This corresponds well with the dominant regimes in the DJIA shown in Fig. 7. To keep things simple we stick to the geometrical Brownian motion presented in Sect. 2.2. For each states, i, this means that the logarithmic changes, ds_i, are normally distributed described by the diffusions

$$ds_i = \mu_i dt + \sigma_i dW_i. \tag{7}$$

The assumptions imply that $\mu_1 = 0$, $\mu_2 > 0$ and $\sigma_1 > \sigma_2$. Every day there is some probability q that the market is in state 1 and a probability $1 - q$ that the market is in state 2. We assume that returns are uncorrelated across states, relaxing the strict non-stationary constraint introduced into the model in Sect. 5.1.

Naively, it appears that this model could never produce GLA. We do not expect any skewness or excess kurtosis, since returns are always normally distributed. Nothing in the model has to do with synchronous, extreme behavior or reflects the presence of correlations. All in all it is as "normal" as can be.

Since we have chosen the simple diffusions above, we can use well-known results to compute the first passage time densities for each state. All we need is to extend (3) to the case with non-zero drift. For state i we find that the density of reaching a return of $+\rho$ after waiting τ is

$$f_i^+(\tau) = \frac{\rho}{\sigma_i\sqrt{2\pi\tau^3}} \exp\left(-\frac{(\rho - \mu_i\tau)^2}{2\sigma_i^2\tau}\right). \tag{8}$$

Similarly, the density for reaching $-\rho$ is

$$f_i^-(\tau) = \frac{\rho}{\sigma_i\sqrt{2\pi\tau^3}} \exp\left(-\frac{(\rho+\mu_i\tau)^2}{2\sigma_i^2\tau}\right) = \exp\left(-\frac{2\rho\mu_i}{\sigma_i^2}\right) f_i^+(\tau). \tag{9}$$

Three things are worth noticing about these densities. First, for each state, the density for losses is a constant times the density for gains. Hence none of the states exhibits GLA, only a constant scaling factor distinguishes the densities. Second, because of this scaling factor, the probability of ever hitting the gain barrier is not the same as hitting the loss barrier, except when the drift is zero. In the case of zero drift, both the gain and loss barrier are hit in finite time with probability 1. However, in the case of positive, non-zero drift, the gain barrier is hit in finite time with probability 1, whereas the loss barrier is hit with the probability $P^- = \exp\left(-2\rho\mu/\sigma^2\right)$. For state 2, with $\mu_2 > 0$, $P^- < 1$. Third, each density has a single peak

$$\tau_1^* = \frac{\rho^2}{3\sigma_1^2}, \tag{10}$$

$$\tau_2^* = \frac{-3\sigma_2^2 + \sqrt{4\rho^2\mu_2^2 + 9\sigma_2^4}}{2\mu_2^2}, \tag{11}$$

where the expression for τ_1^*, (10), is the same as (4). We can show that the peak for state 1 occurs earlier than the peak for state 2 if, and only if, $\sigma_1 > \sigma_2$ and $\rho < 3\frac{\sigma_1}{\mu_2}\sqrt{\sigma_1^2 - \sigma_2^2}$. We assume that this is satisfied.

Summing up, in state 1, there is no GLA and the first passage time densities are identical for gains and losses, with density peaks at τ_1^*. In state 2, there is no GLA, and the first passage time densities peaks at $\tau_2^* > \tau_1^*$. However, whereas the probability of reaching $+\rho$ in finite time is 1, it is only $P^- < 1$ for reaching $-\rho$ in finite time. These observations are conditional on being in either of the two states.

If we do not know which state we are in, the situation is different. Now there is a probability q of being in state 1 and a probability $1-q$ of being in state 2. Assuming that states are persistent, we can neglect the effect of mixed states in the inverse statistics results. We can now calculate an unconditional first passage time density for reaching the gain barrier, $+\rho$, of

$$g_{+\rho}(\tau) = q f_1^+(\tau) + (1-q) f_2^+(\tau). \tag{12}$$

Similarly, the unconditional first passage time density for reaching $-\rho$ is[8]

$$g_{-\rho}(\tau) = q f_1^-(\tau) + (1-q) f_2^-(\tau) = q f_1^+(\tau) + (1-q) P^- f_2^+(\tau) \neq g_{+\rho}. \tag{13}$$

[8] To keep the usual convention within inverse statistics one should normalize (13) by dividing by $q + (1-q)P^-$.

5.2.1 A Numerical Example

We will now illustrate the model with a numerical example. Observing Fig. 7, we choose representative parameter values of $q = 0.5$, $\rho/\sigma = 5$, $\mu_1 = 0$, $\mu_2 = 0.1/250$, $\sigma_1 = 0.15/\sqrt{250}$, and $\sigma_2 = 0.08/\sqrt{250}$,[9] and each state has a persistence of 20 days. In this case, the conditions for $\tau_1^* < \tau_2^*$ are satisfied. For these parameter values we find that $\tau_1^* = 9.3$ days, and $\tau_2^* = 30.6$ days. Furthermore, one finds that the probability of reaching the negative barrier in state 2 is $P^- = 0.21$, so only 21% of down targets are reached in finite time.

The results of simulating the model using these parameter values are shown in Fig. 8. The probability distribution describing losses is seen to peak earlier and with a higher amplitude than the gain distribution. Thus, we have found something very similar to the typical GLA result. Other parameter values would be able to more closely replicate empirical findings.

Varying ρ parallel shifts the two curves on a logarithmic time scale so that peak locations remain roughly in a ratio of 2:1. This is in accordance with the stylized facts of GLA as presented in Sect. 2.3.1.

It is worth noticing that all we do is mix two normal distributions, which have no GLA. Neither do they possess any skewness or excess kurtosis. But for the mixture as a whole, we find a skewness of –0.04 and an excess kurtosis of 0.93. Thus, by simply mixing two normal distributions, we get negative skewness and fat tails. Not to mention GLA.

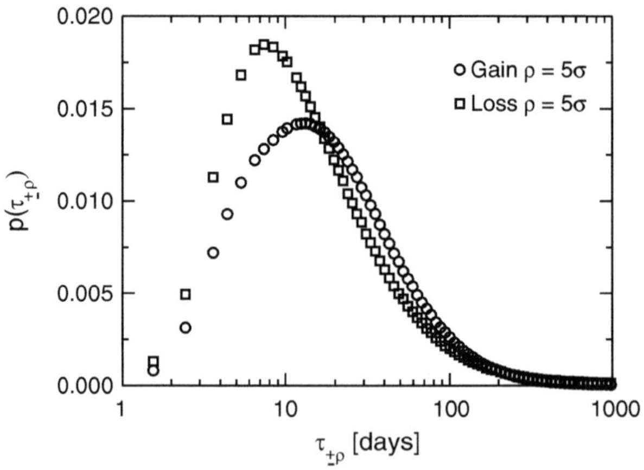

Fig. 8 The figure shows inverse statistics for a Gaussian mixture with two states alternating every 20 days. With *squares*, the loss density curve shows a maximum at ~7 days, whereas the *circles* show the gain density, peaking at ~13 days. Thus, a simple Gaussian mixture model is capable of producing GLA. A better fit would be possible if more than the two states were allowed and parameters were optimized to capture the empirically found GLA curves

[9] Considering the logarithmic price changes without unit, the drift, μ, is measured in units of (day^{-1}), while volatility, σ is measured in units of ($\sqrt{\text{days}}^{-1}$).

5.3 *Empirical Findings for the DJIA*

Everything presented so far in this section relies on the assumption that there are regimes in the series with apparent GLA. It is reasonable to believe that this is actually the case just from at quick look at Fig. 7. Below, we shall put this observation to the test by fitting a Gaussian mixture model to daily Dow Jones price returns since 1928.

5.3.1 Three States that Explain the DJIA

We find that three states best describe historical price returns. The parameters for each state are shown in Table 2. We see that the majority of points are driven by two of the three processes. One flat, $\mu = 0\%$, with middle volatility, $\sigma = 18\%$, and the other with positive drift, $\mu = 15\%$, and low volatility, $\sigma = 8\%$, all values are pr. annum. This is as described in the above models. The third and last process only appears with 5% probability and has large negative returns but also extreme volatility. This state appears to be responsible for the large negative and positive single day returns described as the synchronous fear step incorporated in the fear factor model described in Sect. 2.4.

As above, each mixture component has no skewness or excess kurtosis. However, the mixture as a whole has a skewness of –0.29 and an excess kurtosis of 10.5. by comparison, daily Dow Jones price returns exhibit a skewness of –0.6 and an excess kurtosis of 25.0. This indicates that the mixture model captures most of the realized behavior of Dow Jones.

5.3.2 Comparing Apples and Oranges

We have now validated the assumptions behind the mixture model presented in Sect. 5.2. Besides verifying our work so far they should alert us. Could it be that the GLA and perhaps many other stylized facts of financial time series are simply due to mixing measurements from processes that are really non-comparable? If some observations come from one distribution and others from another, we compare apples with oranges. This can lead to all sorts of spurious conclusions.

Table 2 Empirical parameters for a Gaussian mixture model fitted to the DJIA from 1928 to 2009. States 1 and 2 represent 95% of all dynamics verifying the assumptions underlying the models presented here. Only 5% of all time steps are driven by state 3, i.e. the only state with negative drift. Both drift and volatility are quoted as yearly numbers

	State 1	State 2	State 3
Probability (%)	43	52	5
Mean annualized return (%)	0	15	−54
Annualized standard deviation (%)	18	8	53

To test whether this hypothesis applies to the actual DJIA data, we can find the most likely data generating process responsible for each point. Using the parameter values presented in Table 2, we classify historical daily returns according to which state generated them. We invoke Bayes' theorem to find the state i that most likely generated a given observation x. Bayes' theorem tells us that

$$P(i|x) = \frac{P(x|i)P(i)}{P(x)} \sim N(x, \mu_i, \sigma_i)q_i. \tag{14}$$

We classify each observation according to the highest state probability. We then select the observations that appear to be derived from state 1 and apply inverse statistics on them. We do the same for observations belonging to states 2 and 3. In all cases, there is no GLA. In the series as a whole, there is GLA, but not for data generated by the single components.

What about the fact that GLA disappears when we filter out small time scale behavior? It turns out that all we need to do is introduce ordering of states. This corresponds to state autocorrelation. In a simple example we can neglect state 3, but alternate between states 1 and 2 every six months. Using wavelet trend filtering, we find that GLA disappears just as found empirically.

Thus, a very simple autocorrelation structure in state occurrences can generate results that are fully consistent with empirical findings. As a pleasant side effect, state auto correlation by definition generates volatility clustering, an important stylized fact that is usually taken as a common key feature between financial and turbulent, chaotic systems.

5.4 Correlations

Let us return to the original explanation of GLA being a result of a collapsing market where time varying correlations suddenly approach unity. Since we have shown, Sect. 3.2, that it is not a general empirical fact that individual stocks have insignificant GLA, we should check whether the synchronous behavior incorporated in the fear factor model is enclosed and captured by the mixture setup.

We choose five stocks all being constituents of the DJIA. The stocks are IBM, Johnson and Johnson, 3M, Disney and Intel. We use a subset to clarify results and have chosen specific stocks on the basis that they belong to different industrial sectors. With daily price return series we use an expectation maximization algorithm to find underlying multivariate Gaussian mixtures. The results are shown in Table 3. Again, a subset of processes dominates most of the time. Ninety-six percent of all data points arise from not very significantly correlated processes. Only state 2 is characterized by collective market behavior where correlations are relatively large. The probability of this state is 4%, which corresponds very well with the fear factor model results described in Sect. 2.4. Here, a fear step probability of 5% replicated the empirical DJIA GLA. In this perspective the findings of [13] are enclosed in the

Table 3 Empirical parameters for a Gaussian multivariate mixture model fitted to IBM, Johnson and Johnson, 3M, Disney and Intel from 1986 to 2009. Besides the probability of every state, its drift and the average and maximum of the absolute value of the off-diagonal correlations between the five stocks are shown. Being the only process with negative drift and significant correlations, only state 2 qualifies as a fear scenario. The expected appearance of state 2 corresponds well with the fear factor model, where 5% of all steps should belong to the fear state, in order to fit the GLA empirically found in the DJIA

	State 1	State 2	State 3	State 4
Probability (%)	48	4	44	4
Mean annualized return (%)	−9	−62	26	−32
Average off-diag. corr.	0.36	0.55	0.33	0.17
Max. off-diag. corr.	0.47	0.74	0.48	0.37

empirical results for the fitted mixture model. But as shown in Sect. 5.2 a fear state characterized by large correlations and negative drift is not needed to produce GLA. The only thing needed to observe GLA in a time series is the existence of more than one underlying process.

5.4.1 Other Stylized Facts

Let us finally investigate whether other stylized facts can be replicated by the mere assumption that the driving process can be split into two or more sub-processes. So far we have argued that two Gaussian processes with some transition probabilities are all that is needed to produce skewness, kurtosis, volatility clustering and GLA. What about findings like the power law distribution of returns and the leverage effect?

Returning to the empirical findings from Sect. 5.3.1, we simulate a time series of the same length as the DJIA and with three states identical to the ones found empirically with properties as shown in Table 2. The absolute return distribution is shown in Fig. 9. Also here the Gaussian mixture model produces results very similar to those found for the DJIA itself. Remember that these parameter values are found by fitting a 3-state Gaussian mixture to the DJIA to capture drift and volatility. Clearly, the parameters of the three states can be tuned to better fit the power law distribution. Furthermore, it is a coarse approximation to allow only three states. Nevertheless, the results highly resemble the empirical findings.

We find that the only well-known stylized fact that the 3-state mixture is not capable of reproducing is the leverage effect. Simulations show that very sophisticated correlation structures are necessary to create the specific relation between negative returns and future volatility. This indicates that the leverage effect is a rare stylized fact in the sense that it does not measure the universal character, but instead gives valuable information about the special local dynamics of financial time series.

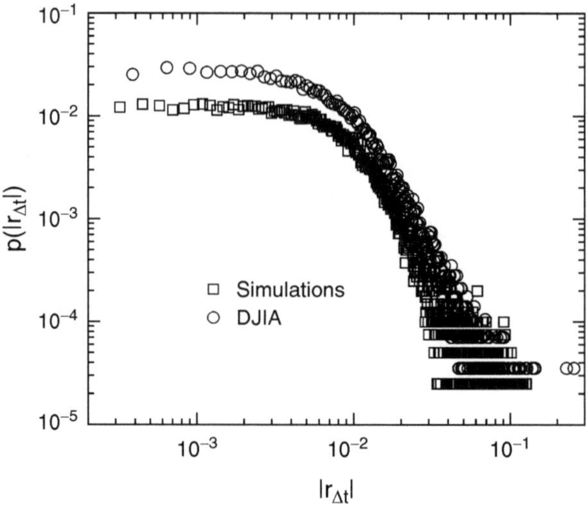

Fig. 9 The absolute return distribution of the DJIA (*circles*) and a simulation (*squares*) of a Gaussian mixture with three states with properties as found empirically for the DJIA and shown in Table 2

6 Conclusion

We have reviewed the existing literature and tested what researchers have not reached consensus about so far. This leads us to the conclusion that emerging markets possess a regular but weak GLA and that single stocks do exhibit GLA. Furthermore we have found that GLA is also a stylized fact of government bond price series. To our knowledge this asset class has not earlier been the subject of inverse statistics. We have shown that a simple data generating mechanism can replicate many stylized facts. Out of distributions with no skewness or excess kurtosis, we get both skewness and excess kurtosis. And although there is no GLA in individual components, there is GLA in the series resulting from mixing the components. Our finding is that GLA could well be an effect of the analysis method alone if multiple data generating mechanisms are present. The same is not necessarily true for skewness and kurtosis, because even a single data generating mechanism can generate skewness and kurtosis. A simple example is the log-normal distribution.

This calls for caution in analyzing and interpreting data. If data are generated by different distributions, they are non-stationary. This is not new knowledge, but it is important to understand the meaning. We should be aware that comparing apples and oranges could lead to artifacts that may be interpreted until doomsday but are ultimately meaningless.

This has important consequences for many financial applications. The empirical multivariate mixture model created from a subset of DJIA constituents, Sect. 5.4, shows that correlations are state dependent with important implications for risk computations, asset allocation and many other financial applications. Similarly, since correlations vary with state, factor analysis may be biased due to market state changes.

We believe that conditions do indeed change but do not insist that components are necessarily Gaussian. For the present work they have served as useful tools explaining why much of the work on financial time series might not be very successful when applied in real life situations. Nevertheless, we find that even the very simple Gaussian mixture model accounts surprisingly well for the many stylized facts known from financial systems.

For the econophysics society, the most important message that follows from this work is that we might be very successful contributors to the financial community if focus turned away from the study of universal properties that still – after many years – constitute the main part of econophysics. Instead, it would not only be instructive but also very useful to learn more about the local properties of the dynamics in these systems. For example, for a financial practitioner, it is useful to know that on a day to day basis, there may be no fat tails, no skewness, no GLA, and no power laws, as long as the regime does not change. However, when the regime changes, all these stylized facts can emerge. It would be even more useful to understand how and why regimes change. If econophysics could even predict regime changes, true value would be added. For now we have to question the methods and stylized facts we use in our daily work.

References

1. Frisch U (1995) Turbulence: the legacy of A.N. Kolmogorov. Cambridge University Press, Cambridge
2. Mantegna RN, Stanley HE (1995) Nature 376:46
3. Doyne Farmer J (1999) Comput Sci Eng 1:26
4. Jensen MH (1999) Phys Rev Lett 83:76
5. Biferale L, Cencini M, Vergni D, Vulpiani A (1999) Phys Rev E 60:R6295
6. Abel M, Cencini M, Falcioni M, Vergni D, Vulpiani A (2000) Physica A 280:49
7. Simonsen I, Jensen MH, Johansen A (2002) Eur Phys J B 27:583
8. Jensen MH, Johansen A, Simonsen I (2003) Physica A 324:338
9. Johansen A, Simonsen I, Jensen MH (2006) Physica A 370:64
10. Jensen MH, Johansen A, Petroni F, Simonsen I (2004) Physica A 340:678
11. Bouchaud J-P, Potters M (2000) Theory of financial risks: from statistical physics to risk management. Cambridge University Press, Cambridge
12. Mantegna RN, Stanley HE (2000) An introduction to econophysics: correlations and complexity in finance. Cambridge University Press, Cambridge
13. Donangelo R, Jensen MH, Simonsen I, Sneppen K (2006) J Stat Mech L11001:1
14. Hull J (2000) Options, futures, and other derivatives, 4th edn. Prentice-Hall, London
15. Karlin S, Taylor HM (1998) A first course in stochastic processes, 2nd edn. Academic, New York
16. Redner S (2001) A guide to first-passage processes. Cambridge University Press, Cambridge
17. Inoue J, Sazuka N (2007) Phys Rev E 76:021111
18. Zhou W-X, Yuan W-K (2005) Physica A 353:433
19. Zaluska-Kotur M, Karpio K, Orłowska A (2006) Acta Phys Pol B 37:3187

20. Karpio K, Załuska-Kotur MA, Orłowski A (2007) Physica A 375:599
21. Lee CY, Kim J, Hwang I (2008) J Kor Phys Soc 52:517
22. Grudziecki M, Gnatowska E, Karpio K, Orłowska A, Załuska-Kotur M (2008) Acta Phys Pol A 114:569
23. Bouchaud J-P, Matacz A, Potters M (2001) Phys Rev Lett 87:228701
24. Ahlgren PTH, Jensen MH, Simonsen I, Donangelo R, Sneppen K (2007) Physica A 383:1
25. Simonsen I, Ahlgren PTH, Jensen MH, Donangelo R, Sneppen K (2007) Eur Phys J B 57:153
26. Siven J, Lins J, Hansen JL (2009) J Stat Mech P02004:1
27. Balogh E, Neda Z, Nagy BZs, Simonsen I Persistent collective trend in stock markets (in preparation)
28. Lillo F, Doyne Farmer J (2004) Stud Nonlinear Dyn Econom 4:3

Communicability and Communities in Complex Socio-Economic Networks

Ernesto Estrada and Naomichi Hatano

Abstract The concept of communicability is introduced for complex socio-economic networks. The communicability function expresses how an impact propagates from one place to another in the network. This function is used to define unambiguously the concept of socio-economic community. The concept of temperature in complex socio-economic networks is also introduced as a way of accounting for the external stresses to which such systems are submitted. This external stress can change dramatically the structure of the communities in a network. We analyze here a trade network of countries exporting 'miscellaneous manufactures of metal.' We determine the community structure of this network showing that there are 27 communities with diverse degree of overlapping. When only communities with less than 80% of overlap are considered we found five communities which are well characterized in terms of geopolitical relationships. The analysis of external stress on these communities reveals the vulnerability of the trade network in critical situations, i.e., economical crisis. The current approach adds an important tool for the analysis of socio-economic networks in the real world.

1 Introduction

There is no doubt that we live in a networked world. We live our lives connected by our family, friend and workmate ties. These social networks connect us to other people in other parts of the world to which we are only a few steps apart. In fact, we live in a "small world." We also live surrounded by infrastructures which form their proper networks, such as supply networks, transportation networks, etc. The

E. Estrada (✉)
Department of Mathematics, Department of Physics and Institute of Complex Systems, University of Strathclyde, 26 Richmond Street, Glasgow G11XQ, UK
e-mail: ernesto.estrada@strath.ac.uk

N. Hatano
Institute of Industrial Science, The University of Tokyo, 4-6-1 Komaba, Meguro-ku, Tokyo 153-8505, Japan
e-mail: hatano@iis.u-tokyo.ac.jp

M. Takayasu et al. (eds.), *Econophysics Approaches to Large-Scale Business Data and Financial Crisis*, DOI 10.1007/978-4-431-53853-0_14, © Springer 2010

economic world in which we also live does not look very much different from this picture. Banks, corporations, industries and even universities and governmental institutions are interconnected forming a complex network of economical and political interrelations. Then, our lives are very much influenced by the structure of these complex networks and by the dynamics of processes taking place on them [1]. In this chapter we are interested in a tiny part of these complex worlds, namely in the analysis of the communicability and community structure of socio-economic networks.

There are different kinds of networks that can be analyzed in a socio-economic context. In general, economics is based on the activity of economic entities in business networks in which such entities are related by business relationships, such as the ones existing between companies and banks [2]. On a macroeconomical scale, world economies are becoming more and more interrelated as a consequence of the globalization, which implies control of capital flow and liberalization of trade policies [3]. Consequently, the study of the international trade network in which countries are represented by the nodes of the network and their commercial trades by the links is an important tool in understanding the interaction channels between countries [4].

The representation of socio-economic systems as networks gives us the possibility of analyzing theoretically how these interrelations are organized and how they influence the dynamics of processes taking place on them. For instance, network analysis permits us to understand how economic perturbations in one country can spread to the whole world. In a complex network the nodes represent the entities of the system and the links their interactions [5]. Then, we consider socio-economic networks represented by simple graphs $G := (V, E)$. That is, graphs having $|V| = n$ nodes and $|E| = m$ links, without self-loops or multiple links between nodes. Let $\mathbf{A}(G) = \mathbf{A}$ be the adjacency matrix of the graph whose elements A_{ij} are ones or zeroes if the corresponding nodes i and j are adjacent or not, respectively. In the next sections we introduce the concept of communicability between the entities in a complex socio-economic network and then analyze how to determine the community structure of such networks.

2 The Concept of Communicability

It is known that in many situations the communication between a pair of nodes in a network does not take place only through the optimal route connecting both nodes. In such situations, the information can flow from one node to another by following other non-optimal routes. Here we consider that a pair of nodes has good *communicability* if there are several non-optimal routes connecting them in addition to the optimal one. The optimal route can be identified in the case of a simple network with the shortest path connecting both nodes. A path is a sequence of different nodes and links in connecting both nodes. Let s be the length of this shortest path. Then, the non-optimal routes correspond to all walks of lengths $k > s$ which connect both nodes. A walk of length k is a sequence of (not necessarily different) vertices $v_0, v_1, \ldots, v_{k-1}, v_k$ such that for each $i = 1, 2, \ldots, k$ there is a link from v_{i-1} to v_i. Consequently, these walks communicating two nodes in the network can revisit nodes and links several times along the way, which is sometimes called "backtracking walks." We assume that the shortest walks are more important than the longer ones [6]:

The communicability between a pair of nodes p, q in the graph is defined as a weighted sum of all walks starting at node p and ending at node q, giving more weight to the shortest walks.

It is well known that the (p, q)-entry of the kth power of the adjacency matrix, $(\mathbf{A}^k)_{pq}$, gives the number of walks of length k starting at the node p and ending at the node q [7]. Then, the communicability function can be expressed by the following formula [6]:

$$G_{pq} = \sum_{k=0}^{\infty} c_k \left(\mathbf{A}^k\right)_{pq}. \tag{1}$$

The coefficients c_k need to fulfill the following requirements: (i) make the series (1) converges, (ii) giving less weight to longer walks, and (iii) giving real positive values for the communicability. For the sake of simplicity we select here $c_k = 1/k!$ [6], which gives the following communicability function:

$$G_{pq} = \sum_{k=0}^{\infty} \frac{\left(\mathbf{A}^k\right)_{pq}}{k!} = \left(e^{\mathbf{A}}\right)_{pq}. \tag{2}$$

The right-hand side (RHS) of the expression (2) corresponds to the non-diagonal entry of the exponential adjacency matrix. Let $\lambda_1 \geq \lambda_2 \geq \cdots \geq \lambda_n$ be the eigenvalues of the adjacency matrix in the non-increasing order and let $\phi_j(p)$ be the pth entry of the jth eigenvector which is associated with the eigenvalue λ_j [7]. Then, using the spectral decomposition of the adjacency matrix the communicability function can be written as [6]

$$G_{pq} = \sum_{j+1}^{n} \phi_j(p)\, \phi_j(q)\, e^{\lambda_j}. \tag{3}$$

In order to understand the communicability function in the context of complex socio-economic networks let us consider the following examples. First, let us consider a series of corporations having business relationships in such a way that they form a linear chain as the one depicted below:

The communicability between the corporations p and q, which are the endpoints of the chain is given by the following mathematical expression:

$$G_{pq} = \frac{1}{n+1} \sum_{j} \left(\cos \frac{j\pi (p-q)}{n+1} - \cos \frac{j\pi (p+q)}{n+1} \right) e^{2\cos\left(\frac{j\pi}{n+1}\right)}, \tag{4}$$

where we have used p and q to designate the number of these nodes in the chain starting by 1. For instance, if the chain is formed by only 4 companies then $p = 1$ and $q = 4$. In this case, the communicability between the corporations p and q tends to zero when the number of corporations is very large. That is, $G_{pq} \to 0$ when $n \to \infty$.

A completely different picture emerges when we consider that all corporations are interrelated to each other forming a compact cluster as the one illustrated below:

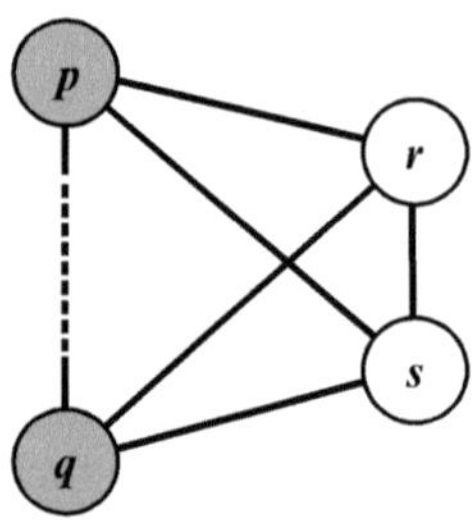

In this case it is easy to show that the communicability between any pair of nodes is given by the following expression:

$$G_{pq} = \frac{1}{ne} \left(e^{n} - 1 \right). \tag{5}$$

Consequently, when the number of corporations is very large the communicability between any pair of them tends to infinity, i.e., $G_{pq} \to \infty$ when $n \to \infty$.

These results indicate that a simple linear interdependence between the corporations gives rise to very poor communicability between the entities involved. On the other hand, a fully interconnected network of corporations is quite effective for the communicability between the entities involved. However, this kind of fully interconnected organizations is very rare at large scales due to infrastructural and cost–benefit reasons. In general, socio-economic networks display complex organizational structures which are neither linear nor fully connected. Then, the communicability function allows us to study the patterns of communication hidden in such structures.

3 Communicability and Socio-Economic Communities

A community in a complex socio-economic network can be understood as a set of entities displaying large internal cohesion. This means that the members of the community are more "tightly connected" among them than with the outsiders of the community. The concept of communicability introduces an intuitive way of finding the structure of communities in complex networks. In this context a socio-economic community is a subset of entities that have larger communicability among the members of the community than with the rest of the entities in the network.

In order to define a socio-economic community we first carry out an analysis of the communicability function (3). The term $\phi_j(p)\phi_j(q)e^{\lambda_j}$ can be positive or negative on the basis of the signs of the pth and qth components of the corresponding eigenvector. A better understanding of this sign pattern can be obtained using the following physical analogy. Suppose that the nodes of the network in question are balls and the links are springs. We argued [6] that the adjacency matrix $\mathbf{A}$ is equivalent to the Hamiltonian of the spring network and the communicability (3) is equivalent to the Green's function of the spring network. Then, the eigenvectors of the adjacency matrix represent vibrational normal modes of the network. The sign of the pth component of the jth eigenvector indicates the direction of the vibration, say up or down. If two nodes, p and q, have the sign for the jth eigenvector it indicates that these two nodes are vibrating in the same direction. In Fig. 1 we illustrate the directions of the vibrational modes for the nodes in a simple network. According to the Perron–Frobenius theorem [7] all the components of the principal eigenvector φ_1 has the same sign. Consequently, it represents the translational movement of the whole network.

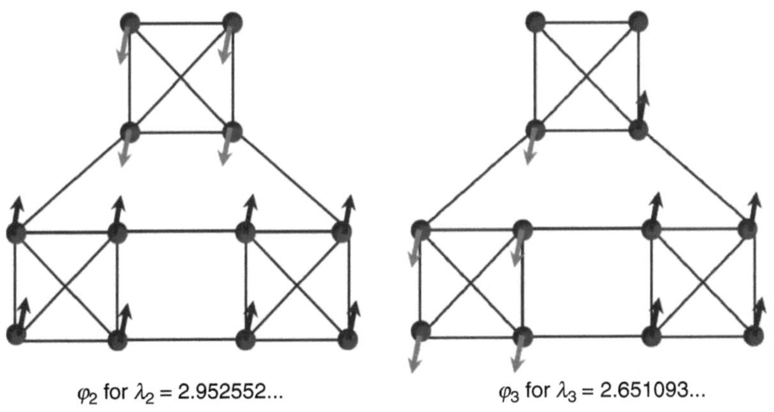

Fig. 1 Illustration of the vibrational modes of the nodes in a network corresponding to the second and third largest eigenvalues

Using this physical analogy we can decompose the communicability function (3) as follows [6]:

$$G_{pq} = \left[\phi_1(p)\phi_1(q)e^{\lambda_1}\right] + \left[\sum_{2\le j\le n}{}^{++}\phi_j(p)\phi_j(q)e^{\lambda_j} + \sum_{2\le j\le n}{}^{--}\phi_j(p)\phi_j(q)e^{\lambda_j}\right] + \left[\sum_{2\le j\le n}{}^{+-}\phi_j(p)\phi_j(q)e^{\lambda_j} + \sum_{2\le j\le n}{}^{-+}\phi_j(p)\phi_j(q)e^{\lambda_j}\right]. \quad (6)$$

The first bracketed term on the right-hand side of the Green's function (6) represents the *movement* of all the nodes (the balls) in one direction after an impact on one node, as if they were part of a giant cluster formed by the whole. In the second bracketed term on the right-hand side of (6), the nodes p and q have the same sign of the corresponding eigenvector (positive or negative); if we put an impact on the ball p, the ball q oscillates in the same direction as the ball p. We thus regard that p and q are in the same cluster if there are more than one cluster in the network. Consequently, we call this second term of (6) the *intracluster communicability*. The last bracketed term of (6), on the other hand, represents an uncoordinated movement of the nodes p and q, i.e., they have different signs of the eigenvector component; if we put an impact on the ball p, the ball q oscillates in the opposite direction. We regard that they are in different clusters of the network. Then, we call this third term of (6) the *intercluster communicability* between a pair of nodes.

As we are interested in the community structure of the network which is determined by the clustering of nodes into groups, we leave out the first term from (6) because we are not interested in the translational movement of the whole network and thereby consider the quantity

$$\begin{aligned}\Delta G_{pq} &= \left[\sum_{2\le j\le n}{}^{++}\phi_j(p)\phi_j(q)e^{\lambda_j} + \sum_{2\le j\le n}{}^{--}\phi_j(p)\phi_j(q)e^{\lambda_j}\right] + \left[\sum_{2\le j\le n}{}^{+-}\phi_j(p)\phi_j(q)e^{\lambda_j} + \sum_{2\le j\le n}{}^{-+}\phi_j(p)\phi_j(q)e^{\lambda_j}\right] \\ &= \sum_{j=2}^{\text{intracluster}}\phi_j(p)\phi_j(q)e^{\lambda_j} - \left|\sum_{j=2}^{\text{intercluster}}\phi_j(p)\phi_j(q)e^{\lambda_j}\right|, \end{aligned} \quad (7)$$

where in the last line we used the fact that the intracluster communicability is positive and the intercluster communicability is negative [6]. Now, we are in conditions of defining a socio-economic community in an unambiguous way:

A socio-economic community is a group of entities $C \subseteq V$ in the network $G = (V, E)$ for which the intracluster communicability is larger than the intercluster one, $\Delta G_{p,q}(\beta) > 0\ \forall (p, q) \in C$.

Using this definition we can generate algorithms that identify the communities in a network without the use of any external parameter. This is the topic of the next section.

3.1 Communicability Graph

In order to find communities in a complex network we need to identify all pairs of nodes having $\Delta G_{p,q} > 0$ in the network. Let us start by representing the values of $\Delta G_{p,q}$ as the non-diagonal entries of the matrix $\Delta(G)$, for which the diagonal entries are zeroes. We are interested only the positive entries of this matrix, which correspond to the pairs of nodes having larger intra- than inter-cluster communicability. Then, let us introduce the following Heavyside function,

$$\Theta(x) = \begin{cases} 1 & \text{if } x \geq 0, \\ 0 & \text{if } x < 0. \end{cases} \tag{8}$$

If we apply this function in an elementwise way to the matrix $\Delta(G)$ we obtain a symmetric binary matrix having ones for those pairs of nodes having $\Delta G_{p,q}(\beta) > 0$ and zero otherwise. This matrix can be represented as a new graph, which we call the *communicability graph* $\Theta(G)$. The nodes of $\Theta(G)$ are the same as the nodes of G, and two nodes p and q in $\Theta(G)$ are connected if, and only if, $\Delta G_{p,q}(\beta) > 0$ in G. Now, a community can be identified as a locally maximal complete subgraph in the network. We recall that a *complete subgraph* is a part of a graph in which all nodes are connected to each other. If this complete subgraph is maximal it is known as a *clique*. The following immediately gives us the method for identifying communities in a complex network [8].

A community is a clique in the communicability graph.

In Fig. 2 we illustrate the communicability graph for the simple network displayed in Fig. 1. There are three 4-nodes cliques and two 2-nodes cliques in this communicability graph, which are also illustrated in this figure.

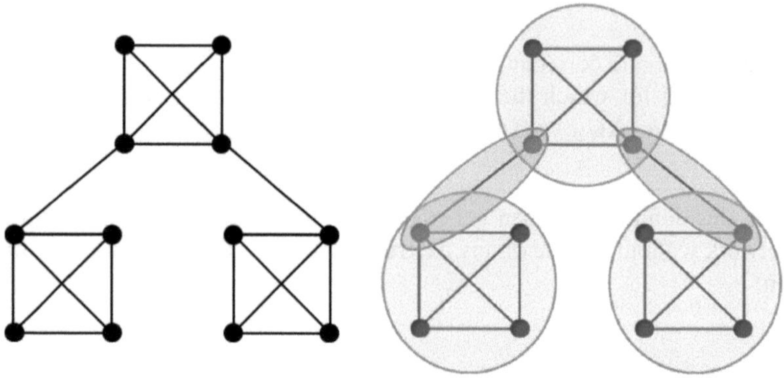

Fig. 2 The communicability graph (*left*) of the network displayed in Fig. 1 and the cliques existing in this communicability graph, which correspond to the communities of the network in Fig. 1

3.2 Overlapping Communities

As can be seen in Fig. 2 there are some nodes that are in more than one community at the same time. Then, the corresponding communities overlap with each other in certain degree. Here we deal with the overlapping of communities and how to merge them to form larger communities.

Two communities are overlapped if they share at least one common node. We can use this information in order to analyze the degree of overlapping between two communities, which can be related to the similarity between the communities in question. Then, we propose the following index as the overlap between the communities A and B in a network [8]:

$$S_{AB} = \frac{2\,|A \cap B|}{|A| + |B|}, \tag{9}$$

where the numerator is the number of nodes in common in the two communities and the denominator gives the sum of the number of nodes in both communities. This index is known in the statistical literature as the Sørensen similarity index [9] and is used to compare the similarity between two samples in ecological systems in particular. The index is bounded as $0 \leq S_{AB} \leq 1$, where the lower bound is obtained when no overlap exists between the two communities and the maximum is reached when the two communities are identical. We can calculate the similarity index S_{AB} for each pair of communities found in the network and represent all results as an overlapping matrix $\mathbf{S}$. If we are interested in identifying only those communities that have an overlap lower than a certain value α, $S_{AB} < \alpha$, we merge those communities for which $S_{AB} \geq \alpha$ into simpler communities.

We have proposed the following general algorithm for the mergence of communities in a complex network [8]:

1. Find the communities in the network following the approach described in the preceding sections
2. Calculate S_{AB} for all pairs of communities found in the previous step and build the matrix $\mathbf{S}$
3. For a given value of α, build the matrix $\mathbf{O}$, whose entries are given by $O_{AB} = \begin{cases} 1 & \text{if } S_{AB} \geq \alpha, \\ 0 & \text{if } S_{AB} \geq \alpha, \text{ or } A = B \end{cases}$
4. If $\mathbf{O} = \mathbf{0}$, go to the end; else go to the step (5)
5. Enumerate the cliques in the graph whose adjacency matrix is $\mathbf{O}$. Every clique in $\mathbf{O}$ represents a group of communities with overlaps larger than or equal to α
6. Build the merged communities by merging the communities represented by the nodes forming the cliques found in the step (4) and go to the step (2)
7. End

4 Socio-Economic Communities Under External "Stress"

Complex networks in general and their communities in particular are exposed to external "stress," which are independent of the organizational architecture of the network. For instance, the web of social relations between actors in a society depends on the level of "social agitation" existing in such society in the specific period of time under study. The network of economic relations between industries and/or banks is affected by the existence of economical crisis, which can change the architecture of such interdependences. The challenge is to capture these external stresses into a quantity that allows us to model the evolution of communities in a complex socio-economic network.

We have proposed to consider the following physical analogy to study this problem [10]. Let us consider that the complex network is submerged into a thermal bath at the temperature T. The thermal bath represents the external situation which affects all the links in the network at the same time. Then, after equilibration all links in the network will be weighted by the parameter $\beta = (k_B T)^{-1}$. The parameter β is known as the *inverse temperature.*

Then, the communicability between a pair of nodes in a network at the inverse temperature β is given by

$$G_{pq}(\beta) = \sum_{j=1}^{n} \varphi_j(p)\varphi_j(q)e^{\beta\lambda_j}. \tag{10}$$

The expression (10) tells us that the structure of the communities depends on the external stress at which the network is submitted. For instance, as $\beta \to 0$ $(T \to \infty)$, the communicability between any pair of nodes in the graph vanishes as

$$G_{pq}(\beta \to 0) = \sum_{j=1}^{n} \varphi_j(p)\varphi_j(q) = 0 \tag{11}$$

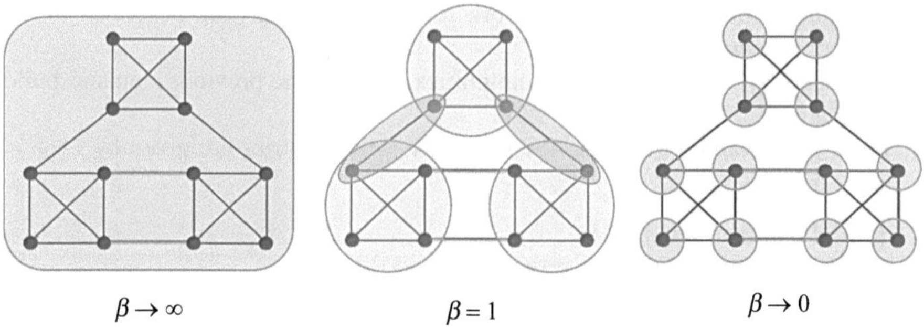

Fig. 3 Illustration of the effect of external stress, accounted for by the inverse temperature, on the structure of communities in a complex network

for $p \neq q$. In other words, when the external stress is very large, such as in a crisis situation there is no communicability between any pair of entities in the network. This situation resembles the case where any individual or corporation is disconnected from the rest in a globally individualistic behaviour.

On the other extreme as $\beta \to \infty$ $(T \to 0)$, the communicability between any pair of nodes in the graph is determined by the principal eigenvalue/eigenvector of the adjacency matrix,

$$G_{pq}(\beta \to \infty) = \varphi_1(p)\varphi_1(q)e^{\beta\lambda_1} \to \infty \tag{12}$$

which means that all entities in the complex network form a unique community. This situation resembles the case where there is an ideal stability in the system which allows all entities to be interrelated to each other forming a unique club in the network. In Fig. 3 we represent these extreme situations for the network represented in Fig. 1.

5 A Social Network Illustration

As an illustration of the concepts explained in the previous sections we consider a friendship network known as the Zachary karate club, which has 34 members (nodes) with some friendship relations (links) [11]. The members of the club, after some entanglement, were eventually fractioned into two groups, one formed by the followers of the instructor and the other formed by the followers of the administrator. This network is illustrated in Fig. 4 in which the nodes are divided into the two classes observed experimentally by Zachary on the basis of the friendship relationships among the members of the club.

The first step in identifying the communities in this network is to obtain the communicability graph. In doing so we need to calculate the matrix $\Delta(G)$ and then dichotomize it using the function $\Theta(G)$. This matrix is the adjacency matrix of the communicability graph, which for the Zachary karate club is illustrated in Fig. 5.

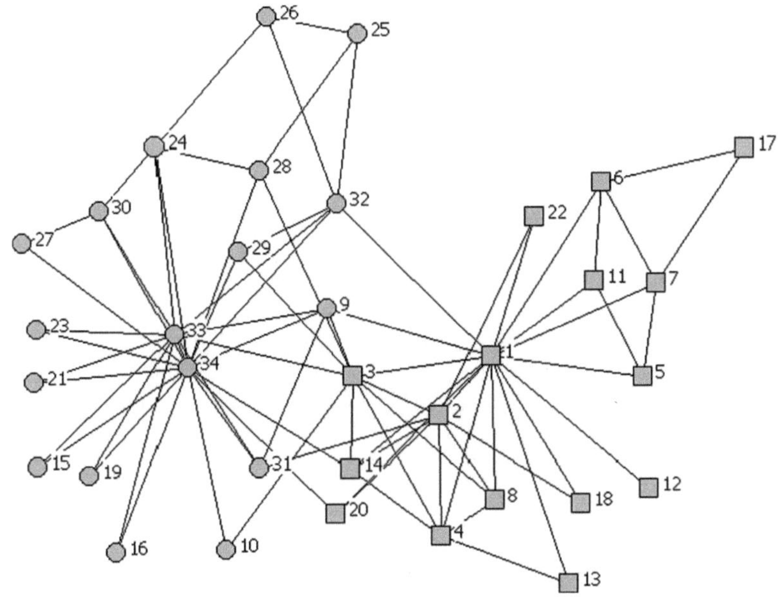

Fig. 4 Network representation of the Zachary karate club with the nodes represented as *squares* and *circles* according to the experimental classification made by Zachary

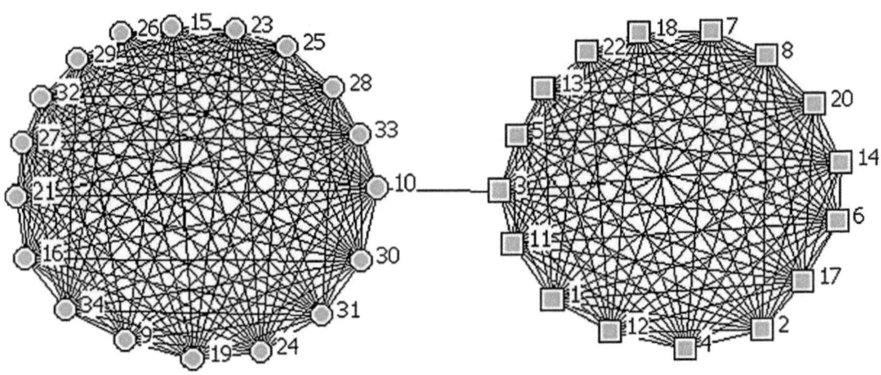

Fig. 5 Communicability graph corresponding to the Zachary karate club network

By finding the cliques in this communicability graph, we detected the following communities [8]:

$$A : \{10, 15, 16, 19, 21, 23, 24, 26, 27, 28, 29, 30, 31, 32, 33, 34\};$$

$$B : \{9, 10, 15, 16, 19, 21, 23, 24, 27, 28, 29, 30, 31, 32, 33, 34\};$$

$$C : \{10, 15, 16, 19, 21, 23, 24, 25, 26, 27, 28, 29, 30, 32, 33, 34\};$$

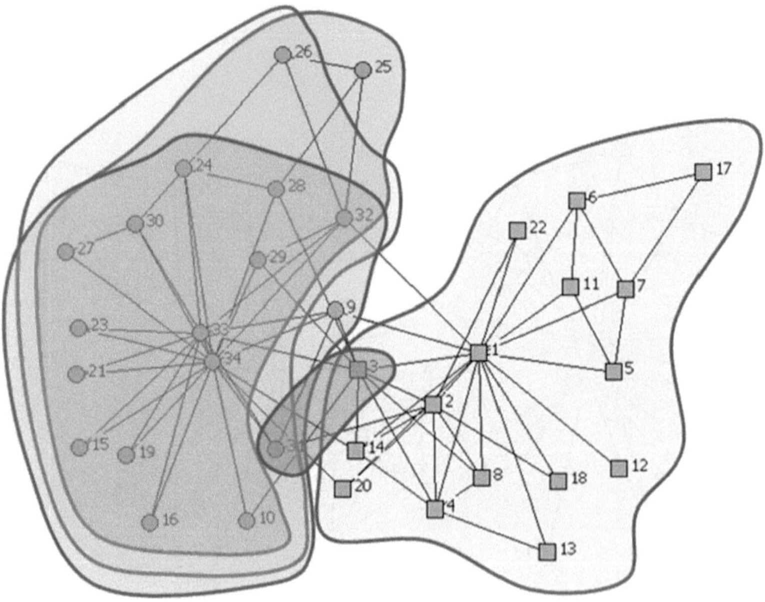

Fig. 6 Illustration of the overlapped communities found in the Zachary karate club network

$$D : \{1, 2, 3, 4, 5, 6, 7, 8, 11, 12, 13, 14, 17, 18, 20, 22\};$$
$$E : \{3, 10\}.$$

These communities display a large overlapping between them as can be seen in Fig. 6.

The large overlapping between these communities can be observed in the community-overlap matrix **S** for this network, which is given below:

$$\mathbf{S} = \begin{bmatrix} 1.000 & 0.938 & 0.938 & 0.000 & 0.111 \\ & 1.000 & 0.875 & 0.000 & 0.111 \\ & & 1.000 & 0.000 & 0.111 \\ & & & 1.000 & 0.111 \\ & & & & 1.000 \end{bmatrix}. \qquad (13)$$

If we consider only those communities having less than 10% of overlap we find only two communities in the network: $C_1 = A \cup B \cup C \cup E$ and $C_2 = D \cup E$. These two communities match perfectly the two factions found experimentally by Zachary in his study for the karate club. These two larger communities are illustrated in Fig. 7 (top-left graphic).

In Fig. 7 we also illustrate the effect of the temperature on the structure of communities in the network. As the temperature increases the communities start to be fractioned until all nodes are forming individual communities. The case $\beta = 0$

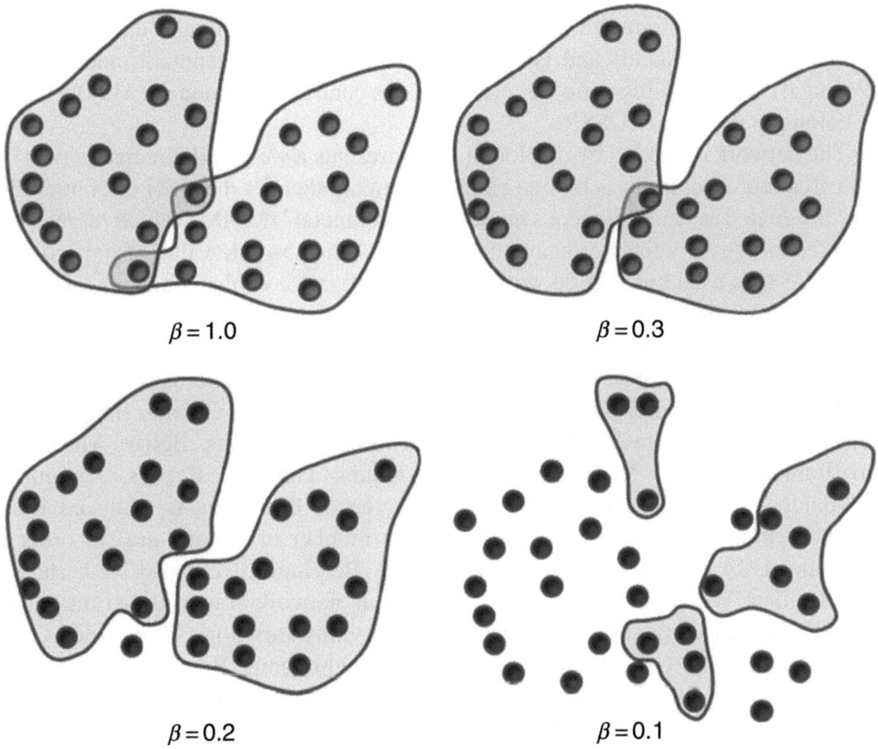

Fig. 7 Illustration of the effect of the temperature on the structure of communities in the Zachary karate club network. The links between individuals have been removed for the sake of simplicity

represents a high level of stress, like a large social agitation. At this temperature the network structure is destroyed and every individual behaves independently. As the value of β increases the stress at which the network is subjected decreases and several organizations of the society start to appear. In an ideal situation of no stress, $\beta \to \infty$, there is only one community in the network. Consequently, the consideration of the parameter β permits us to analyze the characteristics of the community structure of a network under different external conditions by considering that such conditions affect homogeneously to the nodes of the network.

6 Trade Miscellaneous Manufactures of Metal

Here we study a real-world dataset on a trade network of miscellaneous manufactures of metal among 80 countries in 1994. The data was compiled [12] for all countries with entries in the paper version of the Commodity Trade Statistics published by the United Nations. For some countries the authors used the 1993 data (Austria, Seychelles, Bangladesh, Croatia, and Barbados) or the 1995 data (South Africa and Ecuador) because they were not available for the year 1994. Countries which are not sovereign are excluded because additional economic data were not available: Faeroe Islands and Greenland, which belong to Denmark, and Macau (Portugal). Most missing countries are located in central Africa and the Middle East, or belong to the former USSR.

The network compiled by de Nooy [12] represents a weighted directed network. The arcs represent imports by one country from another for the class of commodities designated as "miscellaneous manufactures of metal" (MMM), which represents high technology products or heavy manufacture. The absolute value of imports (in 1,000 US$) is used but imports with values less than 1% of the country's total imports were omitted.

In general, all countries that export MMM also import them from other countries. However, there are 24 countries that are only importers and have no export of MMM at all. They are: Kuwait, Latvia, Philippines, French Guiana, Bangladesh, Fiji, Reunion, Madagascar, Seychelles, Martinique, Mauritius, Belize, Morocco, Sri Lanka, Algeria, Nicaragua, Iceland, Oman, Pakistan, Cyprus, Paraguay, Guadalupe, Uruguay, and Jordan. If the degree of the nodes is analyzed it is observed that the countries having the larger number of exports are (in order): Germany, USA, Italy, UK, China, Japan, France, Belgium/Luxemburg, Netherlands and Sweden. When an undirected version of this network is considered the same countries appear as the larger exporters, with only tiny variations in the order: Germany, USA, Italy, UK, Japan, China, France, Netherlands, Belgium/Luxemburg, Sweden. On the other hand, the in-degree is practically the same for every country. For instance, the average in-degree is 12.475 and its standard deviation is only 3.15. Then, we can consider here only the undirected and unweighted version of this network. Here, the nodes represent the countries and a link exists between two countries if one of them imports miscellaneous manufactures of metal (MMM) from the other. The undirected network is depicted in Fig. 8.

We first calculated the communicability for every pair of countries in the dataset and then computed the $\Delta(G)$ matrix. Using this information we determined the number of cliques in the communicability graph, which correspond to the overlapped communities in the trade network. We found 27 communities with diverse degrees of overlapping. The overlapping matrix is illustrated in Fig. 9.

In order to obtain more valuable information for the analysis of this network we study the communities having less than 80% of overlap. In doing so, we have merged the communities with more than 80% of overlap following the procedure described in Sect. 3.2. Using this approach we reduced the number of communities to only 5, which will be analyzed in the following section.

6.1 Analysis of the Communities

The first community found is formed by 34 countries, 64.7% of which are located in America and 29.4% in Asia. Only two European countries, Germany and Spain, appear in this community. The second community is formed by 32 countries basically

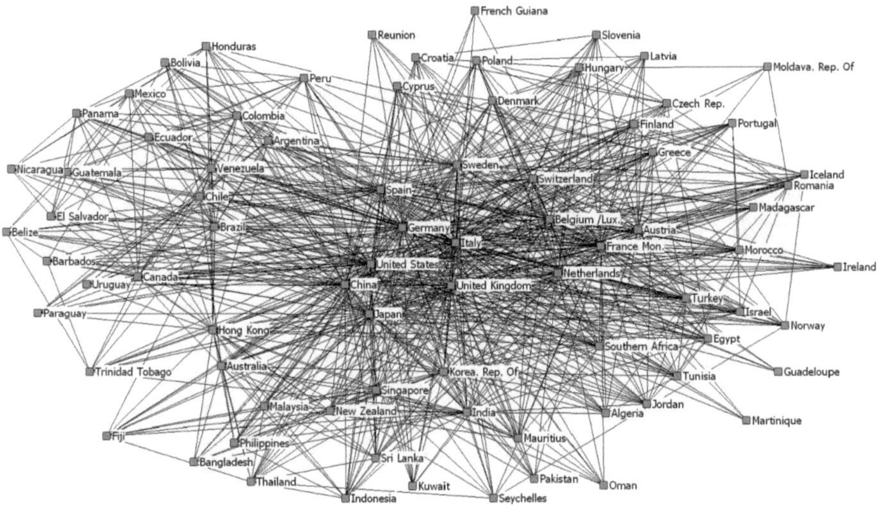

Fig. 8 Trade network of miscellaneous manufactures of metal (MMM). Nodes represent countries and a link exists between two countries if one exports MMM to the other

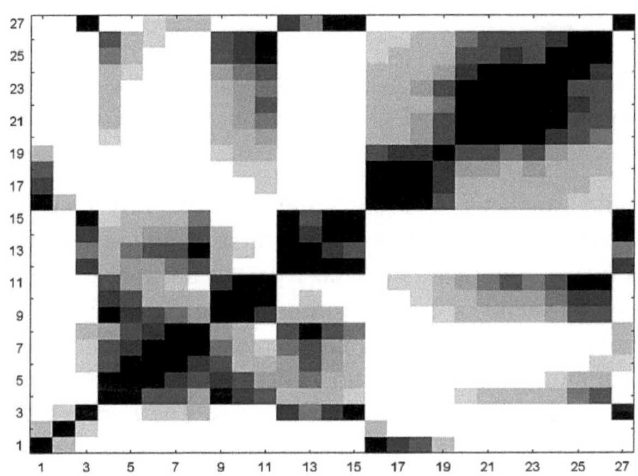

Fig. 9 Overlap matrix among the 27 communities found on the basis of the communicability for the trade network of MMM. Overlap increases from white to black in a scale from 0 to 1

from Asia, as 62.5% of them are in this continent. The third community is formed by 39 countries, 61.54% of them are located in Europe. The fourth community is formed by 21 countries, one third of them are in Europe and the rest in Asia and Africa. The fifth community is the smallest one, which is formed by only eight countries in Europe and the Caribbean (Italy, Greece, Spain, Sweden, Cyprus, French Guiana, Guadeloupe and Martinique). Four of these five communities are represented in Fig. 10.

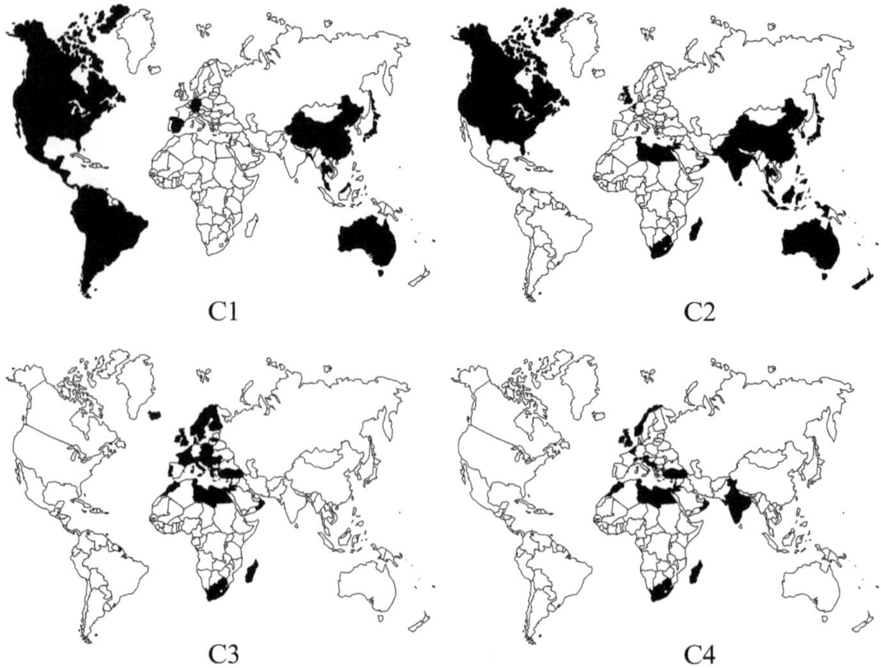

Fig. 10 Graphical representation of the countries forming four of the five communities having less than 50% of overlap in the trade network of miscellaneous manufactures of metal

A more detailed analysis of these communities reveals some interesting facts about the trade between these countries. The first community is dominated by the trade in and between America and Asia. It is not unexpected that Spain appears in this community instead of in the European one due to its historical relations with Latin-America. The reason why Germany is in this community is not clear. The second community reveals the trade between Anglo-Saxon countries (UK, USA, Canada and Australia) with Asian countries. The European community (C3 in Fig. 10) is mainly based on trade within them and with African and Middle-East countries. This community does include neither Spain nor Germany, which are instead included in the America-Asia community. This result reflects clearly the geopolitical nature of the trade clustering between countries. The fourth community reveals the trade between some European countries and some former African/Asian colonies of these European nations.

6.2 *Trade Under External Stress*

We study here the effect of the external stress on the community structure of the trade network analyzed. The external stress is simulated here by the inverse temperature β. At the "normal" temperature $\beta = 1$ we have found 27 communities with

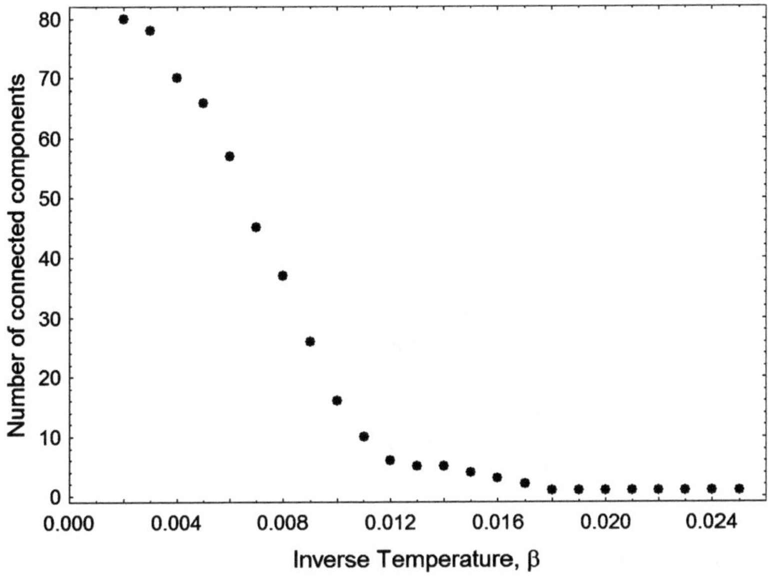

Fig. 11 Effect of external stress, represented by the inverse temperature, on the community structure of the trade network

diverse degrees of overlapping, as described above. These communities form a connected component. That is there is not any community which is isolated from the rest. As the temperature is increased (β decreased) we start to observe the evolution of these communities under the effect of an external stressing factor. At $\beta = 0.017$ the first disconnection occurs in this trade network. The disconnection of other countries from the main connected component is very fast as the temperature increases. For instance, at $\beta = 0.008$ there are 37 isolated clusters of countries, which are disconnected from each other. At this point global trade is impossible and only some local exchanges between very few countries are possible. In Fig. 11 we display the progress of this evolution as the temperature tends to infinity ($\beta \to 0$).

7 Conclusions

We have introduced the concept of communicability in complex socio-economic networks. This concept allows a series of analysis from which we have selected the one related to the communities in socio-economic networks. The method for finding communities based on this concept permits the unambiguous identification of all communities in a network without the use of any external parameter. These communities display different level of overlapping, which can be managed a posteriori to merge communities according to specific necessities of the problem under analysis. The method is also unique in the sense that it permits to study the influence of external factors, like economic crisis and social agitation, by considering a network temperature.

We analyze here a trade network of miscellaneous metal manufactures in 80 countries. The communicability analysis revealed the existence of 27 communities with different level of overlapping. By considering only those communities with less than 80% of overlapping we have identified the existence of certain clusters that reflect the geopolitical relationships in the trade of such manufactures. Another important point is the analysis of the resilience of these communities to external stresses. It has been shown that as the external "temperature" increases there are several countries which separate from the main trading network. This represents the vulnerability of the trade network in situations of critical external influences, such as deep economical crisis. In closing, we hope that the communicability function and the method presented here for studying the identification and evolution of communities can add some valuable information to the study of complex socio-economic networks in the real world.

Acknowledgments EE thanks B. Álvarez-Pereira and K. Deegan for help with the calculations. EE also thanks partial financial support from the New Professor's Fund given by the Principal, University of Strathclyde.

References

1. Jackson MO (2008) Social and economic networks. Princeton University Press, Princeton
2. Souma W, Fujiwara Y, Aoyama H (2003) Physica A 324:396–401
3. Asmin S (1997) Capitalism in the age of globalization: the management of contemporary society. Zed Books, London
4. Serrano MA, Boguñá M (2003) Phys Rev E 68:015101
5. Newman MJE (2003) SIAM Rev 45:167–256
6. Estrada E, Hatano N (2008) Phys Rev E 77:036111
7. Cvetković D, Rowlinson P, Simić S (1997) Eigenspaces of graphs. Cambridge University Press, Cambridge
8. Estrada E, Hatano N (2009) Appl Math Comput 214:500–511
9. Sørensen T (1948) Biol Skr 5:1–34
10. Estrada E, Hatano N (2007) Chem Phys Let 439:247–251
11. Zachary WW (1977) J Anthropol Res 33:452–473
12. de Nooy W, Mrvar A, Batagelj V (2005) Exploratory social network analysis with Pajek. Cambridge University Press, Cambridge

On World Religion Adherence Distribution Evolution

Marcel Ausloos and Filippo Petroni

Abstract Religious adherence can be considered as a *degree of freedom*, in a statistical physics sense, for a human agent belonging to a population. The distribution, performance and life time of religions can thus be studied having in mind heterogeneous interacting agent modeling. We present a comprehensive analysis of 58 so-called religions (to be better defined in the main text) as measured through their number of adherents evolutions, between 1900 and 2000, – data taken from the World Christian Trends (Barrett and Johnson, "World Christian Trends AD 30 – AD 2200: Interpreting the Annual Christian Megacensus", William Carey Library, 2001): 40 are considered to be "presently growing" cases, including 11 turn overs in the twentieth century; 18 are "presently decaying", among which 12 are found to have had a recent maximum, in the nineteenth or the twentieth century.

The Avrami–Kolmogorov differential equation which usually describes solid state transformations, like crystal growth, is used in each case in order to obtain the preferential attachment parameter introduced previously (Europhys Lett 77:38002, 2007). It is not often found close to unity, though often corresponding to a smooth evolution. However large values suggest the occurrence of extreme cases which we conjecture are controlled by so-called external fields. A few cases indicate the likeliness of a detachment process. We discuss a few growing and decaying religions, and illustrate various fits. Some cases seem to indicate the lack of reliability of the data, but others some marked departure from Avrami law. Whence the Avrami evolution equation might be surely improved, in particular, and somewhat obviously, for the decaying religion cases.

We point out two major difficulties in such an analysis: (1) the "precise" original time of apparition of a religion, (2) the time at which there is a maximum number of adherents, both information being necessary for integrating reliably any evolution equation.

M. Ausloos (✉)
GRAPES, Université de Liège, B5 Sart-Tilman, 4000 Liège, Belgium
e-mail: marcel.ausloos@ulg.ac.be

F. Petroni
GRAPES, Université de Liège, B5 Sart-Tilman, 4000 Liège, Belgium
and
DIMADEFA Facoltà di Economia, Università di Roma "La Sapienza", 00161 Rome, Italy
e-mail: fpetroni@gmail.com

M. Takayasu et al. (eds.), *Econophysics Approaches to Large-Scale Business Data and Financial Crisis*, DOI 10.1007/978-4-431-53853-0_15, © Springer 2010

1 Introduction

Barrett and Johnson [1] have surveyed the number of adherents to religions between 1900 and 2000 and publish abundant tables. We have considered that some evolution could be studied according to physics laws [2]. Indeed, religion like sex, age, wealth, political affiliation, language, ...can be considered to characterize a group or an individual status. Whence religion or sex, age, wealth, political affiliation, language distributions can be studied as a function of time and space, as well as auto-correlated, or correlated with any other variable or "parameter" characterizing a population. Thus it may find a role in socio-economic studies pertaining to attitudes, behaviors, opinion formations [3], etc. Several interesting considerations well known in statistical physics can be found in most sociological systems: the role of nucleation, growth, aging, death, criticality, self-organization, epidemic spreading, and subsequent avalanches. If some geometric-like transition or some thermodynamical-like transition exists then some human degree of freedom fluctuations should be observed and discussed.

Recently the dynamics of world's languages, especially their disappearance due to competition with other languages [4] has been of interest [5] in such a respect. A set of similar questions though on religions, within a statistical physics framework, can be raised when attempting to quantify some religion dynamics as measured from individual adherence distribution functions [2]. We emphasize that we are not interested here neither in any religion's fundamental origin or content history nor in finding any hierarchy, but rather in the statistical physics-like aspects of a complex non-equilibrium biological agent based system [6, 7].

Notice the interesting fact when strong fluctuations arise: history is full of examples of individuals or entire groups of people changing their religion, – for various reasons: following the "leader", e.g. Constantinus, Clovis, or not changing at all under "external pressure", leading to martyrdom, or like at inquisition time, or following a fatwah, but also cases of "internal pressure" (Khazars, maybe) or so-called adaptation under proselytism action, e.g. sun worshipping Incas in presence of catholic missionaries, zoroastrian Persians in presence of muslim arabs, etc. Such a competition through agent interactions or under "external field conditions" exist in many cases indeed. Thus, the number of adherents can evolve drastically due to such various conditions [8], independently of the population growth size. We do emphasize that external field conditions can be rather more drastic in the religious domain than in language history.[1] See also Appendix A for some discussion outlining a few aspects, i.e. "differences" between languages and religions, from a physics point of view, perspective or input into modeling such sociological features.

[1] However one can recall the case of "Brugse Metten", May 18, 1302, when flemish "klauwaerts" killed the french "leliaerts", recognized as such, because they could not pronounce correctly "*schild en vriend*". Another case is that of the red khmers, in Cambodia, killing Vietnam educated intellectuals. Finally remember that Gileadites killed Ephraimites, selected because they could not pronounce *Shibboleth*, at a Jordanford (Judges 12:5–6).

In the following we consider as the fundamentally relevant variable the number of adherents of each religion.[2] Only this number is treated as the physics object/measure. Thus a religion is hereby considered as a (socially based) variable, like a language or wealth, to be so studied like any other organizational parameter for defining a thermodynamic-like state. We recognize that a religious state[3] is more individualistic than a linguistic state. Thus, in some sense one can better define the religious adherence of an agent than the linguistic one. Indeed one can hardly be multi-religious but one can be a polyglot. Of course one can switch easily, i.e. through "conversion", from one religious denomination to another, not so in language cases. Thus the observation time of a religious state needs careful attention in surveys.

From another time point of view, one can notice that a religion can seem to appear rather instantaneously, often as a so-called sect, at the beginning, and its number of adherent can grow steadily (see the recent Mormon or Rastafarianism case) or not. The notion of nucleation through homogeneous or heterogeneous fluctuations could be thought of. A religion can also rather quickly disappear, like the Antoinists in some coal mine regions of Western Europe, because its social role or its feeding ground disappears. Both cases though quite interesting are actually outside the realm of this paper. Yet the life time, or aging, of a religion can be studied, through the number of adherents, surely for modern times.

In so doing several pertinent questions can be raised, e.g. from a "macroscopic" point of view: (1) How many religions exist at a given time? (2) How are they spatially distributed ?... From a "microscopic" view point: (3) How many adherents belong to one religion? (4) Does the number of adherents increase or not, and how? And maybe why? (5) Last but not least is there some modelization, through some agent based model possible?

First, let us recognize that the definition of a religion or an adherent (or adept) might not be accepted univocally, but the same can be said about language definition and language knowledge or practice (see Appendix A). We recognize that there are various denominations which can impair data gathering and subsequent analysis; like many, we admit to put on the same footing religions, philosophies, sects and rituals. *Idem*, we do not distinguish between adherents or adepts; there are also agnostics, atheists or "not concerned". In fact, a similar set of considerations exists when discussing languages and dialects, slangs, etc. There are, e.g. three definitions of a language [9]. Similarly one could "weight" the level of adherence to a religion, one could try as for languages to define a religion through its rituals, and quantity of practitioners. Many other indicators are possible (see Appendix B). To consider such variants would lead us too far away from the main stream of the present research and is left for further investigations when possible.

[2] It is sometimes hard to know or to be sure whether an adherent, a disciple, ... is truly a member of a religious denomination or church. However this caveat pertains to usual problems encountered in sociological investigations.

[3] ... Admitting indifference, atheism, agnosticism, ... as a sort of religion, from our point of view.

Thus to address some of these issues, we have followed classical scientific steps as in physics investigations [2]. We have *accepted* as such and subsequently analyzed “empirical” data on the number of adherents of religions. We have discussed in [2] two different freely available data sets. The exactness of both data sets from an experimental (laboratory or naturally based) physics point of view is debatable. Some discussion will be rejuvenated in Sect. 2. Yet, it has been found in [2] that empirical laws can be deduced for the number of adherents, i.e. the *probability distribution function* (pdf). Two quite different statistical models were proposed, both reproducing well the data, with the same precision, one being a preferential attachment model [10], like for heterogeneous interacting agents on evolving networks, e.g. it is more likely that one has the religion of one’s mother or neighbor... (leading to a log-normal distribution), another based on a “time of failure” argument (leading to a Weibull distribution function), as in [11]. Another approach, based on a truly *biological evolution* consideration, has been presented by Hashemi [12].

Moreover, a population growth-death equation has been conjectured to be a plausible modeling of the evolution dynamics in a continuous time framework, i.e. the time evolution of several “main” religions, from a microscopic interpretation is plausible along the lines of the growth Avrami–Kolmogorov equation describing solid state formation in a continuous time framework, the solution of which is usually written as

$$f(t) = 1 - \exp[-Kt^n], \tag{1}$$

where $f(t)$ is the volume fraction being transformed from one phase to another; K and n are adjustable parameters (Fig. 1). For $n = 1$, this equation reproduces the loading of a capacitance in series with a resistance R, for which the differential equation for the voltage $V(t)$ across the capacitance C reads

$$\frac{d}{dt}V(t) = \frac{E - V}{RC} \tag{2}$$

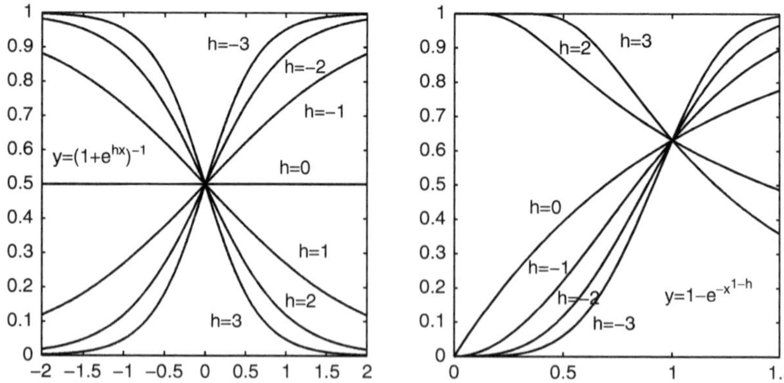

Fig. 1 Logistic map or Verhulst law (2) (*left*) and theoretical behavior of the solution of an Avrami Equation as (5) in reduced units for various typical h values (*right*)

in terms of the electromotive force E, and for which one remembers that one interprets RC as a relaxation time τ. It is also the behavior of the Verhulst logistic map above the inflection point; indicating that this Avrami equation is of interest for so-called late stage growth, i.e.

$$\hat{f}(t) = \frac{1}{1 + \exp[-Kt]}. \tag{3}$$

For $n \neq 1$, (1) can correspond to complex non linear electronic circuits containing an infinity of elements, or also to an infinite combination of springs and damping elements (R. Kutner, private communication after and during an invited talk at FENS07, Wrocław, Poland) in classical mechanics.

A priori, in analogy with crystal growth studies [13, 14], we have considered that a microscopic-like, continuous time differential equation can be written for the evolution of the number of adherents, in terms of the percentage with respect to the world population, of the world main religions, as for competing phase entities in an Avrami sense, i.e.

$$\frac{d}{dt} g(t) = \gamma t^{-h} [1 - g(t)]. \tag{4}$$

It can be solved easily giving the evolution equation for the fraction $g(t)$ of religion adherents

$$g(t) = 1 - \eta \, \exp\left[-\frac{\gamma}{1-h} t^{1-h} \right]. \tag{5}$$

where, adapting to our case (4), η is related to the initial condition, γ is a (positive for growth process) rate (or scaling) parameter to be determined, see discussion in Sect. 3, and h is a parameter to be deduced in each case, measuring the attachment-growth (or death) process in this continuous time approximation. The *relaxation time* τ_n, since $n \equiv 1 - h$, of this stretched exponential growth is

$$\tau_n = \left(\frac{\gamma}{1-h} \right)^{-1/(1-h)} \tag{6}$$

which is markedly rate (K) dependent. For further consideration, let us explicitly write the "very (infinitely) slow" growth case h=1, i.e.

$$\frac{d}{dt} g(t) = \gamma t^{-1} [1 - g(t)], \tag{7}$$

whence

$$g(t) = 1 - \beta t^{-\gamma}, \tag{8}$$

where β, being positive (negative) for a growth (decay) case, is set by initial conditions; for $h = 1$, there is no "relaxation time", but a scaling time $\tau_1 = \beta^{1/\gamma}$, or $\beta = \tau_1^{\gamma}$.

The h-cases which can be illustrated through an Avrami equation are shown in arbitrary time units in Fig. 1 for various h values, for $\eta = 1$ and $\gamma = 1 - h$. They are compared to the (generalized, i.e. $n \neq 1$) logistic map.

What should be emphasized is the fact that religions have appeared at some time t_0 which is somewhat unknown, or to say the least, loosely defined, due to a lack of historical facts but also due to the inherent process of the creation (nucleation, in physics terms) of a religion. This initial time is a parameter intuitively much more important than in crystal growth, but less well defined. Therefore we rewrite the Avrami equation as

$$g(t) = 1 - \exp\left[-\left(\frac{t - t_0}{t_1}\right)^{1-h}\right], \tag{9}$$

thereby allowing also for a time scaling through t_1 related to some growth (or death) rate process. Notice or so that the maximum in such theoretical laws occurs at zero or $+/-$ infinity, – a time information on which there is not much data in the case of religions. Moreover when the number of adherents presents a maximum or a minimum which can be recognized to occur, in the forthcoming analyzed data, during the present or a recent century we will use a second order polynomial like $y = A + Bt + Ct^2$ for the fit.

If $(t - t_0)/t_1$ is much smaller than 1, (9) can be expanded in Taylor series, taking only the first order, and gives

$$g(t) = \alpha + \left(\frac{t}{t_1}\right)^{1-h}, \tag{10}$$

where we have chosen t starting from 0 (instead of 1900, as in our data, being this completely arbitrary and purely conventional) and α representing the initial condition, i.e. the value of the number of adherents for $t = 0$. These are obviously the most simple non linear expressions, i.e. with three unavoidable parameters, which can be used in such a data analysis.

A few examples of religions for which the number of adherents is increasing (e.g., Islam), decaying (e.g., Ethnoreligions) or rather stable (e.g., Christianity and Buddhism) is already shown in Fig. 4 of [2]. In such cases we have found that $h \simeq -1.8$, 6.9, 1.5 and 1.4, respectively in the time range of interest (1900–2050). However in [2] the main denominations were "loosely grouped". To be more specific: Christians in [2] were the results of grouping together 12 denominations; similarly we grouped 15 denominations as "Muslims".

Here we present a more complete and somewhat more detailed analysis of the values of h and its meaning for 58 "time series", where 58 results from: "55" + "1" + "2" (religions). More precisely there are 56 data sets for specific "large" religions, in the World Christian Encyclopedia (WCE) and World Christian Trends (WCT) reference books [1,15], most of them being in the *main denomination* brackets, i.e. in the upper part of the pdf as obtained from the surveys taken between

1900 and 2000. The "1" refers to some data containing 3,000 religions which are put together, as "*other* religions" in the WCT tables. The "2" is due to the set of data about *atheists* and *nonreligious* persons, as mentioned in Tables 1–2 of [1]. Thereafter for conciseness, we will also identify/call these three sets as "religions".

Emphasis will be on distinguishing between growing and decaying cases, discussing our "theoretical" fit, comparing to the forecasting in [1], for 2025 and later, and observing diverse anomalies, thus raising questions to be further investigated.

The remainder of the paper is organized as follows: in Sect. 2 the data bank is briefly discussed, – and criticized, though accepted for further research and subsequent analysis along the theoretical and methodological tools used here which we

Table 1 Values of the parameters h, α, and t_1, used for fitting the data of "increasing religions" with a power law formula; see (10); religions are hereby ranked based on the size of the attachment parameter h which can be negative or positive but ≤ 1; the result of a visual conclusion concerning whether our fit over five data points overshoots (O), undershoots (U) or is approximately the same (=) as the forecasting from WCT is given; reference to the corresponding figure is made

Religion	h	α	t_1	Predict.	Fig.
Shaivites	−5.32	0.032	239	O	2
Hanbalites	−4.66	0.000305	527	O	3
Hanafites	−3.84	0.0629	211	O	2
Zoroastrians	−3.64	3.29e − 005	530	O	3
Kharijites	−2.88	0.000196	1,150	O	3
Afro-Caribbean religionists	−2.75	−5.06e − 007	1,800	O	3
Black Muslims	−2.36	−8.06e − 006	1,110	O	3
Pentecostals/Charismatics	−2.19	−0.00186	208	O	2
Independents	−1.61	0.00427	288	O	2
Shafiites	−1.49	0.024	528	O	2
Afro-American spiritists	−1.32	6.87e − 005	4,990	O	3
Ithna-Asharis	−1.26	0.0137	812	U	4
Afro-Brazilian cultists	−1.21	5.52e − 005	2,610	O	3
Zaydis	−1.10	0.000741	3,450	=	4
Alawites	−1.09	0.000154	7,560	=	4
Ismailis	−1.04	0.00142	1,870	O	4
Yezidis	−1.01	1.84e − 005	2.23e + 004	O	4
High spiritists	−0.83	2.33e − 005	5,690	O	3
Sikhs	−0.792	0.00182	3,130	O	3
Ahmadis	−0.789	4.32e − 005	4,170	=	4
Baha'is	−0.368	4.87e − 006	1.38e + 004	=	4
Druzes	−0.366	4.38e − 005	8.85e + 004	=	4
Neo-Hindus	−0.212	6.19e − 005	1.28e + 004	O	2
Marginal Christians	−0.206	0.000569	1e + 004	=	4
Mandeans	−0.0667	5e − 006	3.17e + 007	U	5
Malikites	0.0566	0.0167	63,803	=	5
Other sectarian Muslims	0.0929	0.000311	2.62e + 006	U	5
Crypto-Christians	0.230	0.0022	1.8e + 004	=	6
Reform Hindus	0.384	0.000154	1.78e + 007	O	6

Table 2 Values of the parameters h, t_0, and t_1 used for fitting the data on "decreasing religions" with (9); h is in this case ≥ 1; the result of a visual conclusion concerning whether our fit over five data points overshoots (O), undershoots (U) or is the same (=) as the forecasting from WCT is given; reference to the corresponding figure is made

Religion	h	t_0	t_1	Predict.	Fig.
Chinese folk-religionists	1.07	$-3.36\text{e}-007$	$3.49\text{e}-015$	O	7
Orthodox	1.14	-0.821	$1.06\text{e}-008$	O	7
Theravada	1.81	-242	3.27	O	8
Mahayana	2.04	-321	16.2	O	8
Karaites	2.09	-99.3	0.00226	=	8
Lamaists	2.77	-614	29.7	O	8

adapt to the considered time series set. The results are largely presented and discussed in Sect. 3 under the form of tables and graphs for various groups of religions, grouping according to the apparent behavior. Some concluding remarks are done in Sect. 4.

2 Data Bank: Theoretical and Methodological Framework

The data analyzed here were taken from the World Christian Trends (WCT) book [1].[4] It is fair to say that this is a remarkable compilation work. Their tables give information on the number of adherents of the world's main religions and their main denominations: 55 specific ("large") religious groups + atheists + nonreligious, plus a set called other religionists made of 3,000 religions which however contains, Yezidis and Mandeans which we consider also, so that we examine $53 + 2 = 55$ (truly recognized) religions. From this data set we have also information on changes during one century of the number of adherents of each religion from 1900 till 2000 (information in the data set are given for the following years 1900, 1970, 1990, 1995 and 2000) – with a forecast for 2025 and 2050. Let us point out that it is not understood (or barely understandable) how such a forecast is made in the data bank.

[4] Data Source Information: The sources used in the WCT database were so numerous and diverse that we only mention here few of them, for a more exhaustive discussion the readers are referred to the WCE. The major physical collections of data built up may be summarized here: around 5,000 statistical questionnaires returned by churches and national collaborators over the period 1982–2006; field surveys and interviews on the spot in over 200 countries conducted by the authors, who over the years 1965–2006 visited virtually every country in the world; the collection of 600 directories of denominations, Christian councils, confessions and topics; a collection of 4,500 printed contemporary descriptions of the churches, describing denominations, movements, countries and confessions; officially published reports of 500 government-organized national censuses of population each including the question on religion, in over 120 countries, covering most decades over the period 1900–2005; bibliographical listings from searches (including computerized enquiries on key-words) in a number of major libraries including those of the British Library (London), Library of Congress (Washington), Propaganda (Rome), Missionary Research Library (New York), and a score of universities.

A critical view of this data has to follow: we have already [2] noticed a break at 10^7 adherents, in the pdf, indicating in our view an overestimation of adepts/adherents in the most prominent religions, or a lack of distinctions between denominations, for these, – as can be easily understood either in terms of propaganda or politics, or because of the difficulty of surveying such cases precisely. Yet one paradoxical surprise stems in the apparent precision of the data. E.g., the data in [1] seems to be precise, in several cases, up to the last digit i.e., in mid-2000, there are 1,057,328,093 and 38,977 Roman Catholics and Mandeans respectively. In strong contrast there are 7,000,000 and 1,650,000 Wahhabites and Black Muslims respectively, numbers which are quite well rounded. Thus a mere reading of the numbers warns about the difficulty of fully trusting the data. Nevertheless the analysis is pursued bearing this *caveat* here below.

3 Results

Results of the h-fit to Avrami equation of the WCT surveys [1] are summarized in Tables 1, 2 and 3: the 58 "denominations" of interest are given. The parameters are obtained by a least-square best fit of the data (not considering the WCT forecast) to the equations mentioned in each table caption for the various cases. The ranking in the tables is according to the fit parameter h or A.

The parameter h values and their meaning deserve some short explanation and discussion here. According to the standard growth (Avrami) process h should be positive and less than 1, since $n \equiv 1 - h$; if it is greater than 1, this is indicating the possibility for *detachment*. We consider that if $|h|$ is outside the (0, 1) interval, we have to imagine that the nucleation growth process is heterogeneous and/or conjecture that it is due to *external field* influences. Moreover notice that when h is greater than 1, the Avrami equation solution decays, ...from a maximum at the time t_0. However it is hardly difficult to know when a religion has attained its maximum number of adherents. Thus the time scale or the initial appearance time of a religion are questionable points. Another point is obvious from Fig. 1. The theoretical expressions do not allow a fit in the vicinity of either a maximum or a minimum. We should expect deviations, if such a case occurs, whence other empirical functions should be considered to be of interest.

3.1 *Intermediary Comments*

In order to read the figures, let us point out the way we have here chosen for the displays. It seems somewhat obvious, from a mathematical or physics point of view that one should consider (1) strictly increasing or decreasing cases, (2) cases of decay after a maximum, or (3) of growth after a minimum. This hints also to consider in future work the curvature as a relevant indicator as for other (financial) time series [16]. Therefore we have grossly ranked the figures and data according to whether the number of adherents seems to be increasing (Table 1), with $h \leq 0$, starting from the lowest value and increasing, in a power law fit. Next we display the "decreasing" cases along an Avrami law, again ranking in order of increasing h, corresponding to fits with parameters given in (Table 2).

Sometimes it is readily observed from the WCT tables that there are "presently growing" religions but for which a minimum is observed during the twentieth century, or a few are decaying after some recent maximum. For such "religions" the number of adherents can be in a first approximation fitted with a second order polynomial $y = A + Bx + Cx^2$, as mentioned in the Introduction, for which the parameters are given in Table 3.

In all cases, each fit has been made over 5 points, represented by dark squares in the figures. The WCT forecasts are indicated by a square with a cross inside.

Table 3 Values of the parameter used for fitting data on 12 "decreasing" and 11 "increasing" religions with the polynomial equation $Cx^2 + Bx + A$; for a warning on the six "central" (in the table) religions, see text; the result of a visual conclusion concerning whether our fit over five data points overshoots (O), undershoots (U) or is the same (=) as the forecasting from WCT is given; reference to the corresponding figure is made

Religion	C	B	A	Predict.	Fig.
Nonreligious	−2.61e − 005	0.103	−102	U	12
Atheists	−1.31e − 005	0.0514	−50.3	U	12
Unaffiliated Christians	−4.38e − 006	0.017	−16.5	U	12
Roman Catholics	−4.2e − 006	0.0165	−16	U	12
New-Religionists (Neoreligionists)	−3.88e − 006	0.0153	−15	U	12
Shamanists	−4.87e − 007	0.00185	−1.74	U	12
Confucianists	−2.11e − 007	0.000831	−0.815	U	12
Wahhabites	−5.53e − 008	0.000215	−0.208	U	12
Taoists	−4.4e − 008	0.000174	−0.171	U	12
Other religionists (in 3,000 religions)	−3.81e − 008	0.00015	−0.148	U	13
Ashkenazis	−2.46e − 008	4.26e − 005	0.0149	U	11
Oriental Jews	−3.23e − 009	1.22e − 005	−0.0112	O	11
Samaritans	7.14e − 012	−3.01e − 008	3.18e − 005	U	14
Sefardis	6.52e − 010	−2.8e − 006	0.00315	O	14
Jains	8.97e − 009	−3.61e − 005	0.0371	U	14
Shintoists	2.05e − 007	−0.000836	0.853	=	14
Saktists	2.12e − 007	−0.000824	0.806	O	9
Protestants	7.66e − 007	−0.00306	3.11	U	10
Anglicans	9.75e − 007	−0.00386	3.83	O	9
Vaishnavites	1.25e − 006	−0.00487	4.82	O	9
Sufis	2.34e − 006	−0.00921	9.11	=	10
Animists	2.77e − 006	−0.0111	11.2	O	9
Evangelicals	5.95e − 006	−0.0233	22.8	O	9

3.2 Growing Religions

In this subsection, we show cases of small or large size religions which are strictly increasing (Figs. 2–6). The illustrations are sufficiently readable and understandable that we do not convey much more hereby than in the figure captions. The fits over 5 points are usually rather good (Figs. 2–4), but we emphasize that the extrapolation either often overshoots or sometimes underestimates the WCT forecast for 2025 and thereafter, except for several cases in Fig. 4, cases for which $0 \geq h \geq -1$. This disagreement is partially due to the lack of saturation implied by the Avrami model.

We observe (Fig. 5) that there are three cases where the growth appears to be linear; in these cases we underestimate the WCT forecast. We emphasize that growth does not mean lack of saturation, as it should be appreciated: this is illustrated by two cases in Fig. 6, i.e. Crypto-Christians and Reform Hindus. Notice that for these last two cases the slope at the *origin*, i.e. looks infinite, and h is ≤ 1. However one should recall that the infinite slope is occurring during the existence of the religion. The number of data points does not allow us to observe the date of the religion birth.

We suggest to the reader to compare the figures with the h values in the tables, and observe that the y-scales are evidently quite different from figure to figure depending on the rank of the religion. Notice that t_1 ranges from 200 to ca. 2.0×10^7. Negative α values naturally result from the fit lack of precision.

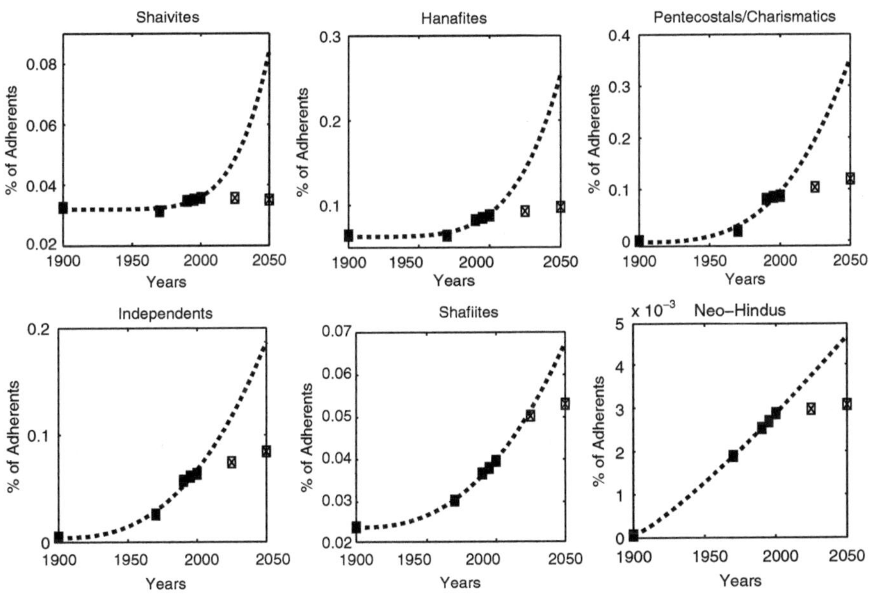

Fig. 2 Six illustrative cases of actually increasing religions, ... with $h \leq 0$. Observe the overshooting in the forecast with respect to WCT in all cases

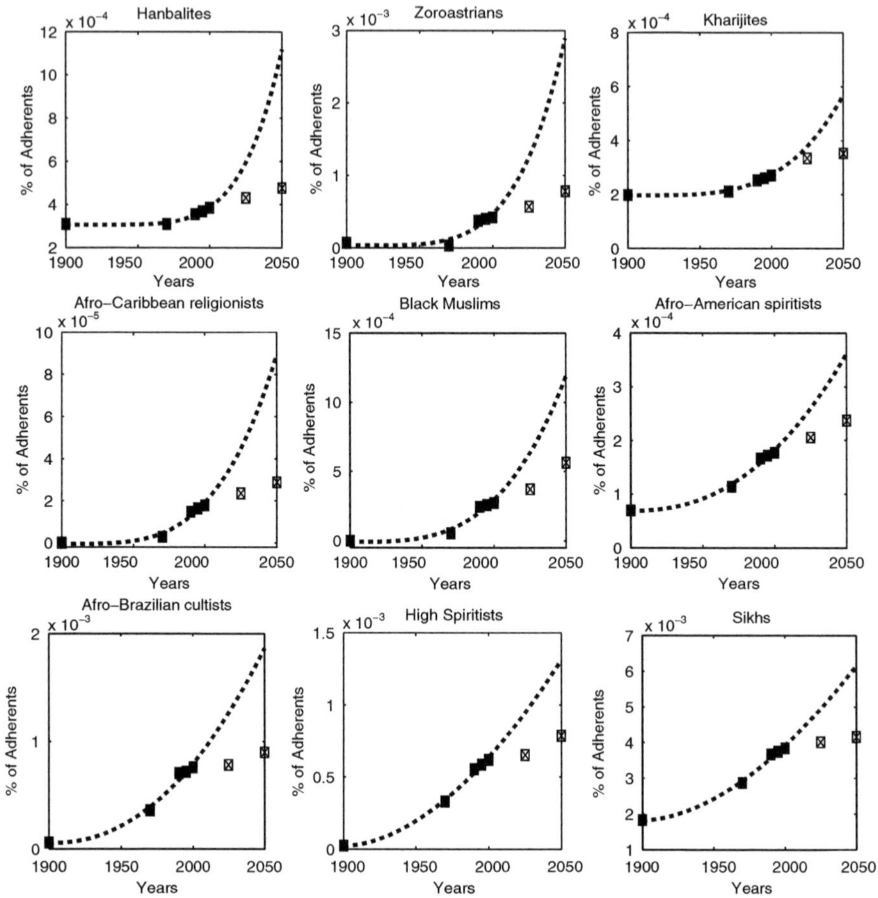

Fig. 3 Six illustrative cases of actually increasing (small size) religions ... with $h \leq 0$; our empirical law does not confirm the WCT forecast in the next years, but overshoots the WCT value

3.3 Decaying Religions

Turning to the apparently decaying religions (Figs. 7–8), the same classification can be made: either there are strongly or smoothly decaying cases, with remarkable fits, over 5 points, even though there is no a priori knowledge of the time when the number of adherents is maximum. The empirical forecast resulting from an extrapolation of the fit over 5 points overshoots the WCT prediction in the twenty-first century in almost all cases. This is again due to the lack of saturation implied by the Avrami model. Notice the very small values of t_1, and the negative ones for t_0, pointing out to the presence of a maximum before 1900. The sharp decay of the cases shown in Fig. 7 will be commented upon below.

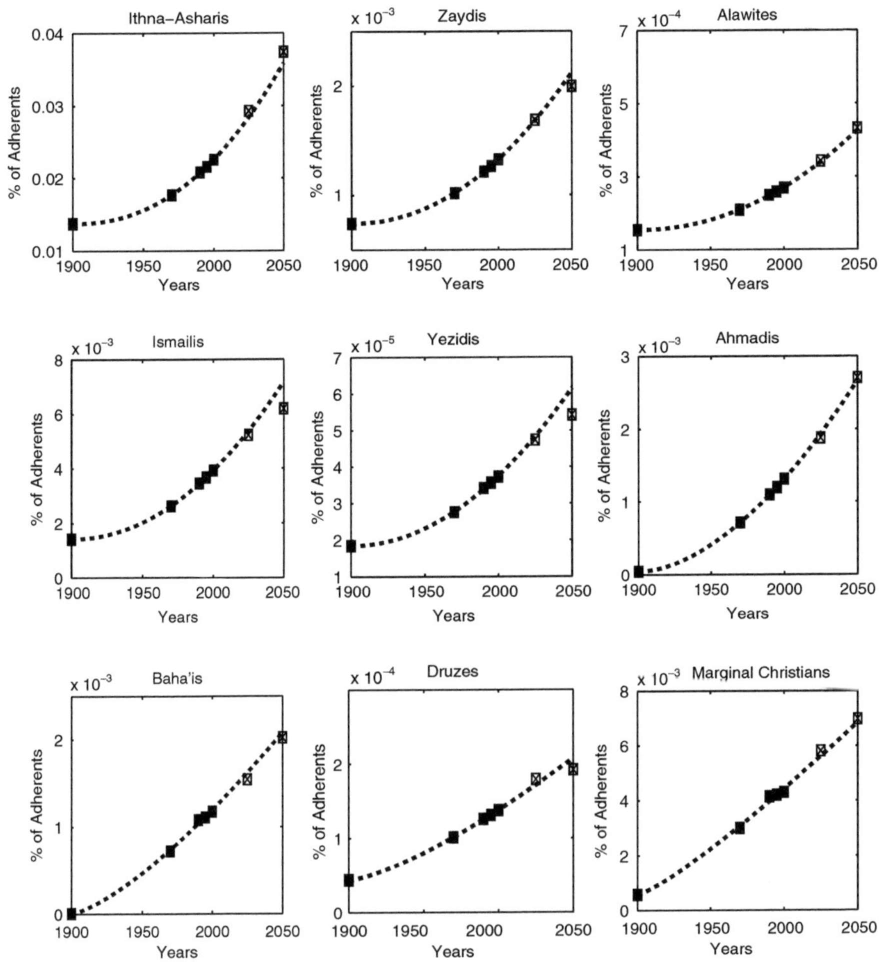

Fig. 4 Six illustrative cases of actually increasing (small size) religions …; our empirical law confirms the WCT forecast

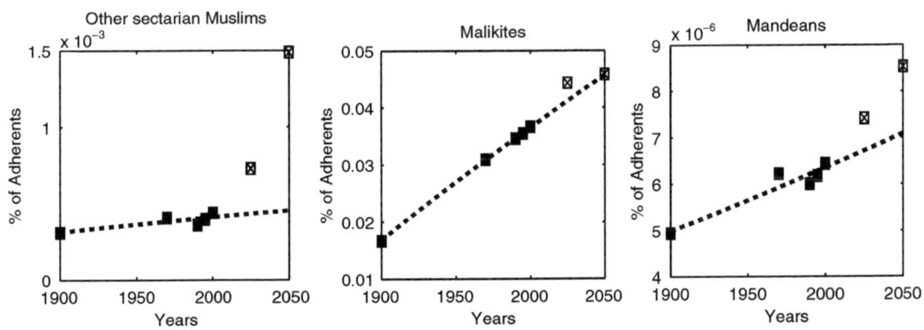

Fig. 5 Three small size religions, with increasing number of adherents; decreasing h from left to right, with h close to 0; see Table 1; our empirical law underestimates the WCT forecast

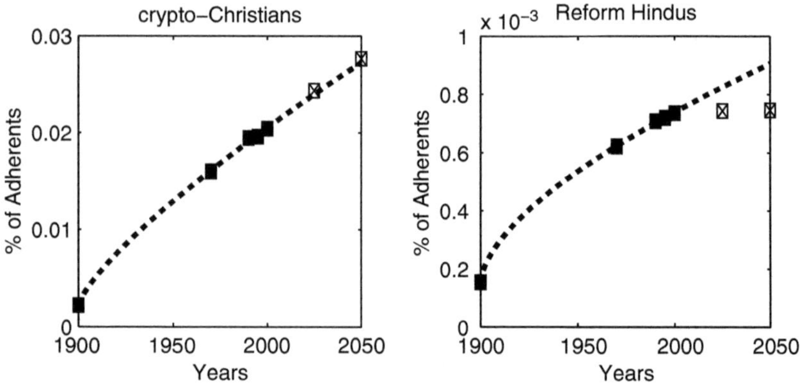

Fig. 6 Two illustrative cases of actually increasing (indicated) religions, with saturating like forecast. Observe the rather good fits, parameters in Table 1, h positive and ≤ 1, and even a rather good confirmation of the WCT forecast

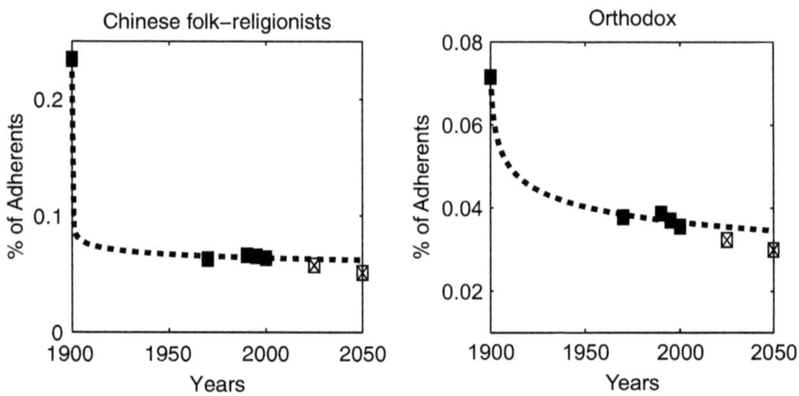

Fig. 7 Two sharply decaying religions, with very small t_1 and $h \geq 1$

3.4 Cases with Presently Observed Extremum

Sometimes there are "presently decaying" ("growing") religions for which a maximum (minimum) is observed during the present centuries (Figs. 9–14). Notice that these are rather large size religions. For such "religions" the number of adherents can be fitted with a second order polynomial $y = A + Bx + Cx^2$; the relevant parameters for these cases are given in Table 3. For the presently decaying (increasing) number of adherents, A and C are negative (positive), B being positive (negative), since for this fit the origin of time is in 0, at the beginning of the christian era.

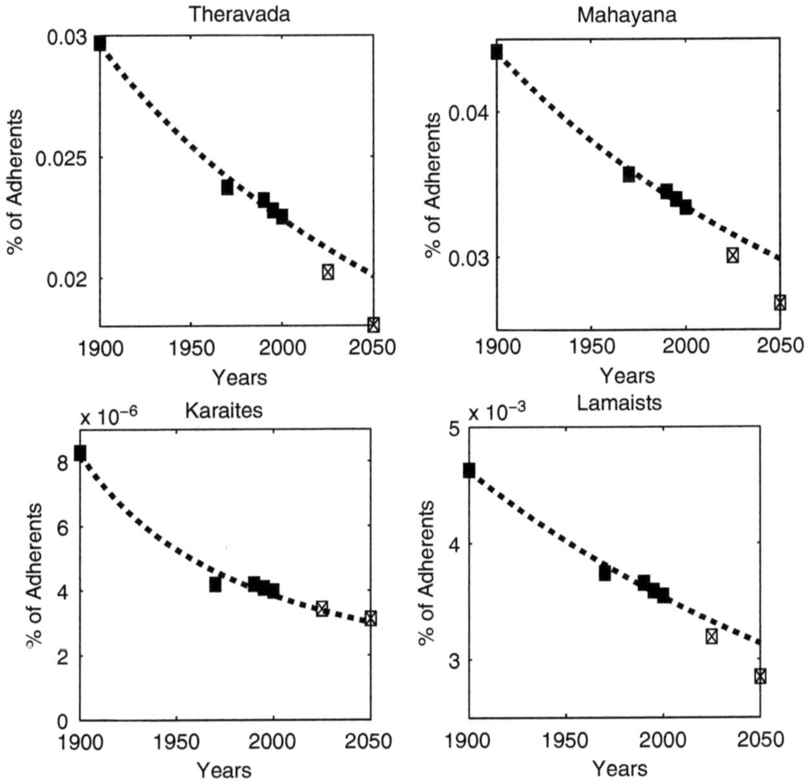

Fig. 8 Four smoothly decaying religions with h much larger than 1; our forecast being similar to that predicted in WCT

In the presently increasing cases we overshoot the WCT forecast (Fig. 9), or not (Fig. 10). For the presently decaying cases as in:

1. Figure 11, the maximum occurs much before 1900; apparently there is a strong prognosis for the collapse of these two religions, Oriental Jews and Ashkenazis, in contrast to the WCT forecast.
2. Figure 12, we underestimate the WCT forecast; the latter being thus more optimistic than the twentieth century data evolution indicates.

In some (3) cases (Fig. 12) the collapse seems rather obvious, though the tail of the evolution law has a (sometimes large) error bar, not shown for data readability. Observe that the "other religions", containing 3,000 or so smaller denominations (Fig. 13) present also a parabolic convex shape. As for the Roman Catholics (see Fig. 12a) and the Jains (Fig. 14a), the WCT is forecasting a marked turn over in the twenty-first century, i.e. somewhat expecting an increase in "small denominations", though it is unclear whether their forecast includes an increase in the *number* of religions.

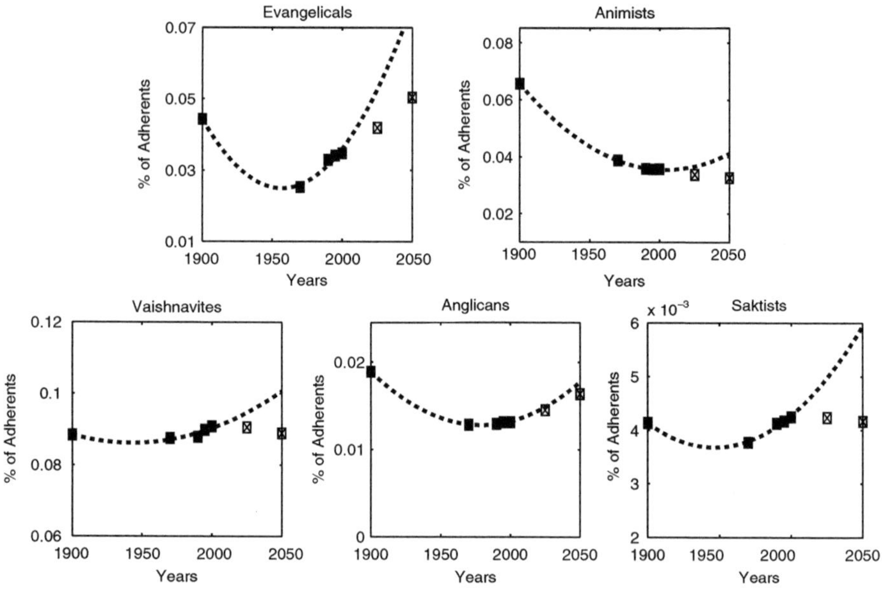

Fig. 9 Five large size religions indicating a turn over with a minimum in the twentieth century; theoretical forecast with respect to WCT is debatable though our fit slightly overshoots the WCT data

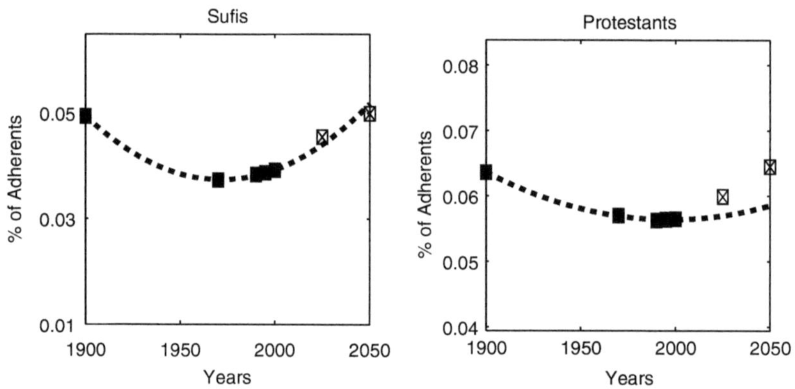

Fig. 10 Two large size religions indicating a turn over with a minimum in the twentieth century; our theoretical forecast slightly underestimates the WCT data

Religions with extreme values of parameters, see Table 3, are illustrated in Fig. 14. they indicate complex situations. It might happen that such religions will disappear, but whether it might be through a long tail or sharp cut-off evolution remains a question. More data should be of interest in such cases.

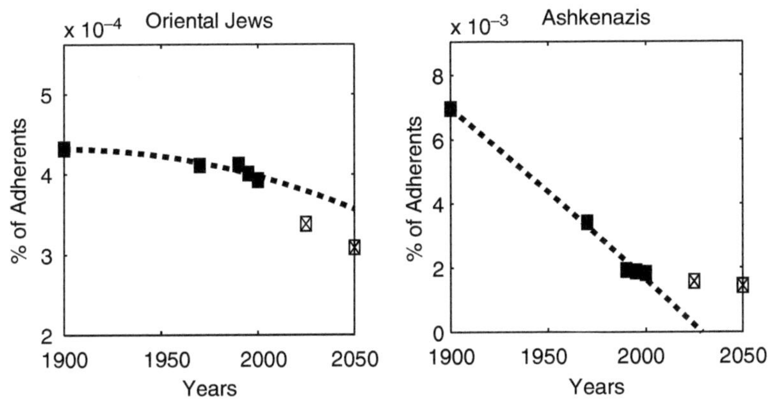

Fig. 11 Two cases of religions having a markedly predicted collapse after having had a maximum in the nineteenth century

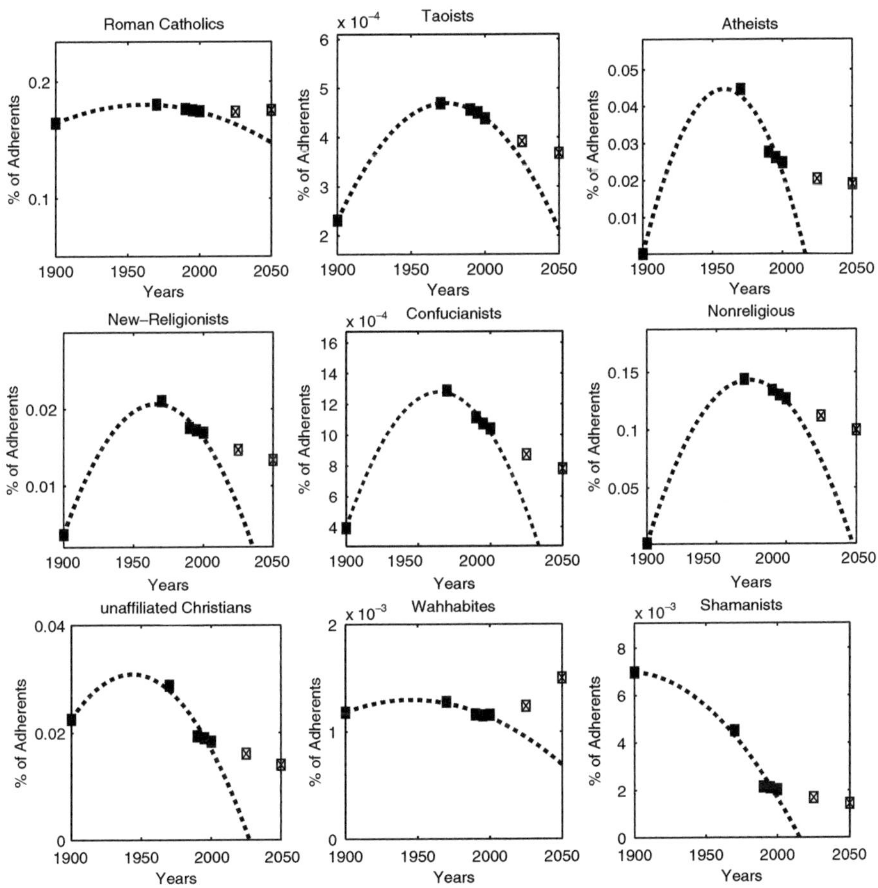

Fig. 12 Nine religions having had a maximum during the twentieth century; the parabolic forecast undershoots the WCT expectation

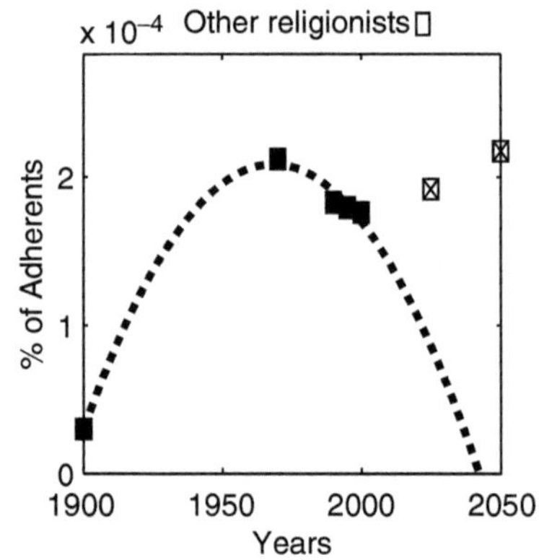

Fig. 13 Case of so-called 3,000 other religions for which a decreasing behavior is observed; notice the marked underestimate of our forecast with respect to WCT – predicting an increase in this twenty-first century

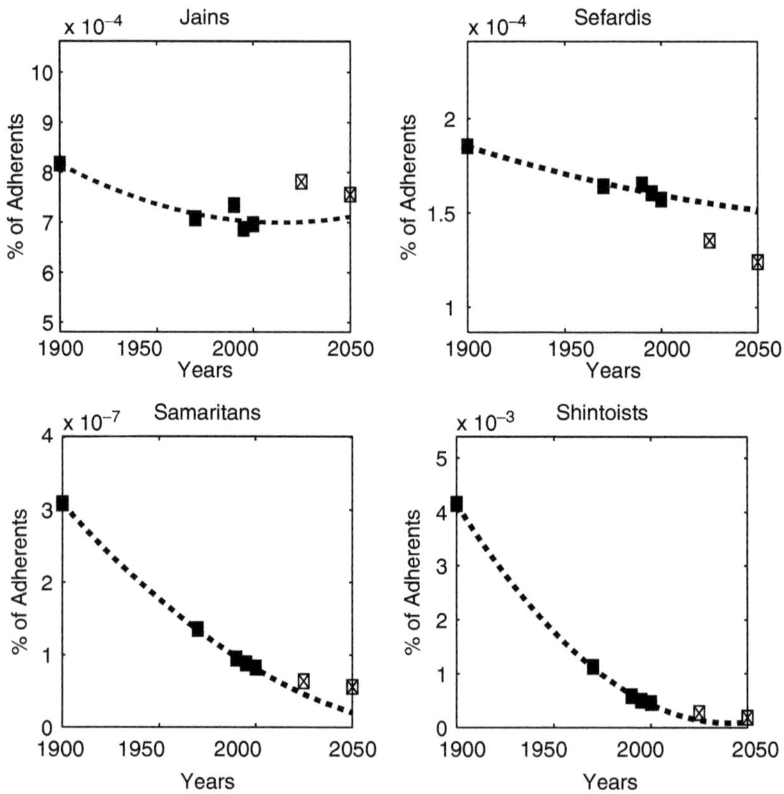

Fig. 14 Four cases of equivalent size, but rather small, religions having a relatively complex behavior, apparently decaying during the twentieth century, but with debatable forecasting with respect to WCT for the twenty-first century

4 Discussion and Conclusions

First, let us warn that we consider that the religious practice is more likely more diverse than WCE and WCT surveys indicate. The data is over a limited quite finite time span. Yet understanding the difficulty of such surveys, and accepting the interest of the data *per se* we suggest to let religious adherence to be a degree of freedom of a population, and follow through with some statistical physics considerations for our enlightenment. Therefore we have analyzed 58 cases of growing and decaying (so-called) religions, observing several groups through their analytical behaviors. We indicate that with an Avrami law the fit can be quite often good, in particular for the growing cases. Physically speaking that gives some support to the conjecture of religions growing like crystals (http://physicsweb.org/articles/news/11/1/1/1). However we cannot expect that the Avrami equation holds true for ever; the system should saturate at some point, except if only a few religions are excessively predominating, and not "allowing" the probability of existence (in a thermodynamic sense) of others. The evolution of the number of adepts over the life time of a religion cannot be expected to be so smooth either. Indeed most of the fits seem often to indicate an exponential behavior. This is clearly wrong on the long term. Whence it would be of interest to develop an alternative set of fits and considerations through the logistic map approach, recalled in the introduction (Fig. 1). In this way the growth of the population due to the inherently limited resources would be more realistically taken into account. However this introduces an extra parameter. In the same self-criticism of our approach and recognizing that more sophisticated ones can be though of, let us mention that beyond the simple analytical forms here above presented, one can imagine describing the number of adherents evolutions as in a Lotka–Volterra approach [17–21]. However when complexifying the description, see also [22, 23], it is not obvious that one can easily quantify or calibrate the parameters.

The same is true for the parabolic fit, which either indicates a quite quickly forthcoming disappearance of a religion or allows for infinite growth. We recognize that these are approximations.

It seems that we often overestimate/underestimate the WCT theoretical trend in the decaying cases and in several growing cases, though we sometimes agree in the latter cases. Again we claim that the WCT trends can be quite arbitrary, supposedly predicting a linear evolution from the last three data points in the surveys. It might be interesting to use other types of statistical analysis to conclude whether the forecasts so much differ from one another. ... As well perform detailed analyses taking into account error bars in the original data. For example, the case of Zoroastrians (in Fig. 3) indicates an anomalous point corresponding to 1975, while other cases seem to indicate major (but unknown) error bars (like in Fig. 5) on the data from the surveys. To resolve such questions is outside the scope of this report.

Turning to the data displayed on different figures, a "high" growth is seen for Hanfites, Shafiites and Malikites which are all Sunnists. Maybe we should not need to add a comment based on "political considerations" here, but we may consider that the meaning of h makes sense again. In fact this is emphasized when considering two of the highest growth rates, i.e. as found for Charismatics and Independents, though a strict *late growth stage theory* might be debated upon. One case where one case trust the data points is likely that of the black muslims (Fig. 3) since they hardly existed before 1900, whence for which an Avrami equation would hold. It would be very interesting to check soon the number of adherents in such a case.

Finally observe that the "non religion" adherent data finds a remarkable position as the fourth growing "denomination". Observe the maximum in the number of "adherents" in such a case near 1970, rendering the theory (or the data!) to be debated upon.

Last but not least, it might considered that the growth in many cases only reflect the population growth in several countries (a comment by J.J. Schneider and D. Stauffer). It is true that it would be interesting to recount the number of adepts, and their rate of adherence, per religion *and* per country, and to correlate the evolution to the birth rate in the examined country with the religion(s) growth rate, and in particular the attachment parameter value. The latter might not be a constant as assumed here above. Such an approach is presently attempted [21].

In conclusion, here above we have shown that we can attempt to make a statistical physics like analysis of the number of adherents in religions, going beyond our first paper [2] on the subject. However the data seem sometimes barely reliable.

Nevertheless one can, expecting better surveys, at a more limited scale, suggest further lines of research. One could suggest agent based models like for languages, including the role of external fields. One could try to have a Langevin equation connexion to Avrami equation; of course we need to define a hamiltonian H and a current: that implies interactions thus competitions between entities; what we do not see here yet. However the hamiltonian can be obtained following standard ideas, like turning over the pdf into its log and defining some temperature. Religions seem to be an interesting field of study for statistical mechanics!

Appendix A: Languages Vs. Religions

Through this Appendix A we wish to outline what we consider are a few interesting aspects, i.e. "differences", between languages and religions, from a physics point of view, perspective or input into modeling their sociological features. In particular one could consider their origin or mode of nucleation [24–29], number (see the orders of magnitude), variety and types of agents, define their opinion leaders, observe the range of time scales, and the relevance of applied fields; see Table 4, as a summary of considerations of interest.

We insist that in physics one should study the response of the system to intrinsic or extrinsic fields. Recall that we may describe the population of agents through a free energy F, Hamiltonian formalism or Langevin equation. In such a way, one would develop the quantity of interest as a series in terms of clusters, e.g., ordered along the increasing size of the cluster according to the number of spin S_i in the cluster $< \ldots >$, as in

Table 4 Comparison: similarities and differences between languages and religions seen from a statistical physics point of view

	Languages	Religions
Origin	[24–26]: Physiology	[27–29]: Myths
Number	More than 6,000	More than 3,000
Huge variety	Dialects, slangs	Denominations, sects
Speaker distribuion	Log-normal	Log-normal (?)
Agents	Multilingual (frequent)	Polyreligious (rare)
Opinion leaders	Authors, teachers	Priests, witches, shamans
Time scales	Nucleation: slow	Nucleation: fast
	Growth: slow	Growth: fast (avalanches)
	Decay: fast	Decay: slow
Tags	Grammar, vocabulary	Images, rituals
Impact factor	Libraries	Worship sites
Diffusion	Slow	Fast
Applied fields	Rare	Many, strong

$$\exp\left[-\frac{F-F_0}{kT}\right] = -\Sigma_i H_i S_i - \Sigma_{<ij>} J_{ij} S_i.S_j - \Sigma_{<ijk>} K_{ijk} S_i.S_j.S_k - \cdots, \tag{11}$$

in obvious notations, i.e. each spin representing an adept in an external field H_i and interacting with another adept through some interaction J_{ij}, etc., or similarly

$$\frac{\partial \Phi_i}{\partial t} = A_{ij}\Phi_j + B_{ijk}\Phi_j\Phi_k + C_{ijkl}\Phi_j\Phi_k\Phi_l + \cdots, \tag{12}$$

for some information flux Φ_i. A vector generalization is immediately thought of. In the same spirit one can write down many terms and differential equations in a Lotka–Volterra approach, taking into account some interaction and competition between religions [21].

Appendix B: Indicators of Religion Status

The time dependence of the number of adherents can be considered to be a very restrictive way to "measure" the evolution of a religion. One could also "weight" the level of adherence to a religion. For example, one could try as for languages to define a religion through its quantity of practitioners, rituals, ... Many other indicators are possible. One can measure diverse quantities related to the religious effect. As in physics one can search for the relation between causes and effects, the response to internal or/and external fields.

As there are several definitions of a language [9], similarly one could also define what a religion "is" in different ways [30].

First let us list a few definitions of religions from the conventional literature, see http://www.religioustolerance.org/rel_defn.htm:

1. Barns & Noble (Cambridge) Encyclopedia (1990): "...no single definition will suffice to encompass the varied sets of traditions, practices, and ideas which constitute different religions."
2. The Concise Oxford Dictionary (1990): "Human recognition of superhuman controlling power and especially of a personal God entitled to obedience."
3. Webster's New World Dictionary: Third College Edition (1994): "any specific system of belief and worship, often involving a code of ethics and a philosophy."
4. Merriam-Webster's Online Dictionary: http://www.m-w.com/dictionary/religion: "a cause, principle, or system of beliefs held to with ardor and faith."
5. Christian Apologetics & Research Ministry (CARM), http://www.carm.org/: "An organized system of belief that generally seeks to understand purpose, meaning, goals, and methods of spiritual things. These spiritual things can be God, people in relation to God, salvation, after life, purpose of life, order of the cosmos, etc."

In fact, we can admit that:

> Religion is any specific system of belief about deity, often involving rituals, a code of ethics, a philosophy of life.

as in http://www.religioustolerance.org/var_rel.htm. Notice that a notion about the worldwide view seems presumptuous or exaggeratedly aggressive. We are accepting that those not included in the above are "non religious"; among them, atheists, agnostics, non-interested ones, etc., can be distinguished. On the other hand, we could distinguish between *adepts* and *adherents*. An adept definition can be found in http://en.wikipedia.org/wiki/Adept:

> An adept is an individual identified as having attained a specific level of knowledge, skill, or aptitude in doctrines relevant to a particular (author or) organization.

As W. Gibbs wrote/said that *Mathematics is a language*, one could conclude from the above that *Physics is a religion*. It is indeed clear that a religious adherent instead of being an analog of an up or down spin, is rather a vector for which each element can be a quantity measuring some value like one of those considered in sociology, i.e. a "quality". The adept attaining an extreme value of one or several vector components. Next we may imagine Potts vector or ferroelectric type (Hamiltonian) models for describing an ensemble of religious agent state or evolution; recall Appendix A. Quantitative and qualitative dynamical evolutions of agents and groups ("denominations") can also find some theoretical source in many competition and organization physics models.

Moreover, we recommend that one should consider religions from another ensemble of point of views (also sometimes called indicators; going to press we found recent and similar considerations, see [31]) among others:

1. Number of groups, sects
2. Number of parishes
3. Number of chapels, churches, worship sites
4. Number of "priests", (clergy)

5. Number of adherents, like "believers", taking into account sex, age, wealth, language, ...
6. Intensity of participation, in rituals, and in practicing the principles
7. Wealth and financing
8. Type of hierarchy, ...

No need to say that physicists are not the first ones to reflect on variability in religion distribution or adherence level. We may find already such considerations in books and papers by specialists of the history or sociology of religions [30].

Acknowledgements The work by FP has been supported by European Commission Project E2C2 FP6-2003-NEST-Path-012975 Extreme Events: Causes and Consequences. Critical comments by A. Scharnhorst have to be mentioned. Moreover this paper would not have its form nor content without comments and constructive criticisms by J.J. Schneider and D. Stauffer whom we gladly thank. Beside COST Action MP0801, MA thanks FNRS FC 4458 - 1.5.276.07 project having allowed some stay at CREA and U. Tuscia.

References

1. Barrett D, Johnson T (2001) World Christian trends AD 30 – AD 2200: interpreting the annual Christian megacensus. William Carey Library, Pasadena, CA
2. Ausloos M, Petroni F (2007) Statistical dynamics of religions and adherents. Europhys Lett 77:38002 (4 pp)
3. Holyst J, Kacperski K, Schweitzer F (2000) Phase transitions in social impact models of opinion formation. Physica A 285:199–210
4. Abrams DM, Strogatz SH (2003) Modelling the dynamics of language death. Nature 424:900
5. de Oliveira VM, Gomes MAF, Tsang IR (2006) Theoretical model for the evolution of the linguistic diversity. Physica A 361:361–370
6. Kaneko K, Tsuda I (1996) Complex systems: chaos and beyond. A constructive approach with applications in life sciences. Springer, Berlin
7. Nicolis G, Prigogine I (1989) Exploring complexity. W.H. Freeman, New York
8. Ormerod P, Roach AP (2004) The medieval inquisition: scale-free networks and the suppression of heresy. Physica A 339:645–652
9. Klinkenberg JM (1994) Des langues romanes. Duculot, Louvain-la-Neuve
10. Barabási AL, Albert R (1999) Emergence of scaling in random networks. Science 286:509–512
11. Rotundo G, Scozzari A (2009) Co-evolutive models for firms dynamics. Lecture Notes in Economics and Mathematical Systems, vol 613. Springer, Berlin, pp 143–158
12. Hashemi F (2000) An evolutionary model of the size distribution of sect/congregations. J Evol Econ 10:507–521
13. Cloots R, Vandewalle N, Ausloos M (1996) Simulations of the kinetic growth of $YBa_2Cu_3O_{7-d}$ grains. J Cryst Growth 166:816–819
14. Gadomski A (1996) Stretched exponential kinetics of the pressure induced hydration of model lipid membranes. A possible scenario. J Phys II France 6:1537–1546
15. Barrett D, Kurian G, Johnson T (2001) World Christian encyclopedia, 2nd edn. Oxford University Press, New York
16. Ivanova K (1999) Ausloos M Low order variability diagrams for short range correlation evidence in financial data: BGL-USD exchange rate, Dow-Jones Industrial Average, Gold ounce price. Physica A 265:279–286
17. Lotka AJ (1925) Elements of physical biology. Williams & Wilkins, Baltimore

18. Volterra V (1931) Leçons sur la théorie mathématique de la lutte pour la vie. Gauthier-Villars, Paris
19. Hayward J (1999) Mathematical modeling of church growth. J Math Sociol 23:255–292
20. Hayward J (2005) A general model of church growth and decline. J Math Sociol 29:177–207
21. Vitanov NK, Dimitrova ZI, Ausloos M (2010) A model of ideological struggle (submitted for publication)
22. Johnson TM, Barrett D (2004) Quantifying alternate futures of religion and religions. Futures 36:947–960
23. Shy O (2007) Dynamic models of religious conformity and conversion: theory and calibrations. Eur Econ Rev 51:1127–1153
24. Ruhlen M (1987) A guide to the world's languages. Stanford University Press, Stanford
25. Diamond J (2006) The third chimpanzee: the evolution and future of the human animal. Harper Perennial New York, pp 141–167
26. Johansson S (2005) Origins of language – constraints on hypotheses. John Benjamins, Amsterdam. http://www.arthist.lu.se/kultsem/pro/SverkerJohanssonsem.pdf
27. Durkheim E (1912/1968) Les Formes élémentaires de la vie religieuse; transl: the elementary forms of religious life. Les Presses universitaires de France, Paris
28. Eliade M (1978) A history of religious ideas, vol I, From the Stone Age to the Eleusinian mysteries. University of Chicago Press, Chicago, IL
29. Dubuisson D (2003) The Western construction of religion: myths, knowledge, and ideology. Johns Hopkins University Press, Baltimore
30. Dennett DC (2006) Breaking the spell: religion as a natural phenomenon. Penguin, New York
31. Herteliu C (2007) Statistical indicators system regarding religious phenomena. J Study Relig Ideol 16:111–127

Index

MIX
Papier aus verantwortungsvollen Quellen
Paper from responsible sources
FSC® C105338

If you have any concerns about our products,
you can contact us on
ProductSafety@springernature.com

In case Publisher is established outside the EU,
the EU authorized representative is:
Springer Nature Customer Service Center GmbH
Europaplatz 3, 69115 Heidelberg, Germany

Printed by Libri Plureos GmbH
in Hamburg, Germany